SCIENCE

SCIENCE
for first examinations

DON MARCHANT
formerly Principal Lecturer Kesteven Training College

Holmes McDougall Ltd · Edinburgh

Holmes McDougall Ltd
Allander House
137-141 Leith Walk
Edinburgh EH6 8NS

ISBN 0 7157 1991-2

© Holmes McDougall Ltd. 1981

Design: Sydney McK. Glen
Drawings: Nancy Bryce
Sub-editor: Lyn Stubbs
Picture Research: Procaudio Ltd.
Printing: Holmes McDougall, Edinburgh

Note to teachers

Some teachers may not find it easy to adopt
the sequence of subject matter of this book.
They may prefer, or be constrained, to treat
the material in a more conventional way.
Because of the arrangement in Topics, this
alternative treatment is not difficult.

Without implying that there *is* an orthodox
sequence in syllabuses, some examples can
be given to show the flexibility and adap-
tability possible:

Physics	Topic
Volume, mass, density	1,3
Force	9,10
Friction	14,15
Floating, hydrometers	18
Fluid pressure	19
Atmospheric pressure	28
Barometers	29,30
Energy forms	57
Conservation	58

Chemistry	Topic
Physical and chemical change	35
Elements, compounds & mixtures	3
Atoms, molecules	6
Atomic structure	56
Air composition	31,32
Rusting of iron	36
Water	17
Solution	21

In fact, in the extreme, it is possible to
abstract material suitable for an elementary
course in one or more of the separate sciences. If it is felt that the introduction of
electricity should come earlier in the course,
this too can be arranged:

	Topic
Electric current	67,68
Electrolysis	53
Magnetic effect of current	91,92
AC	94
House wiring	113

However, the author hopes that teachers
will teach the sequence of topics as set down.
He hopes that they will find the synthesis
refreshing and stimulating.

PREFACE

This book provides a simple basic course in science. It covers the requirements of the syllabuses for CSE General Science; typical examination questions from CSE examinations are included. It also follows quite closely the syllabus for the O level General Science for Nurses Examination of Oxford Delegacy for Local Examinations.

This book will appeal to many readers besides CSE candidates. The subject matter is presented as an undifferentiated whole, treated in topics grouped under twelve major themes. There is no conscious division into biology, chemistry and physics. Those who are curious about the behaviour of natural things are not at first aware of the various branches of science.

There is a general need, at present, for greater scientific literacy. How often do we find that contestants in popular quiz programmes can answer correctly such questions as 'Who hung the sword over the head of Damocles?', or 'Who, according to Shakespeare, was a "sweeper-up of unconsidered trifles"?', or 'What have Vaughan Williams' Second Symphony and Haydn's 104th in common?', but many of these same contestants cannot give the formula for common salt or say what is measured in amperes.

Although it may not be fair to assume that quiz programmes provide an accurate index of culture, it is clear that ignorance of basic science is less culpable in a cultured citizen than unfamiliarity with details of mythology, literature, art or music. Science is not embraced so joyfully as the other disciplines.

But science, inasmuch as it is the outcome of many centuries of human thought and endeavour, is an important element of our culture. Convincing the student of this must, in the final analysis, remain the responsibility of the teacher, by choice of experiment and discussion. This book encourages the reader towards this end.

I am very much indebted to Dr. E. Halfpenny and Mr. M. Atherton who read through the typescript and made numerous criticisms and constructive suggestions, also to Nancy Bryce who drew the line drawings from my rough sketches. Finally I should like to thank my publishers, in particular Mr. T. R. Horne, for much valued advice and guidance.

September 1980 D.M.

CONTENTS

PATTERNS & STRUCTURES

There are many different things in the world around us. Write down a list of the first ten things you think of. The list might be: the school, a cow, a washing-machine, the moon, a butterfly, a frying-pan, a radio, a cabbage, a diamond, an oak-tree. These things differ in ways that may be used to sort them out. For example: some, like the moon and the cabbage, are natural things; others, like the radio and the washing-machine, are man-made. Some things are living, e.g. cow and cabbage, and others are non-living, e.g. diamond and frying-pan. We could also arrange them under the headings:

Animal   cow, butterfly
Vegetable (or plants) cabbage, oak-tree
Mineral   (or made from mineral things) washing-machine, radio or diamond.

They can be placed in order of size, as they differ in both volume and mass.

Both living and non-living things are made up of smaller units. A building like a school is made up of bricks, windows and so on. Animals and plants are made up of units called **cells**. Rocks, minerals and chemicals are made up of **atoms** and **molecules**.

The earth itself is a collection of **minerals** and **rocks**. You will study one rock in particular. It is found as limestone in the Pennines, and as chalk in the chalk-downs of the South. The chemical name of limestone and chalk is calcium carbonate.

Windscale nuclear power station

Size 1

Animals differ greatly in size. A whale can be as long as 30 metres, but a mouse is only a few centimetres long [100 centimetres (cm) equals 1 metre (m)]. Length alone does not give a good idea of bulk. You need to measure the **volume** of a thing to find out how much space it takes up. Volume is measured in cubic centimetres (cm³) or in cubic decimetres (dm³). A cubic decimetre is also called a litre (l) (1 dm³ equals 1 000 cm³).

If the object being measured is something like a brick, you can find its volume by multiplying length by breadth by height.

In the case of an irregular-shaped thing, a good method is to put it into water in a can with a spout and see how much water flows out, or put it under water in a measuring cylinder and see how far the water rises (see diagrams below).

1 Write down the numbers of the **animals** shown in the drawing below. Name as many as you can.
2 Write down the numbers of the **plants** in the drawing below. Can you name any of them?
3 What is the **volume** of the stone in diagram **b** below?
4 What is the **volume** of the block pictured here?

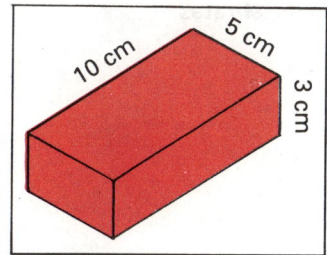

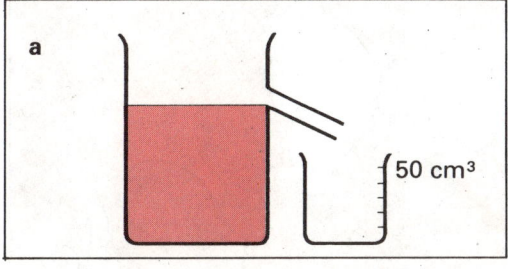

a

50 cm³

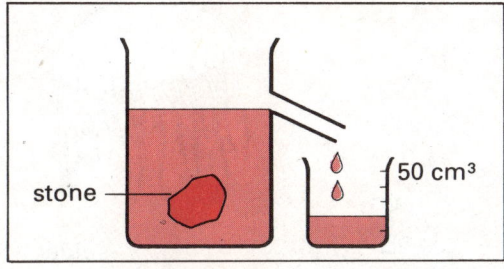

stone

50 cm³

Measurement of volume
a The object is inserted into a spouted can and the volume of water that flows out is noted
b The rise in the water level is noted after the object is inserted into the measuring cylinder

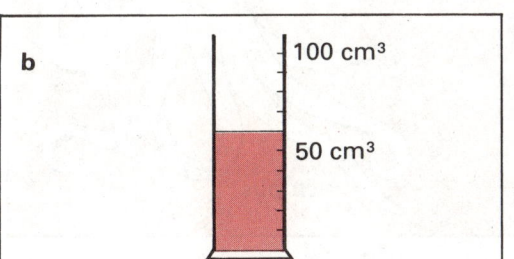

b

100 cm³

50 cm³

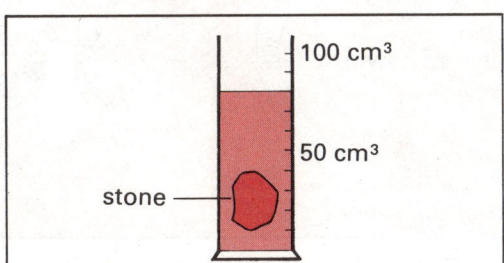

100 cm³

50 cm³

stone

Plants and Animals

All the animals shown so far have a head, four legs and a backbone. They are called **vertebrates**. The bones that make up the backbone are called **vertebrae**. Man is a vertebrate. He seems different because he stands only on two limbs and uses the other two limbs as arms.

The animals shown here are different. They are **invertebrates**. You can see how they are different by looking at an earthworm. It is easy to find earthworms in the garden. Put some soil containing earthworms on a sheet of brown paper and study them closely.

Invertebrates

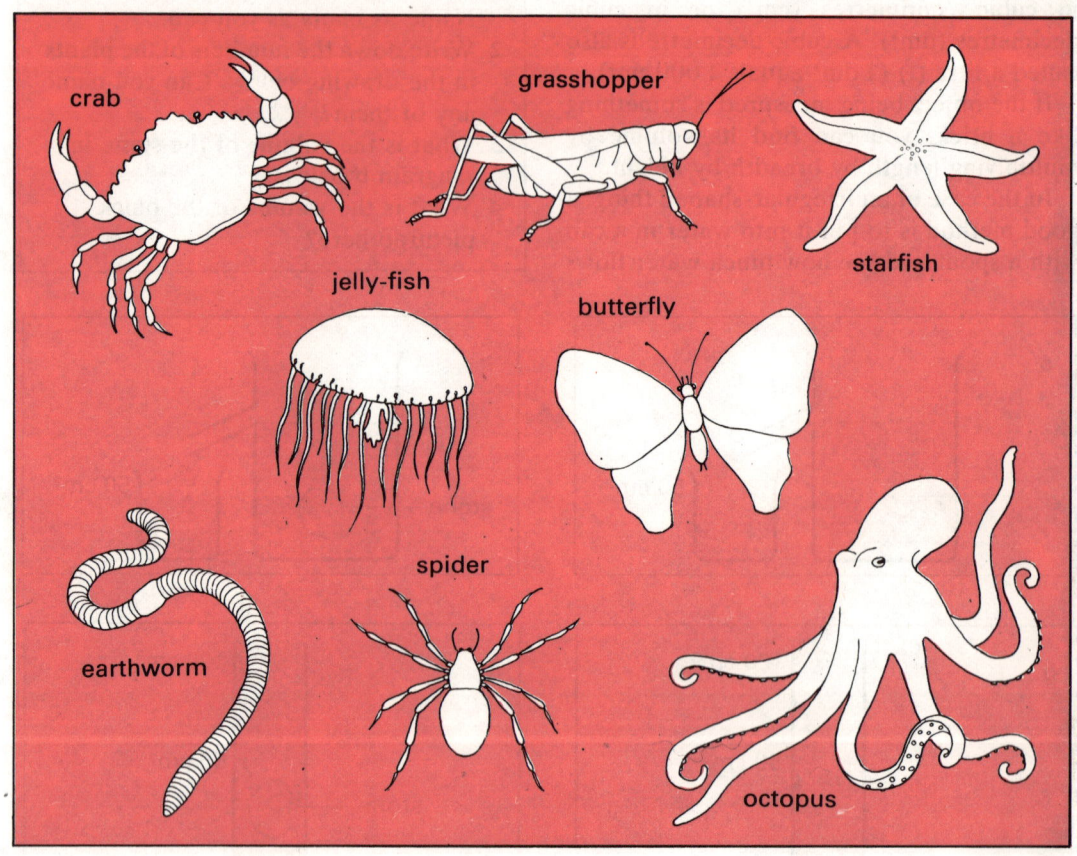

crab

grasshopper

starfish

jelly-fish

butterfly

spider

earthworm

octopus

Earthworm

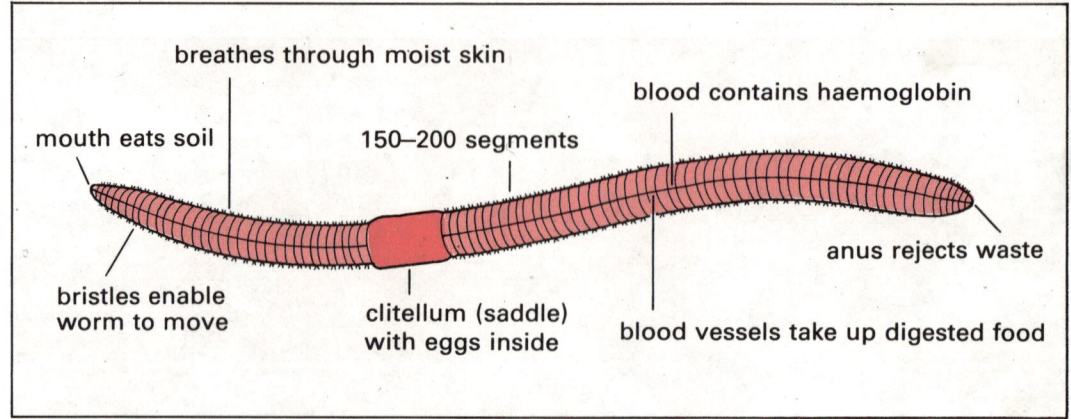

breathes through moist skin

blood contains haemoglobin

mouth eats soil

150–200 segments

anus rejects waste

bristles enable worm to move

clitellum (saddle) with eggs inside

blood vessels take up digested food

If you walk heavily on a lawn at night, the worms that are on the surface quickly go back into their burrows. This is because they feel vibrations. The earthworm is **irritable**, which means that it reacts to an outside stimulus (your feet walking on the lawn).

Most animals can do these things: they move, feed, breathe, reproduce, excrete and show irritability.

Vertebrates

The main classes are:

Fish Have gills for breathing

Amphibians Smooth moist skin, e.g. frog, newt

Reptiles Scales on outside, e.g. crocodile, tortoise

Birds The only animals with feathers

Mammals Have hair; young develop inside body, e.g. man, monkey, kangaroo, bat, squirrel, rabbit, rat, elephant, whale

Invertebrates

These include many animals, e.g. worms, jelly-fish, starfish, snails and oysters. But the most important group is the **Arthropods**. This group is divided into the classes:

Crustaceans They have many pairs of legs, all different, e.g. crab, lobster, shrimp

Spiders Have four pairs of legs

Centipedes and millipedes These have several pairs of legs, mostly all alike

Insects This class includes more species than all other groups together. An insect has three pairs of legs

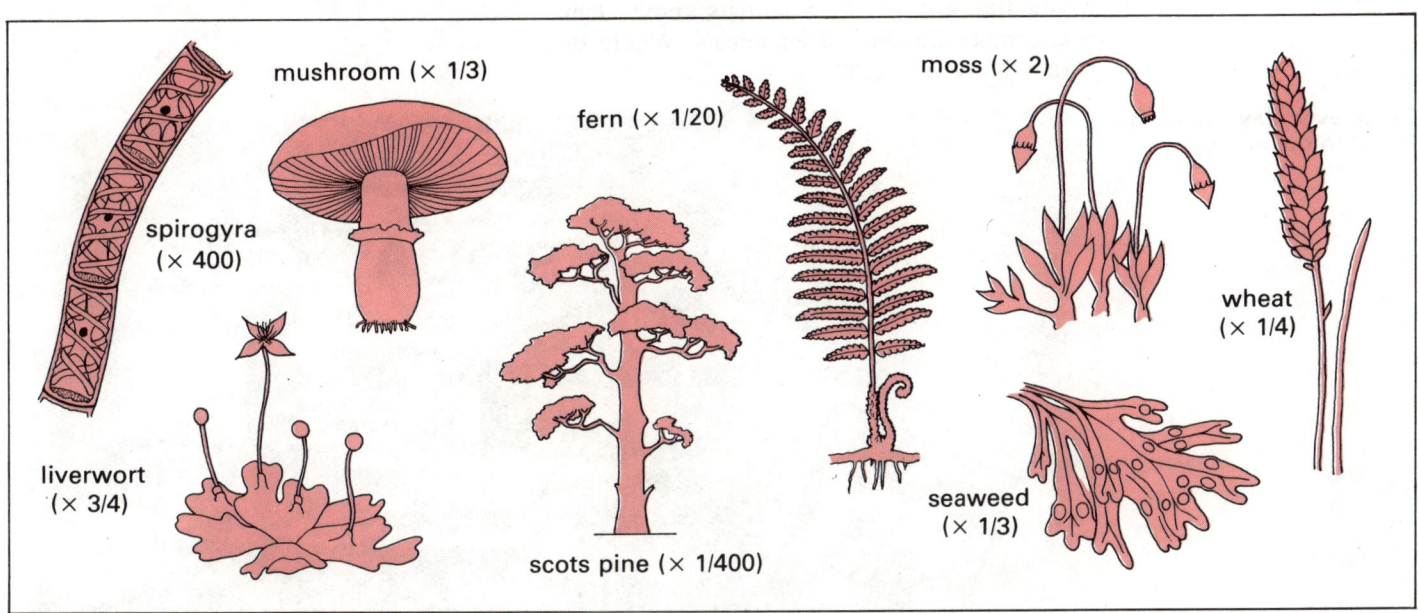

mushroom (× 1/3)

fern (× 1/20)

moss (× 2)

spirogyra (× 400)

wheat (× 1/4)

liverwort (× 3/4)

scots pine (× 1/400)

seaweed (× 1/3)

Plants

Not all plants have roots, stem, leaves and flowers. Besides flowering plants we have:

Algae These are quite simple, e.g. sea-weed

Fungi Produce spores, e.g. mushroom

Liverworts and mosses These are leaf-like structures

Ferns Also produce spores

Cone-bearing trees e.g. firs

Mass

The volume of a thing can be misleading about the amount of matter in it. A pillow filled with feathers could easily be thrown about. The same volume of lead would be much heavier. We call the amount of matter in a thing its **mass**. Mass is measured in grams (g) or kilograms (kg); 1 gram is the mass of 1 cm³ of water; 1 kg equals 1 000 g, and is about 2.2 pounds.

You find the mass of an object by weighing it. The greater amount of matter (mass) in anything, the larger the force required to hold it up (its **weight**).

Masses of animals compared

Mass	Animals
0.1 to 1 g	housefly, ants
1 to 10 g	earthworm
10 to 100 g	mouse, sparrow, frog
100 to 1 000 g (1 kg)	rat, pigeon, herring
1 kg to 10 kg	cat, rabbit
10 to 100 kg	man, sheep
100 to 1 000 kg	horse, cow
1 000 to 10 000 kg	elephant
10 000 to 100 000 kg	some sharks
Just over 100 000 kg	largest whales

Plants

1 What do you understand by 'vertebrates'?

2 How many legs does an insect have? 2, 6, 4, 8 or 10.

3 Which of these is a mammal? Salmon, rat, hen, frog, lizard.

4 State two important differences between sea-weed and a buttercup.

5 Which of the plants shown above has the greatest mass and the smallest mass?

11

Chemicals

You can find many chemicals in your home. In the kitchen you find flour, vinegar and salt. On the dining room table sugar, milk, butter. There might be aspirin and magnesia in the medicine cabinet. There is petrol and oil in the car and perhaps ammonium sulphate and weed-killer in the garden shed.

You must not forget water from the tap and, of course, the air that we breathe. Can you find any more? Scientists know of an enormous number of chemicals. Where do they all come from?

Organic substances

The chemicals that are made from minerals are called inorganic substances. But many chemicals come from living things; sugar is made from sugar cane or sugar beet; citric acid is present in lemons. Such compounds are called organic substances. They mostly contain carbon, hydrogen and oxygen and perhaps nitrogen and sulphur. They rarely contain metals. Inorganic compounds often do.

Some everyday chemicals found in the home

Minerals

Many chemical substances are found in the earth as **minerals**. Sometimes these minerals are pure substances, e.g. common salt, sulphur, copper and gold. Others must be purified. Chemists can change these raw materials into many other substances. Large masses of minerals are found below the top layer of soil. They form the crust of the earth. These are called rocks. Rocks can be divided into three classes according to how they were formed.

Igneous rocks were formed by the cooling of molten masses (a good example is lava from volcanoes)

Sedimentary rocks have been blown or washed down by water from other places (examples are sand, sandstone and limestone)

Metamorphic rocks have been altered from their original form by heat or pressure or both (slate is a good example, it has come from a kind of clay)

Galena

Iron pyrites

Quartz

Compounds and elements

Pure chemicals may be made up of different substances joined together. Thus common salt, which is sodium chloride, is a **compound** of sodium and chlorine. Sodium is a soft metal which almost explodes when put in water. Chlorine is a poisonous greenish-yellow gas. Yet when these two (sodium and chlorine) join together chemically, they make a white powder —. common salt. Neither sodium nor chlorine can be split into anything different. Substances that cannot be divided into different substances are called elements. Sodium and chlorine are therefore **elements**. Some other elements are carbon, iron, copper, oxygen and sulphur.

Sodium chloride is a chemical compound. Some substances we have mentioned are really mixtures. Milk for instance, is not a single chemical substance: it is a mixture of proteins, fat, sugar and water.

Although millions of chemical substances are known, there are only 103 chemical elements. (Eight of these do not occur naturally — they have been made by man.)

Heaviness

An elephant is heavier than a cat. But is iron heavier than cork? A large cork raft is heavier than an iron nail! Obviously you must compare the same volumes. 1 cm³ of iron weighs about 8 g; 1 cm³ of cork weighs about 0.3 g. So iron is **denser** than cork.

$$\text{Density} = \text{mass of unit volume} = \frac{\text{mass}}{\text{volume}}$$

We sometimes compare the density of an object with the density of water.

Relative density (or specific gravity)

$$= \frac{\text{density of substance}}{\text{density of water}}$$

$$= \frac{\text{mass of substance}}{\text{mass of same volume of water}}$$

1 What is the density of aluminium, if 10 cm³ has a mass of 27 g?
2 What would 10 cm³ of lead weigh if the density of lead is 11 g/cm³?
3 You find a golden-coloured nugget in a piece of rock. Is it gold? It has a mass of 20 g. When it is lowered into a measuring cylinder, the water level rises to 4 cm³. Use relative densities to decide if the nugget is gold (gold relative density is between 15 and 20; iron pyrites about 5; copper about 4).

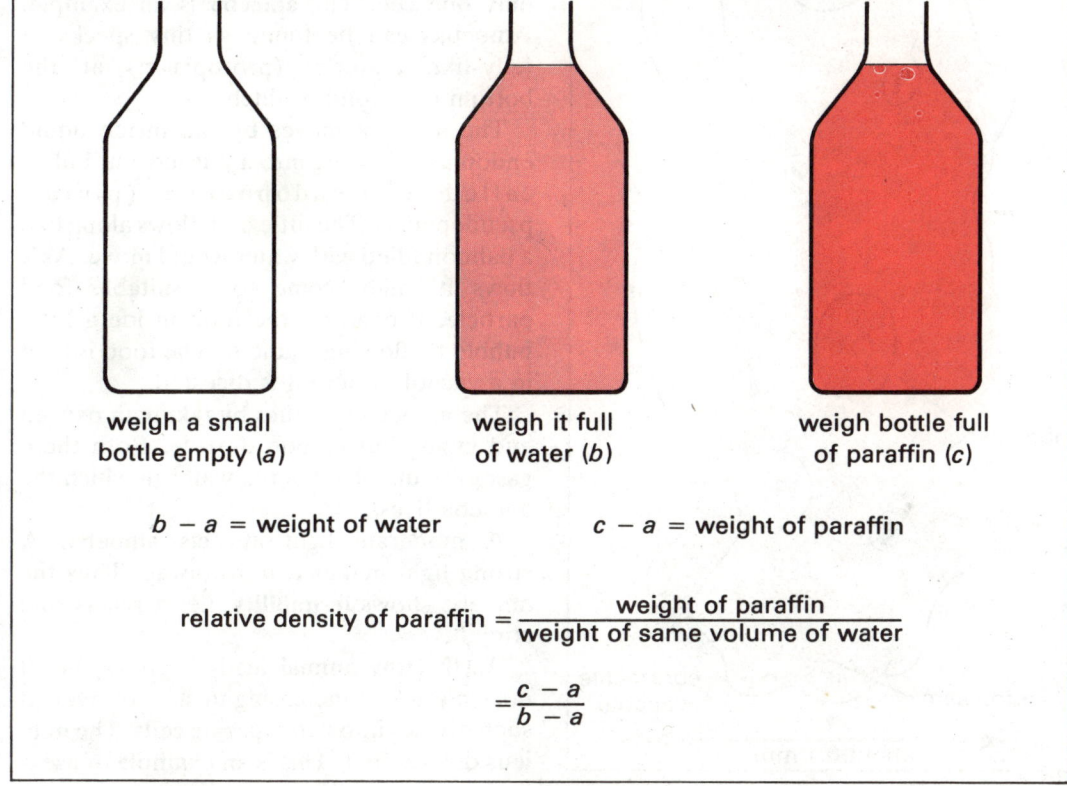

Finding the relative density of paraffin

weigh a small bottle empty (a)

weigh it full of water (b)

weigh bottle full of paraffin (c)

b − a = weight of water

c − a = weight of paraffin

$$\text{relative density of paraffin} = \frac{\text{weight of paraffin}}{\text{weight of same volume of water}}$$

$$= \frac{c - a}{b - a}$$

13

Detailed Structure

Cells

When small pieces of plant or animal tissue are looked at under a microscope, they are seen to be made up of 'units'. These units are called **cells**. Plant cells have a firm outline. They have cell walls made of cellulose. Animal cells are bounded only by a thin membrane.

Nucleus

In most cells there is a part that stains easily when dyes are put on them. This is the **nucleus**. The fluid around the nucleus is called the **cytoplasm**. The nucleus seems to control the life of the cell — a kind of 'brain'. Spaces filled with clear liquid can sometimes be seen in cells, these are called **vacuoles**.

Plant cells (left)
Cells from onion skin

Animal cells (right)
Cells rubbed off inside of the cheek

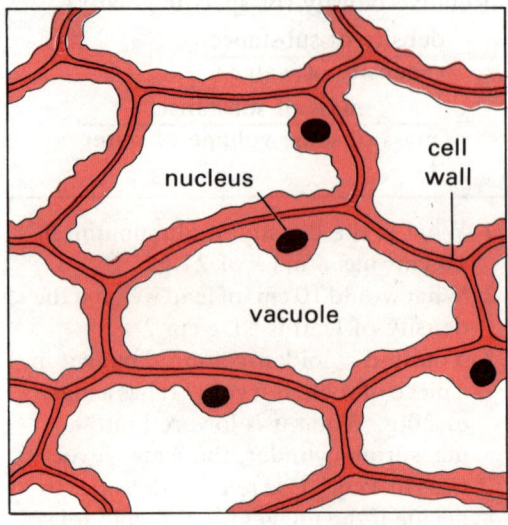

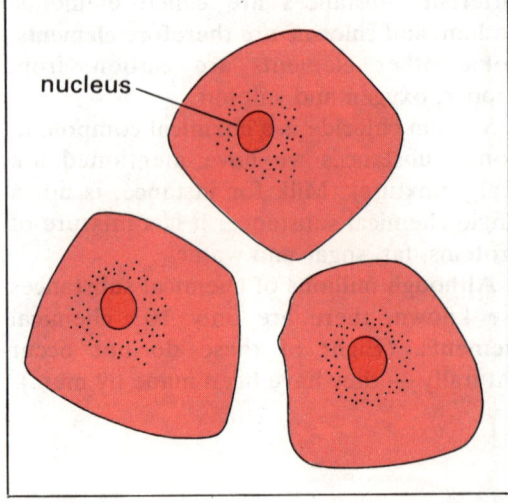

Amoeba

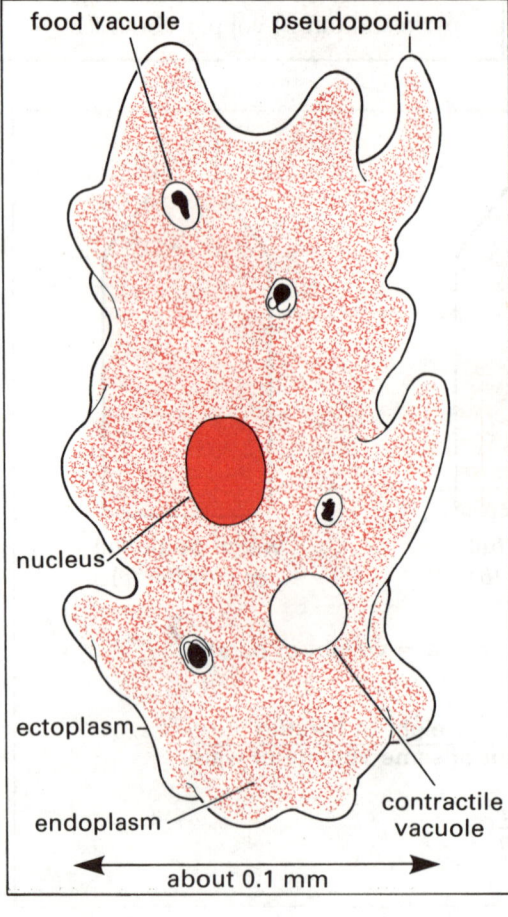

about 0.1 mm

Amoeba

Most animals and plants consist of many cells. But there are some organisms that have only one cell. The **amoeba** is an example. Amoebas can be found as tiny specks of jelly-like material (protoplasm) at the bottom of a pond or ditch.

The amoeba moves by the inside liquid endoplasm flowing into a pushed-out bulge, called a pseudopodium (plural: pseudopodia). The little blob flows along like a balloon filled with water would move. As it flows it might come to a suitable food particle. It captures the food inside a little bubble by flowing round it. The food is held in a vacuole where it is digested.

The amoeba breathes by taking in oxygen and giving out carbon dioxide. Both these gases are dissolved in the water in which the amoeba lives.

A moderate light attracts amoeba. A strong light makes it move away. Thus the amoeba shows **irritability**, i.e. it reacts to a stimulus.

As the tiny animal feeds it grows, but it does not go on increasing in size for ever. It soon divides into two separate cells. The nucleus divides first. This is an example of asex-

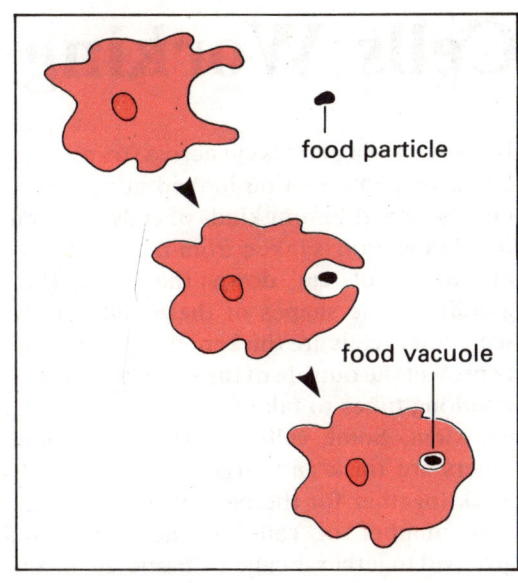

Amoeba feeding

food particle

food vacuole

ual reproduction. This is when offspring are produced from only one parent. All the protoplasm of the parent is found in the bodies of the young amoebas. If the pond dries up, the amoeba can survive by withdrawing the pseudopodia and forming a cyst. When the dry period is over, the amoeba comes out of the cyst. Sometimes a number of tiny amoebas come out.

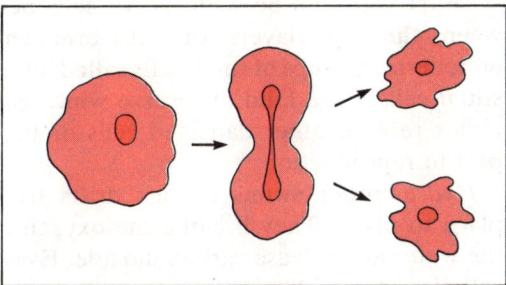

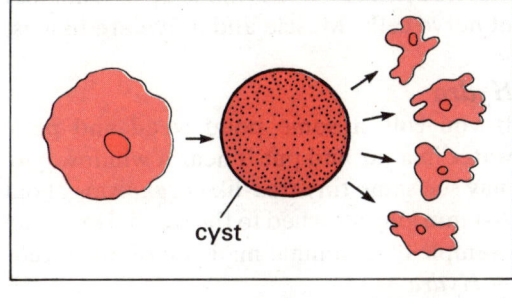

Asexual reproduction of amoeba

cyst

Paramecium

This is another one-celled animal. It has a definite shape, something like a slipper. It can move through water by waving tiny hairs covering the body. It has a double nucleus and its method of reproduction is much more complicated.

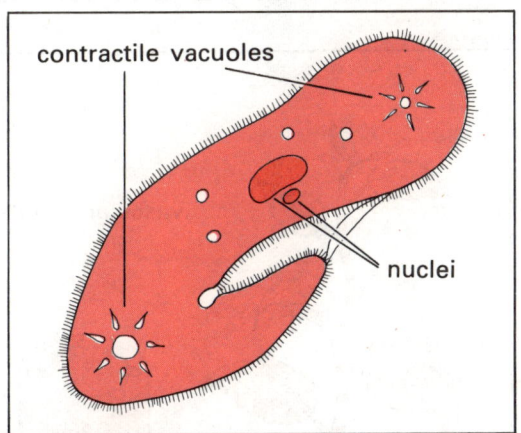

Paramecium

contractile vacuoles

nuclei

Pleurococcus viridis

This long name belongs to a single-celled plant. It is found as a green powder on the north side of tree trunks. If some of this powder is scraped off gently into a drop of water and looked at under the microscope, it is seen to be tiny green cells. Some of the cells are linked together to form colonies. The cell has a nucleus and a cell wall. There is a green basin-shaped 'body' inside. This is called a chloroplast.

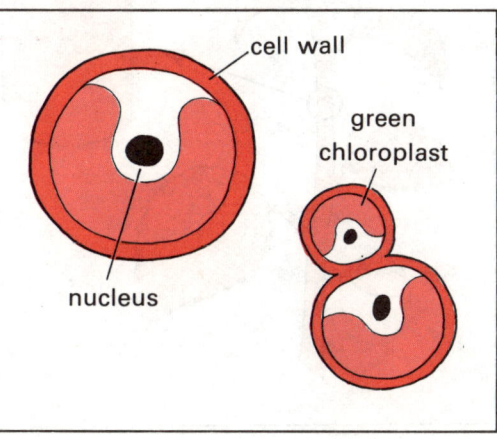

Pleurococcus viridis

cell wall

green chloroplast

nucleus

1 Describe, with the aid of diagrams how a single-celled organism, such as the amoeba, reproduces.
2 Draw two fully labelled diagrams to show the differences between an animal cell and a plant cell.
3 'If a cell has a cellulose wall it is a * cell.' What is the missing word?

Cells Working Together

If the stem of a plant is cut across (transverse) and a very thin section looked at under the microscope, different kinds of cells are seen. If a thin section is taken from the same piece of stem, but cut down the stem (longitudinal), the shapes of these cells can be seen. Some cells are thickened on the outside to protect the outside of the stem. Others link into long tubes to take food, in solution, up the stem. Some cells are thin and small, others are thick and large, but all the cells work together for the benefit of the plant.

A number of cells of the same kind grouped together is called a **tissue**. In an animal, a muscle is made up of a number of muscle cells. A nerve is made up of a number of nerve cells. Muscle and nerve are tissues.

Hydra

If you collect some pond weed and pond water in a jar and put it near a window, you may see some tiny tree-like organisms about 2–4 mm long attached to the weed. Here is an example of an animal made up of many cells — *Hydra*.

A section across the body wall shows that there are two layers of cells. The outer layer (ectoderm) is for support and protection. These outer cells have little 'tails' which, by contracting, make the *Hydra* shorter — a kind of primitive muscle action. The inner layer (endoderm) cells digest small organisms, e.g. water-fleas, that are taken in through the mouth into the gut (enteron) by the waving of the tentacles.

The capture of food is helped by special 'trigger-cells' called cnidoblasts, found in the tentacles. When the little trigger (cnidocil) is touched, a thread (nematocyst) is released and this stuns or poisons or wraps around the prey.

Other cells can be seen: nerve cells between the two layers of ectoderm and endoderm. Groups of small cells called interstitial cells are a kind of reserve which can either replace other damaged cells or take part in reproduction.

Hydra either swims or just drifts from place to place. They breathe the oxygen in the water and release carbon dioxide. Every cell can breathe because all the cells are in direct contact with water.

Cell types from a plant stem

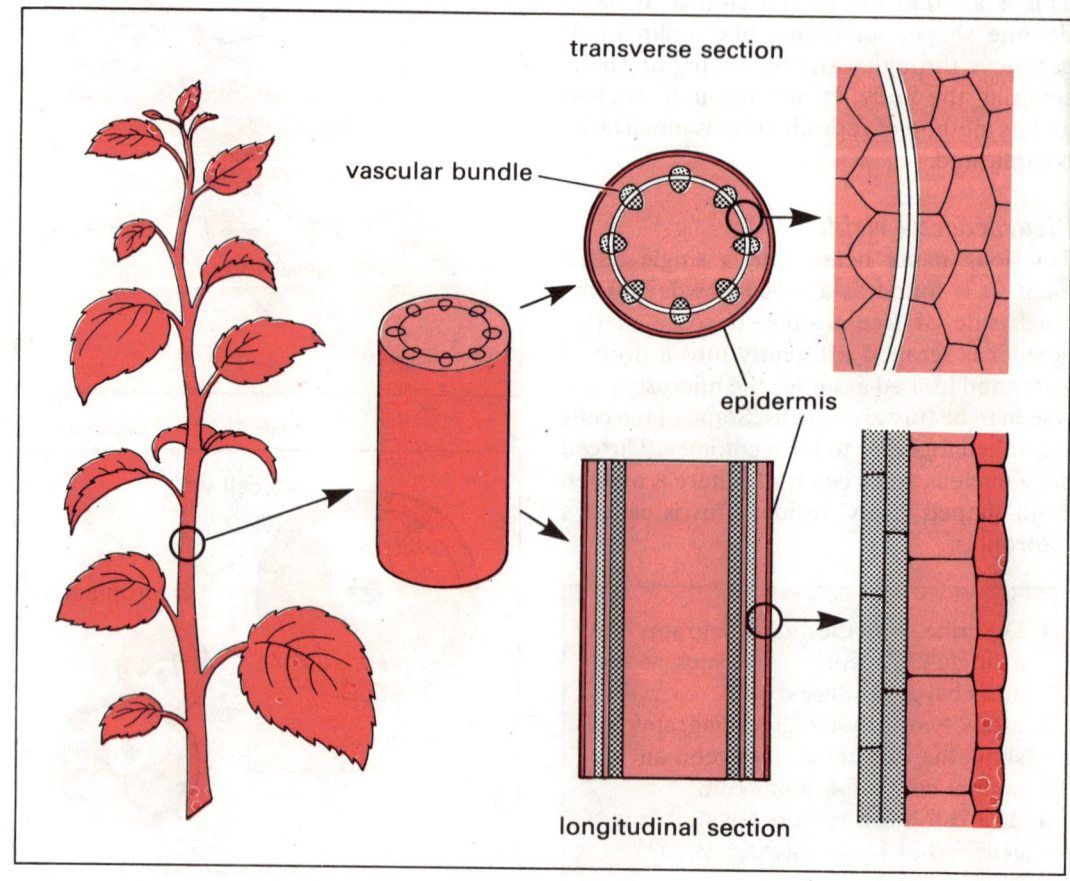

transverse section

vascular bundle

epidermis

longitudinal section

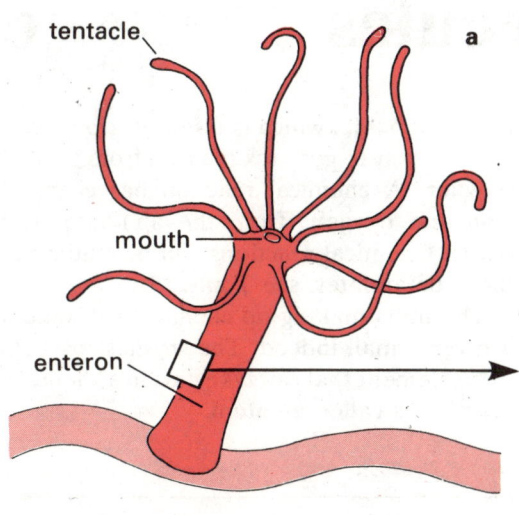

tentacle
mouth
enteron

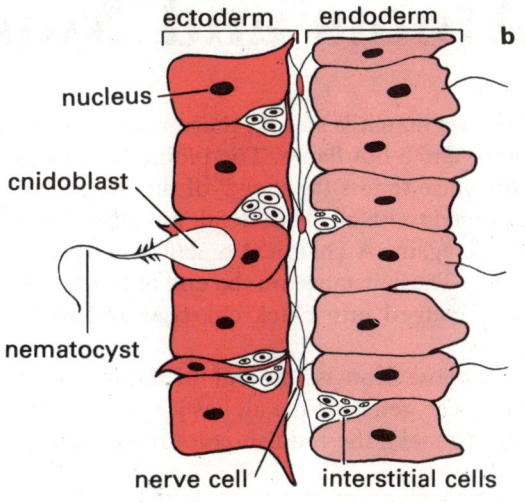

ectoderm | endoderm
nucleus
cnidoblast
nematocyst
nerve cell | interstitial cells

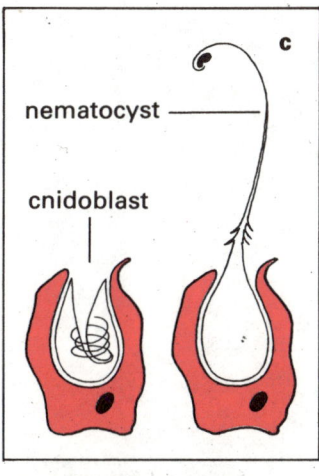

nematocyst
cnidoblast

Structure of Hydra
a Whole organism
b Detail of body wall
c Detail of cnidoblast
  catching food

Reproduction

(a) *Asexual*. A bud forms on the body and grows into a small *Hydra*. Later this daughter *Hydra* separates from the parent.

(b) *Sexual*. Both male and female organs can form on the same *Hydra*. (A bisexual animal, like this, is called **hermaphrodite**.) The male organ, the testis (plural: testes) forms near the tentacles. A number of cells with tails develop inside. These are called sperm. The ovary is the female sex organ. It forms at the base of the trunk and may be as wide as the *Hydra* itself. Inside, one large cell, the ovum (egg), forms.

When they escape from the testis, the sperm swim to the ovary of another *Hydra*. The ovum is not usually mature at the same time as the testis of the same *Hydra*. The sperm are attracted chemically towards the ovum. One sperm buries itself in the ovum and loses its tail. The contents of the sperm and ovum fuse together, forming a zygote which is the start of another *Hydra*.

This process of fusing is called **fertilisation**. It is the joining together of the nucleus of a male cell with the nucleus of a female cell.

Hydra viridis is a green species and at first sight it might be thought to be a plant. But we know it is an animal, and it catches food and eats it. The cells also have no cellulose walls.

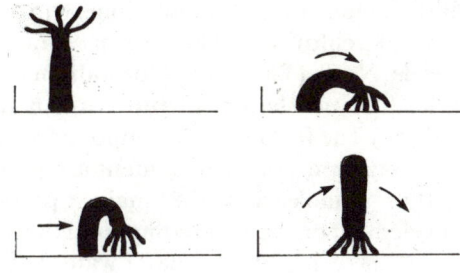

Movement of Hydra

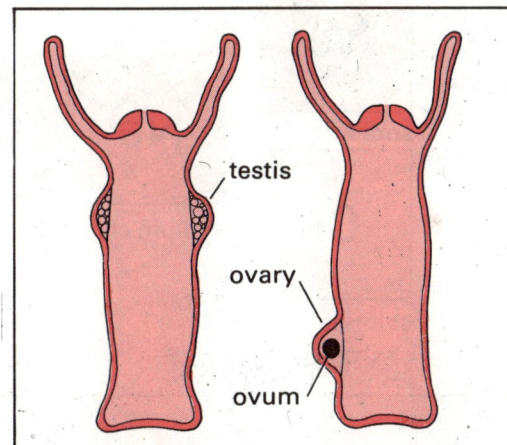

testis
ovary
ovum

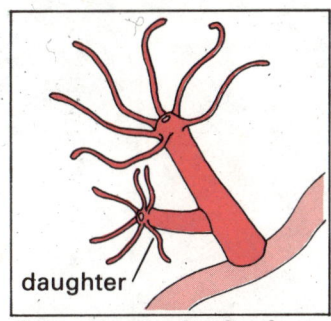

daughter

Asexual reproduction in Hydra (above)

Sexual reproduction in Hydra (left)

1 The diagram shows a living cell. Write down the names of the parts A,B,C,D. Is it a plant cell or an animal cell?
2 How does the *Hydra* feed?
3 Write down the names of the following:
  (a) the inner layer of cells of *Hydra*;
  (b) the organ that produces sperms;
  (c) the organ in which the egg (ovum) develops.

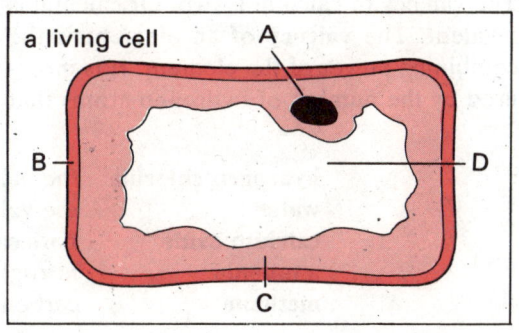

a living cell
A
B
C
D

—B

Atoms and Molecules

When a match is struck, there is a crackling noise and a hot flame. The blob of chemical substance that is the 'head' of the match has changed to black. The 'dead' match will not strike again. A chemical reaction has taken place. The substance on the end of the match has changed into black charcoal and some gases.

Suppose iron is left out in damp air. It rusts. Oxygen from the air has joined with the iron to make the reddish-coloured rust. It is now iron oxide, which is different from iron. It is not easy to get back the iron from the red powder. A chemical reaction has changed iron and oxygen (from the air) into iron oxide. Chemical reactions will be studied in more detail later, see Topic 35.

The units making up chemical substances are very small indeed. The smallest particle of an element that can take part in a chemical reaction is called an **atom**.

In sodium chloride, sodium is joined with chlorine. One atom of sodium joins with one atom of chlorine. This forms sodium chloride, NaCl. (The symbol for sodium is Na — the first two letters of the Latin name, Natrium.) The formula of a compound shows how many atoms of each element are joined together. A **molecule** is the smallest particle of an element or chemical compound that can exist on its own. A molecule of water is represented as H_2O. Even when atoms of the same element combine, the result is called a molecule, e.g. a molecule of oxygen gas is O_2.

A good way to understand how atoms join together is to make a set of cards for the elements. For a sodium atom you cut a card 2 cm by 2 cm. The card for a chlorine atom is the same size. The sodium atom and the chlorine match together to form sodium chloride. Elements do not always join together in the 1 to 1 way (singly). A calcium atom, Ca, joins with two chlorine atoms. The card for calcium is made 2 cm by 4 cm, and it needs two chlorine atom cards to match the calcium card.

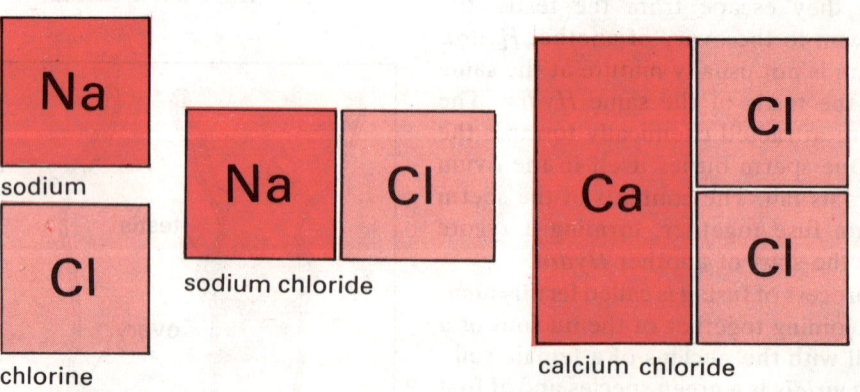

sodium

chlorine

sodium chloride

calcium chloride

Valency

The valency of calcium is two. Or calcium is bivalent. The **valency** of an element is the combining power of the element. It is measured by the number of hydrogen atoms that one atom of the element will combine with or displace. So the formula for a compound depends on the valencies of the elements. The diagrams show this in the cases of:

hydrogen chloride	the valency of chlorine is one
water	the valency of oxygen is two
calcium oxide	both calcium and oxygen are bivalent
ammonia	nitrogen has a valency of three
methane	carbon has a valency of four

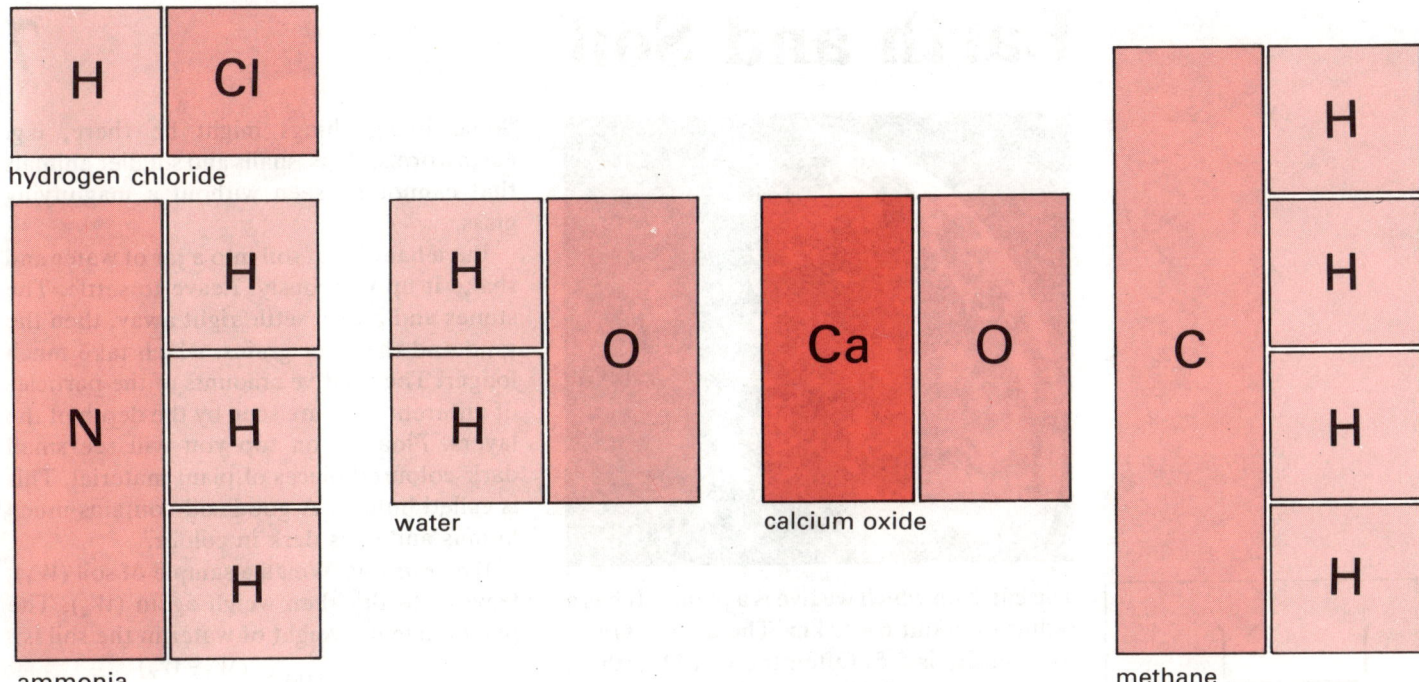

hydrogen chloride

ammonia

water

calcium oxide

methane

To make a set of valency cards, you have to note the valencies of the elements. When an element has more than one valency the number in brackets shows the valency in use. These cards should be kept safely because they will be useful later.

Valencies and symbols of some common elements

Valency	1	2	3	4
Size of card	2 cm × 2 cm	2 cm × 4 cm	2 cm × 6 cm	2 cm × 8 cm
	Hydrogen, H	Calcium, Ca	Aluminium, Al	Carbon, C
	Chlorine, Cl	Iron(II), Fe(II)	Iron(III), Fe(III)	Lead(IV), Pb(IV)
	Bromine, Br	Copper(II), Cu	Nitrogen, N	
	Iodine, I	Magnesium, Mg	Chromium, Cr	
	Fluorine, F	Oxygen, O		
	Sodium, Na	Sulphur, S		
	Potassium, K	Zinc, Zn		
	Silver, Ag	Nickel, Ni		
	Gold, Au	Lead(II), Pb(II)		

The diagram shows a few more formulae.

Examples of formulae

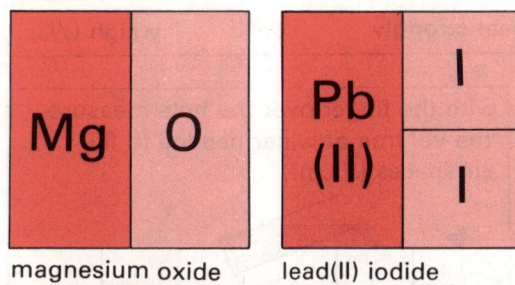

magnesium oxide

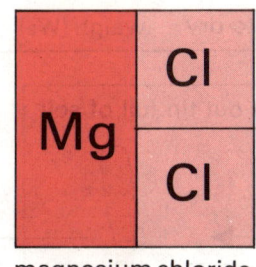

lead(II) iodide

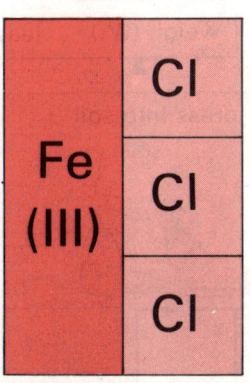

magnesium chloride

iron(III) chloride

1 With the help of the valency cards, write down the formulae for copper(II) oxide, hydrogen sulphide, aluminium chloride, aluminium oxide (you will need two Al cards here!), carbon chloride.

Earth and Soil

The earth on which we live is a planet. It has a radius of about 6 400 km. The average relative density is 5.5. Often the word 'earth' is used to mean soil. The top soil in which most plants' roots grow is only a thin layer. Below this is sub-soil, then a mass of rocks.

Soil is made up chiefly of broken pieces of these rocks. Various forces wash away or chip off pieces of the main mass of rock. So some soils may be the same as the rock underneath, but other soils have been blown or washed from other places.

Soil

To study soil put a spadeful of garden soil on a sheet of brown paper. Note particles of different sizes: stones, gravel, sand and some pieces of decaying plant and animal remains.

Some living things might be there, e.g. earthworms, slugs, snails and smaller animals that cannot be seen without a magnifying glass.

Put a handful of soil into a jar of water and shake it up vigorously. Leave to settle. The stones and gravel settle right away, then the sand and the finer grains, which take much longer. The relative amounts of the particles of different sizes are seen by the depth of the layers. Floating on top you will see small dark-coloured pieces of plant material. This is called humus. A good soil contains much humus and so is dark in colour.

Water in soil. Weigh a sample of soil (W_1), leave it to dry then weigh again (W_2). The percentage by weight of water in the soil is:

$$100 \times \frac{(W_1 - W_2)}{W_1}$$

This might be large if there had been rain but even 'dry' soil contains some water.

Humus. Take the dried soil, after weighing, then make it red-hot by heating it in a bunsen flame. Cool and weigh again (W_3). This gives us the amount of humus in the soil, because the humus has now been burnt off. What would the percentage of humus be?

Air in soil. Here you can see how a tin may be used to find the percentage by volume of air in soil. Most people are surprised to find that even the soil from a well-trodden path can contain 15–20% of air.

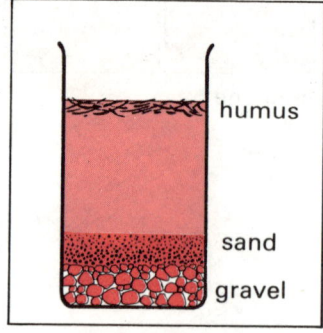

Soil composition
Layers of different components after shaking

Experiment 1
Finding the weight of water and humus in soil

Experiment 2
Finding the percentage of air in soil

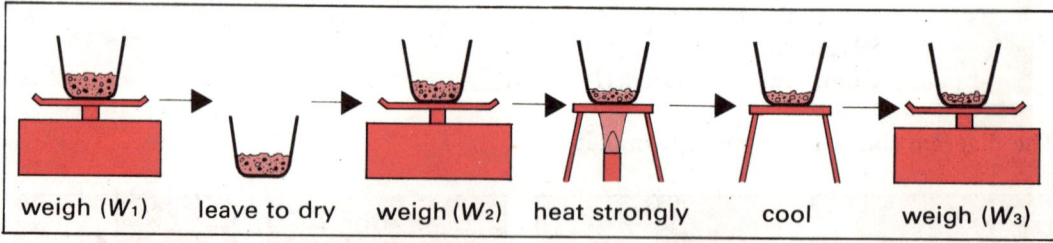

weigh (W_1)   leave to dry   weigh (W_2)   heat strongly   cool   weigh (W_3)

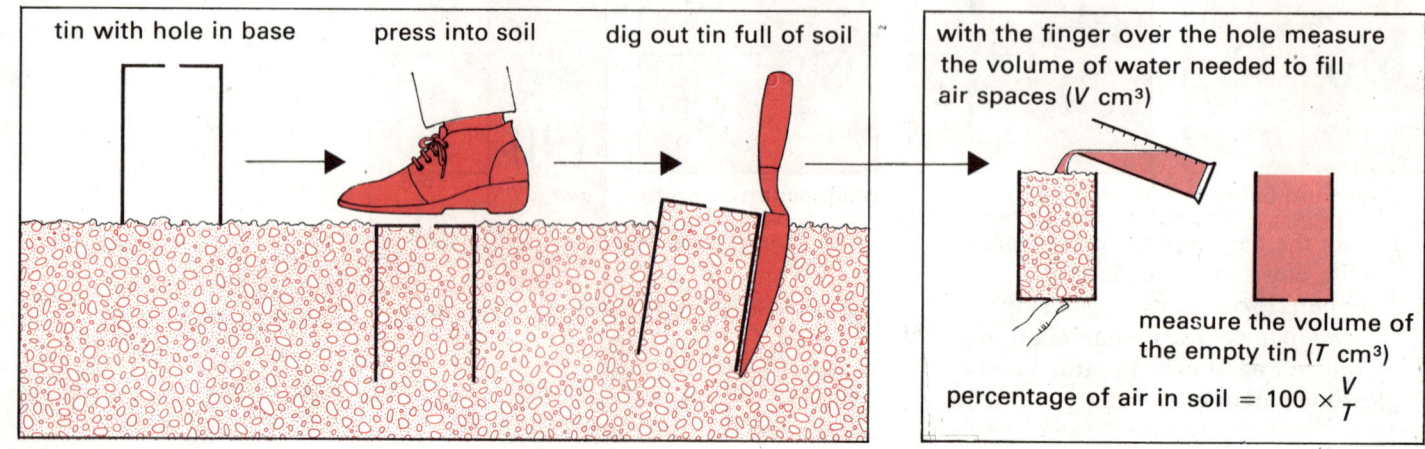

tin with hole in base   press into soil   dig out tin full of soil   with the finger over the hole measure the volume of water needed to fill air spaces (V cm³)

measure the volume of the empty tin (T cm³)

percentage of air in soil $= 100 \times \frac{V}{T}$

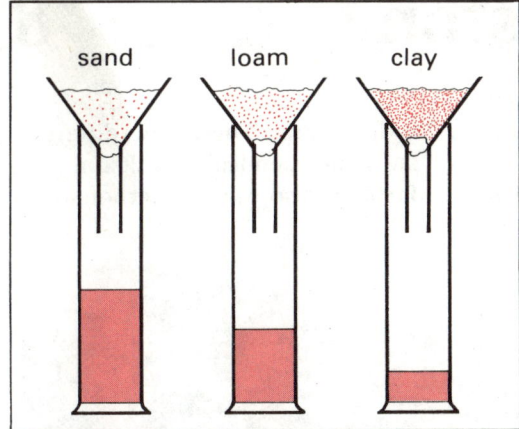

Drainage
How to show the difference
of drainage in soil

Drainage

Pour equal volumes of water into funnels containing sand, loam and clay. The soil in each funnel is held in position by a loose cotton-wool plug. Sand allows water to drain through it faster than the others.

Capillarity

Place dry samples of sand, loam and clay in tubes, plugged at the bottom, and stand the tubes in water. The water rises best in clay. Clay has a greater capillary effect. So loam is a good soil. It is a mixture of sand and clay. It allows water to drain through and not become water-logged, unlike clay. Yet loam does not become so dry in hot weather that it prevents water from reaching the roots of plants — as can happen with sand.

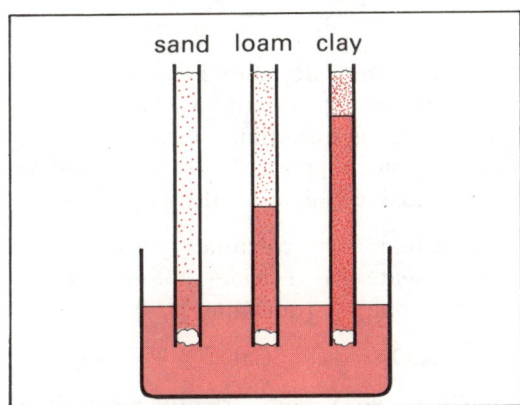

Capillarity effect

Mineral salts

Certain substances must be present in soil. If they are not present the farmer must provide them by adding fertilizer. The most important elements needed can be remembered by the catch phrase:

Charlie HOPKiNS — Mighty good CaFe

Animals in soil

Ordinary soil contains a remarkable variety of animals. Some are harmful, others useful. There can be about 700 000 worms and over 600 million insects in a hectare of soil (1 hectare is about 2.5 acres). Earthworms burrow into the soil and open up spaces for air and water. Some worms bring soil to the surface as wormcasts. Earthworms are known to bring 25 tonnes of soil per hectare to the surface during a year. They aid the gardener by letting air into the ground and turning it over. Also they drag leaves down to line their burrows and this makes the soil more fertile.

Insects

Many different kinds of insects are found in the soil. Some of them, such as ants and beetles, increase the amount of humus present. This makes the soil more fertile. Many insects go through one or more stages of their lives underground. Larvae ('grubs') are found in the soil and many moths pupate underground. The larvae of the Click beetle ('wireworms'), of the Cockchafer beetle and of the Crane fly ('Daddy-long-legs'), are examples of destructive larvae.

Centipedes and millipedes

These have some features in common with insects but are not insects. The name 'millipede' is misleading. The record number of legs on any one of them is 750, and that millipede is not found in Britain!

Bacteria

These are the most important of all the living organisms in the soil. This subject is dealt with in Topic 99.

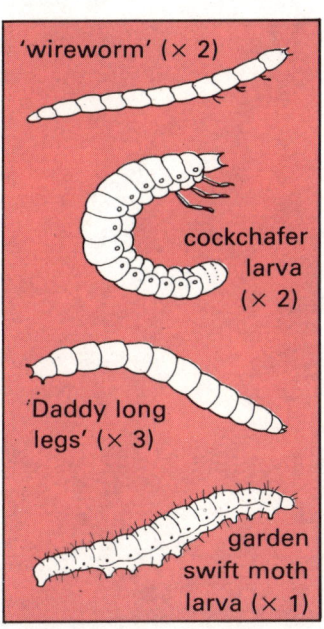

'Underground' stages of insects

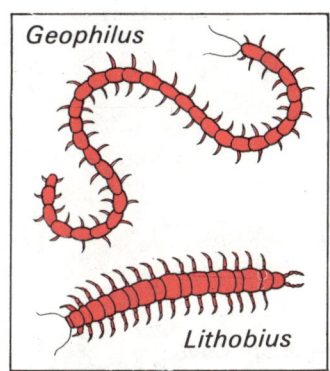

Centipedes

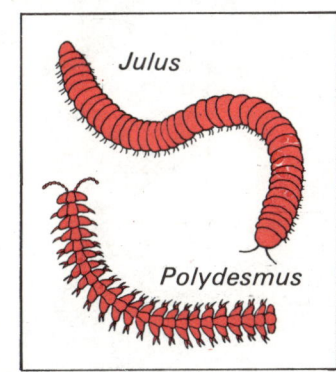

Millipedes

1 How do millipedes differ from insects?
2 Identify the following soils:
   A is sticky when wet, cracks during hot spells, has poor drainage and is difficult to work;
   B does not retain water, contains little natural plant food, is easily worked and warms up quickly.
3 Why are earthworms regarded as gardeners' friends?
4 Name three elements that a good fertilizer should provide.

Chalk and Limestone

Calcium carbonate

Calcium carbonate ($CaCO_3$) is found in many parts of Britain. It is the limestone of the Pennines, the edge of the Lake District and Derbyshire. It occurs as chalk in the Downs in the south. These are different forms of the same chemical — calcium carbonate. Marble is also calcium carbonate.

Malham Cove
This is typical limestone scenery

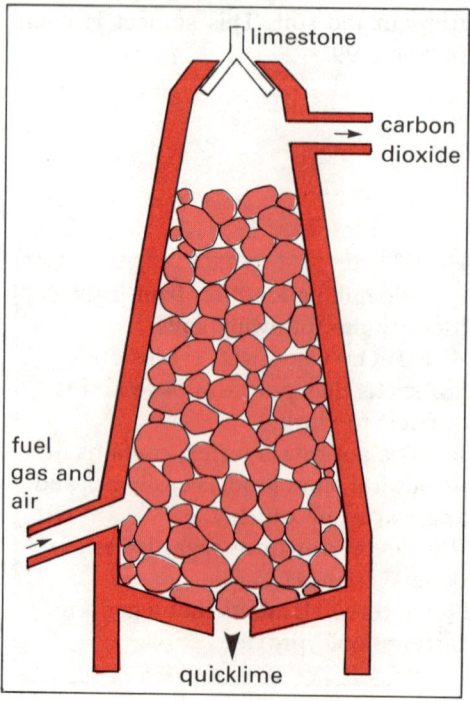

Diagram of a lime kiln

Lime

When calcium carbonate is very strongly heated, it decomposes into a white powder. This powder is quicklime or calcium oxide (CaO). What happens is that a gas called carbon dioxide is driven off by the heat:

calcium carbonate (limestone)	→	calcium oxide (quicklime)	+	carbon dioxide
$CaCO_3$	→	CaO	+	CO_2

Quicklime glows very brightly when it is strongly heated. In the early days of the cinema, before electric light, a flame directed on to a cylinder of quicklime gave the 'limelight' for showing films. Lime is such an important chemical that several million tons are produced each year. The process is carried out in limekilns.

Slaked lime

If a few drops of water are dropped on a cold lump of quicklime a hissing noise is heard, steam is given off and the lump gets very hot. It crumbles to a dry white powder. The water has joined chemically with the quicklime. It forms a different substance slaked lime, which is calcium hydroxide:

$$CaO \quad + H_2O \rightarrow \quad Ca(OH)_2$$
quicklime $\qquad\qquad$ slaked lime

Lime-water

When slaked lime is shaken with water, a little of it dissolves. The solution is called lime-water. This solution is used to test for the gas carbon dioxide.

When this gas is bubbled into lime-water it turns 'milky' or 'cloudy'. A precipitate is formed. Another chemical change has taken place and we are back to calcium carbonate:

calcium $\quad$ carbon $\quad$ calcium
hydroxide + dioxide → carbonate + water

$$Ca(OH)_2 + CO_2 \rightarrow CaCO_3 + H_2O$$

But if too much carbon dioxide is bubbled into lime water, a clear solution is obtained. The precipitate disappears. This is because calcium hydrogencarbonate has been formed:

calcium $\quad$ carbon $\qquad$ calcium
carbonate + dioxide + water → hydrogen
$\qquad\qquad\qquad\qquad\qquad$ carbonate

$$CaCO_3 + CO_2 + H_2O \rightarrow Ca(HCO_3)_2$$

(Calcium hydrogencarbonate used to be called calcium bicarbonate.) When carbon dioxide is bubbled into lime water for a long time it can be shown as:

$$Ca(OH)_2 + CO_2 \rightarrow Ca(HCO_3)_2$$

(Do not be afraid of these symbol equations. When you get used to them they help you to understand the chemical changes that take place.)

If the solution of calcium hydrogencarbonate is boiled, the chalk precipitate is formed again:

calcium
hydrogen → carbonate + dioxide + water
carbonate

calcium $\quad$ calcium $\quad$ carbon
hydrogen → carbonate + dioxide + water
carbonate

$$Ca(HCO_3)_2 \rightarrow CaCO_3 + CO_2 + H_2O$$

These reactions are important for when we study the Hardness of Water (Topic 27).

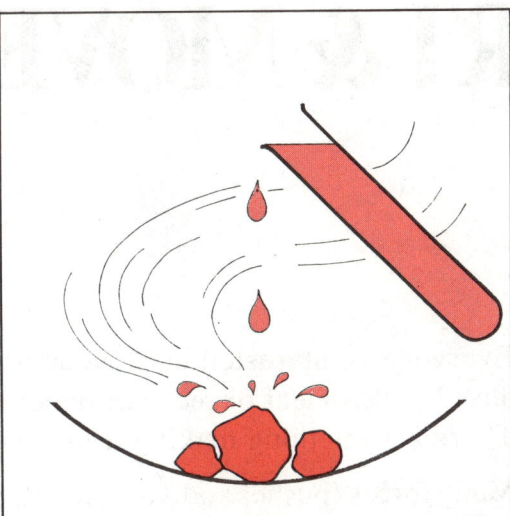

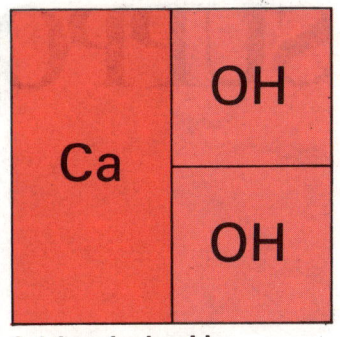

Calcium hydroxide

Adding water to quicklime

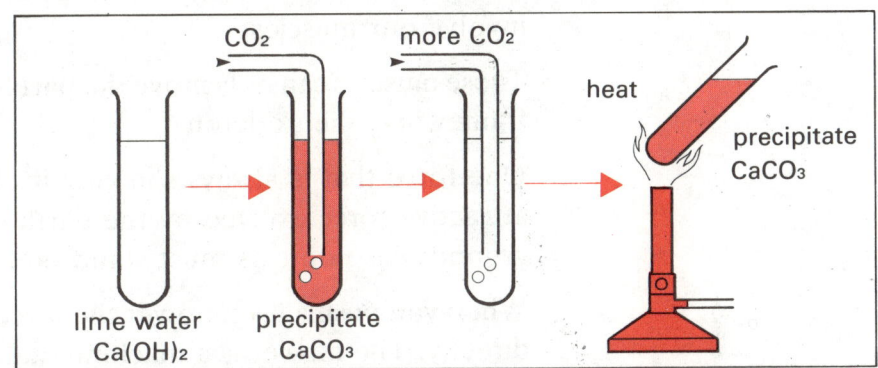

lime water $\qquad$ precipitate
$Ca(OH)_2$ $\qquad\quad$ $CaCO_3$

Lime water
The lime water turns milky as the carbon dioxide is bubbled into it

Mortar

Slaked lime has many industrial uses. When mixed with sand and water, it forms mortar used in bricklaying. The mortar hardens as the water evaporates and slowly the calcium hydroxide reacts with the carbon dioxide in the air to form calcium carbonate.

Large amounts of slaked lime are also used in agriculture to treat acid soils.

Cement

This is made when calcium carbonate is roasted with clay for some hours in a horizontal rotary kiln. The kiln is heated internally by a blast of burning coal dust. A mixture of calcium silicate and calcium aluminate is made. When this is mixed with sand, pebbles and water, it sets to a hard mass known as concrete.

Rotary kiln for cement manufacture

1 Calcium oxide is the chemical name for which of the following: quicklime; chalk; limestone; slaked lime; washing soda.
2 State the two substances from which cement is made. What solids are added to cement to make concrete?

SUPPORT & MOVEMENT

Everyone is interested in movement and speed. If a thing moves must it be alive? A dead leaf or piece of paper blowing about in the wind is not living. There is something making it move – a force is acting.

Many forces (pushes and pulls) are to be seen in everyday life. Opening a door, lifting something, kicking a ball, pushing a lawn mower. All of these actions involve our muscles.

These muscles can only move the parts of our body because they are joined to a framework, the skeleton.

One force that is always showing itself is the force of **gravity**. This is the attractive force exerted by the earth on all objects. Things fall towards the ground. Also objects must stand on a firm base if they are not to fall over.

When you push a lawnmower the force you use does not actually cut the grass directly. The mower is a machine that lets us cut grass more easily. Machines when working get hot because of friction between the moving parts. This wastes some of the effort you put in. You can reduce this loss of energy by lubricating the machine.

Cable windlass
Gears transmit the forces needed to pull cables wrapped around the drums

Forces and Movement

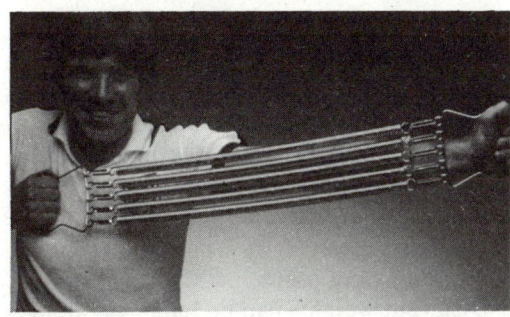

Measuring forces

If you want to develop your muscles you can try to pull out a set of springs. The harder you pull (the greater the force you use) the further you stretch the springs.

Hang up a single spiral spring with a balance pan on the lower end. Measure the length of the spring between two marks X and Y. Now put a load of 20 g in the pan; measure the distance between X and Y again.

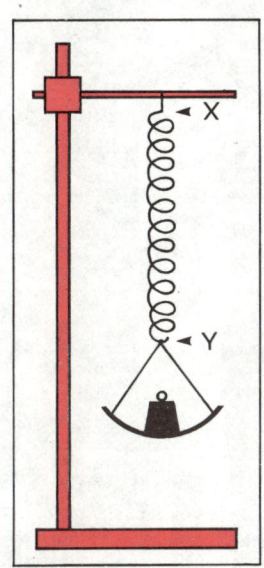

Take off the 20 g weight and XY should now be the same as at the start. The spring is elastic and recovers after being stretched. Next put 40 g on the pan and measure XY. Repeat with 60 g on the pan and so on. Note how far each load extends the spring. You can put the results in the form of a graph, plotting extension against load. The graph is a straight line, AB. So the amount by which the spring is stretched is proportional to the load. This is known as **Hooke's Law**. Extension is proportional to load.

The weights put in the pan are really forces pulling the spring down, so this is a method of measuring forces. This method is used in weighing with the spring balance. A pointer attached to the lower end of the spring shows the weight put on the spring.

If you add more and more weights to the spring, there is a point when the spring will not go back to its original length when the weights are taken off. The spring is now distorted and no longer elastic. The point at which this happens (C on the graph) is the elastic limit. If you add yet more weights, you could reach the breaking point, D. A length of rubber or elastic gives results similar to those obtained with the spring.

Load (g)	Length (cm)	Extension (cm)
0	10	0
20	12	2
40	14	4
60	16	6
80	18	8
100	20	10

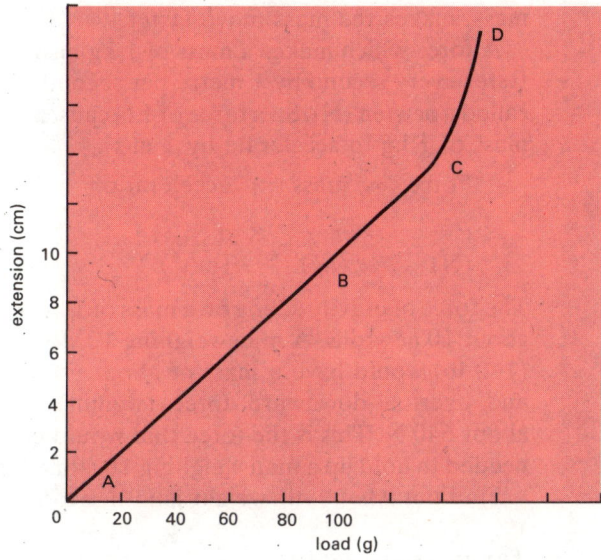

Graph of extension against load

Inertia

As you have seen, an object has mass. Some objects are much more difficult to move than others; they have greater inertia. They stay in the same place and do not move unless some force makes them do so. Try this trick: put a coin on a piece of card which is itself put on a tumbler. Now with a quick flick of the finger horizontally, push the card away so that the coin falls into the tumbler. Its inertia stops it from moving with the card because the force you use is so sudden.

A moving thing also has inertia; any moving object keeps on moving unless a force stops it. This is seen in a car collision when the car is suddenly stopped, the driver, because of his inertia, goes on moving. He will go through the windscreen if he is not held securely by his seat belt. One kind of seat belt does in fact depend on inertia. When the driver is thrown forward suddenly the belt is jerked but it does not unwind from the heavy roller on which it is coiled because of inertia. So the driver is held back by the belt.

Mass is a measure of inertia; bodies of great mass have great inertia.

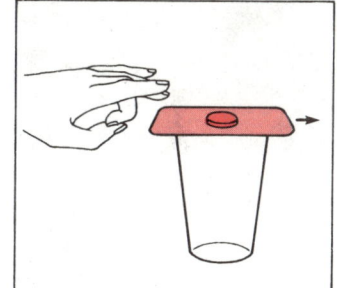

A trick with inertia

Forces in Action

Mass and weight

Any two pieces of matter attract each other. This force of attraction is very small for most things, but the earth being large attracts an object with a force that we call the weight of the object. Mass is an amount of matter; weight is a force.

Mass and acceleration

When things fall downwards they accelerate. This is true of all forces. A force acting on a mass, makes the mass move faster.

A force which makes a mass of 1 kg move faster every second by 1 metre per second is called a **newton** (N). So a force of 1 N causes a mass of 1 kg to accelerate by 1 m/s².

$$\text{Force} = \text{mass} \times \text{acceleration}$$

$$F = m \times a$$
$$\text{(N)} \quad \text{(kg)} \quad \text{(m/s}^2\text{)}$$

The force of gravity acting on a mass of 1 kg is about 10 newtons. A man weighing 10 stone (140 lb) would have a mass of about 64 kg and exert a downward force (weight) of about 640 N. This is the force that would be needed to hold up a man weighing 10 stone.

Find out what your weight is in newtons.

Terminal velocity

An object falling in air does not accelerate indefinitely. It reaches a steady velocity because of the resistance of the air. A parachutist takes advantage of this. When he opens his parachute, the resistance on the parachute keeps him falling at a steady rate.

Forces balancing

A force may not cause acceleration because it is opposed by another force. When a man is sitting in a chair his weight is a force acting downwards. But he does not go through the chair because the chair pushes upwards with a force equal to his weight.

A horse pulls a cart along, but the cart pulls the horse back with an equal force. This is an example of action and reaction on different bodies. The reason why the cart moves is because there are other forces at work, between the horse's hooves and the ground, and between the cartwheel and the ground. In firing a rifle, the explosive charge forces the bullet forward, but the bullet reacts on the rifle and there is a recoil. When a rocket is fired, the exploding gases are forced out, but the reaction forces the rocket upwards.

Turning forces

Some forces we use make things turn round, e.g. opening a door or pedalling a bicycle. Try to close a door by pushing, not on the handle, but at a point near the hinge. It is more difficult. So the effectiveness of a turning force depends on how far it is applied from the pivot or turning point.

Different weights can be balanced along a metre rule, suspended by a pin passing through its midpoint.

Here are some typical results:

Left hand side			Right hand side		
Weight	Distance from pivot	Distance × weight	Weight	Distance from pivot	Distance × weight
0.5 N	0.1 m	0.05 N-m	0.2 N	0.25 m	0.05 N-m
0.4 N	0.2 m	0.08 N-m	0.5 N	0.16 m	0.08 N-m
1.0 N	0.05 m	0.05 N-m	0.5 N	0.1 m	0.05 N-m

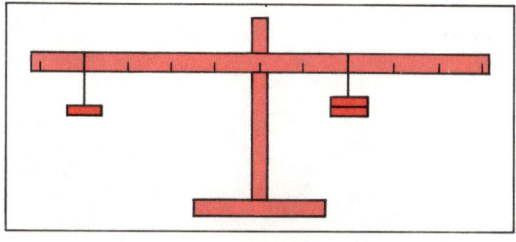

The weight multiplied by the distance from the pivot (or fulcrum) is called the **moment** of the force.

In all the above cases the moment of the weight on the left hand side equals the moment of the weight on the right hand side.

Example. A girl has picked some berries but has no machine for weighing them. She has balanced the bowl of berries against a 1 kg bag of sugar. With the sugar at 1 m from the fulcrum, the bowl must be placed 40 cm from the fulcrum.

The bowl when empty put at 1 m from the fulcrum balances the sugar at 50 cm. The diagram shows how the mass of the berries (2 kg) can be found.

Balance

In a chemical beam balance and also in household scales, the two scale pans are usually set at equal distances from the pivot. So the downward force of gravity on an object placed in the left hand balance pan (its weight) will be equal to the downward force on standard weights put in the right hand pan.

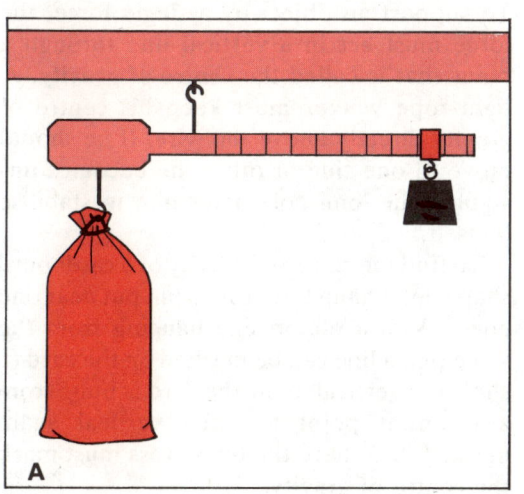

A

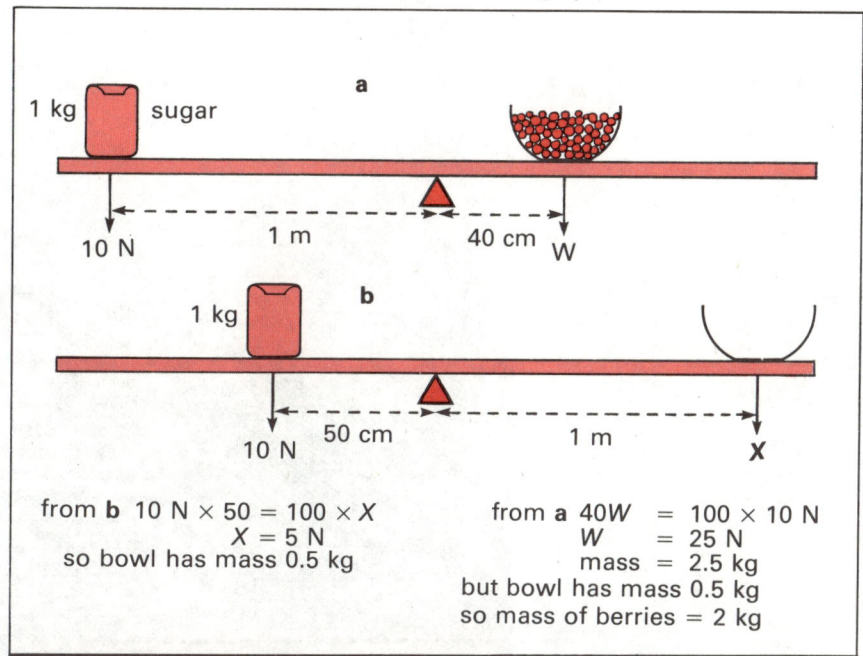

from **b** 10 N × 50 = 100 × X
X = 5 N
so bowl has mass 0.5 kg

from **a** 40W = 100 × 10 N
W = 25 N
mass = 2.5 kg
but bowl has mass 0.5 kg
so mass of berries = 2 kg

Example
How to find the weight of some berries without a weighing machine

1 The diagram (B) below shows a windlass. Hanging on the end of the rope is a mass which gives a downward force of 100 N. What force (F) should be exerted on the handle to raise the load?

2 The diagram (A) below shows a type of balance. How does it differ from the household scales.

Windlass

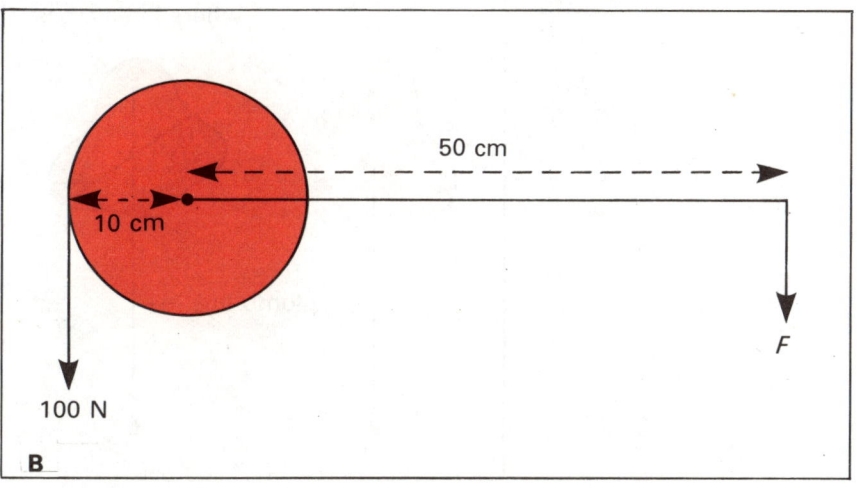

B

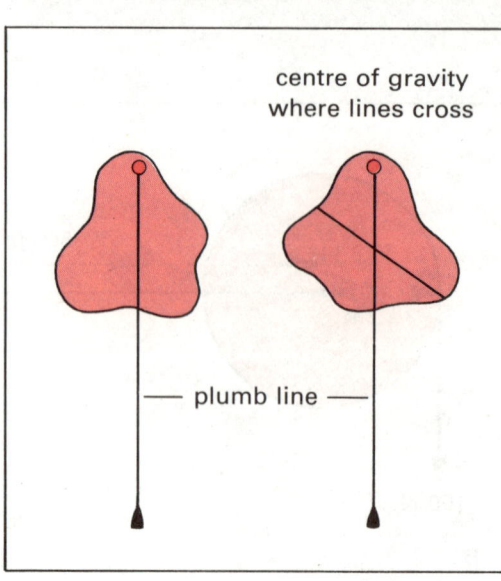

Centre of gravity is where the two lines cross

centre of gravity
where lines cross

— plumb line —

Centre of gravity

To support an object by a single force, the force must act in a vertical line through a point that is called the **centre of gravity**. A tight-rope walker must keep his centre of gravity directly above the wire. If he should move to one side or other, he becomes unstable. The long pole helps him to stabilise himself.

To find the centre of gravity of a cardboard shape, let it hang free from a pin put near one edge. With a plumb line hanging from the same pin, a line can be marked on the card to show the vertical. Now the card is hung from a different point and the vertical again marked in. Where the lines cross must mark the centre of gravity.

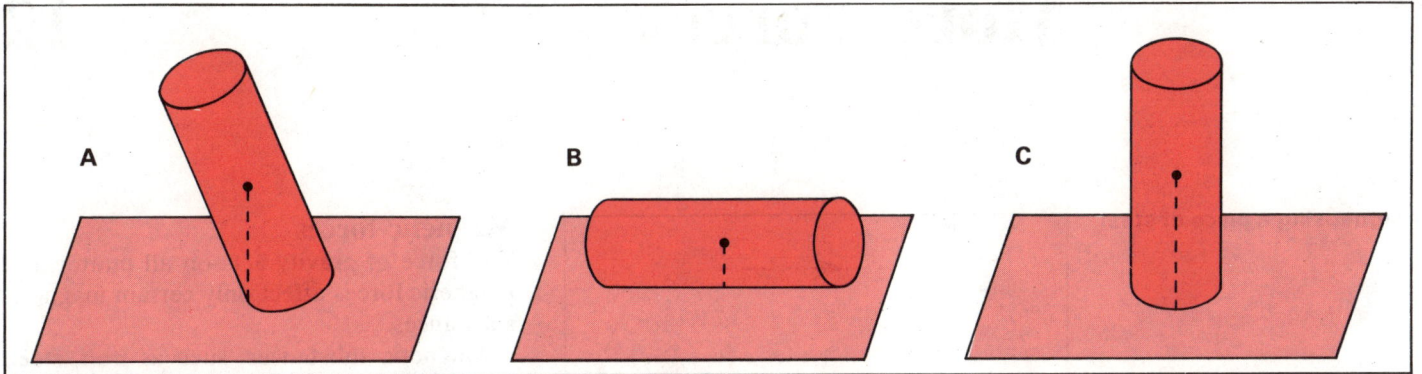

Stability

An object keeps its position so long as the vertical line through its centre of gravity falls inside the base on which it rests. But if the centre of gravity falls outside this boundary line, the object is unstable and will topple. This is because the centre of gravity always moves lower if it can.

An object can be in any one of three equilibrium positions. In A, the balance is critical – the least push sideways and the object topples. The push lowers the centre of gravity. This is unstable equilibrium.

In B a sideways push moves the object but the centre of gravity remains at the same height, so the object is still stable. This is a case of neutral equilibrium.

Now look at C. A slight sideways push will raise the centre of gravity higher, so when released the object will fall back into its first position. This is stable equilibrium. The object is in a stable position unless it is disturbed so far that its centre of gravity moves beyond the perpendicular line marking the edge of its base.

The centre of gravity of a bus should be as low as possible. This will reduce the chance of the bus toppling over, e.g. when cornering, especially if most of the passengers are on the upper deck.

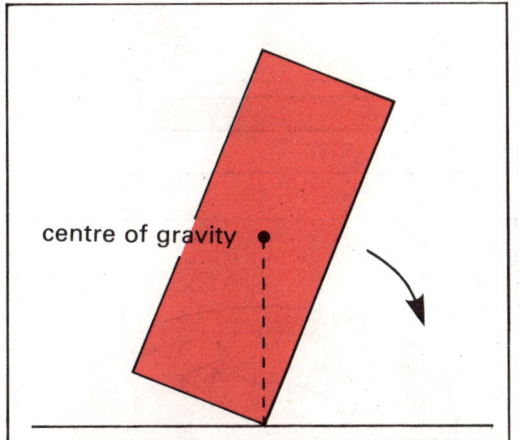

centre of gravity

Stability

An object is unstable if a line downwards from the centre of gravity falls outside the edge of the base

1 In the following diagram which would be least likely to topple over: **a, b** or **c**? X marks the position of the centre of gravity. For each of these cases, measure the angle through which the vehicle could be pushed over without toppling.
2 You are given a large wheelbarrow to transport a very heavy bag of sand and ten house bricks to a neighbour's house. Draw a diagram to show where the load would best be placed. Give your reason for this.

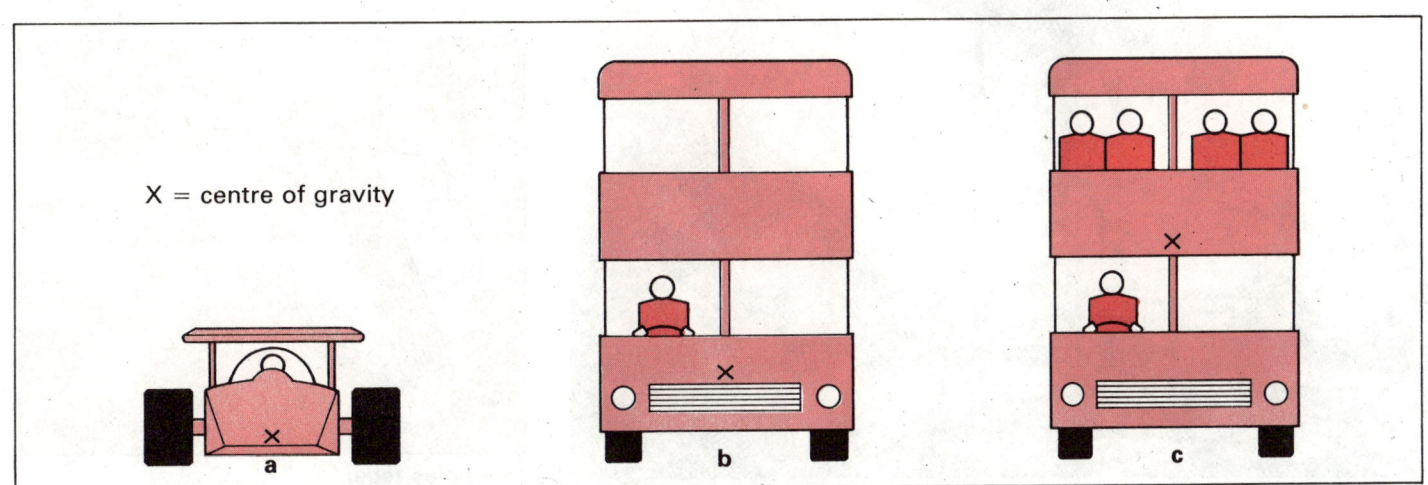

X = centre of gravity

More Forces

Magnetising a piece of steel

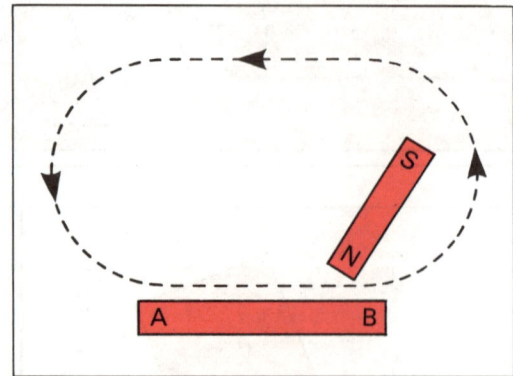

The magnetised steel can become a compass

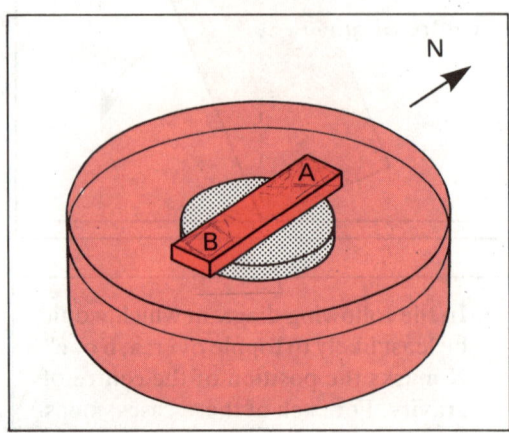

This magnet is used to lift scrap iron

Magnetic forces

The force of gravity acts on all matter, but magnetic forces affect only certain magnetic substances.

Magnetic substances, such as iron, steel, cobalt, nickel and a few special alloys can be magnetised. Copper, lead and aluminium are not magnetic substances.

An iron nail, for instance, is attracted by a magnet and can be picked up by it.

Making a magnet

A piece of steel AB is magnetised by stroking it a few times with the north end of a magnet following the path shown by the dotted line. If the piece of steel is then put on a large cork, which is then floated in water, it comes to rest in a north–south direction, with end A pointing towards the north. So a north-seeking pole has been made at A and a south-seeking pole at B. The north-seeking pole of a magnet is called the north pole for short.

Another way of making a magnet is by using an electric current flowing in a coil around the piece of steel (see Topic 90).

Attraction and repulsion

With your magnet still floating on the cork, bring near to it the north pole of another magnet. Putting this north pole near each of the floating magnet's poles in turn you will find that the north pole pushes away (repels) another north pole, but attracts a south pole. A south pole repels another south pole but attracts a north pole.

So **like poles repel, unlike poles attract.**

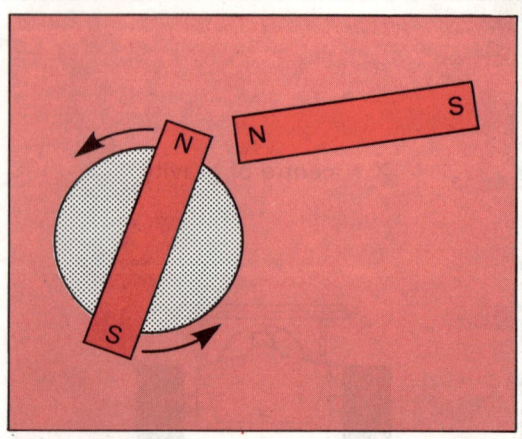

Like poles repel

Steel and soft iron

Pick up as many steel nails as you can with a magnet. Holding the top nail, take the magnet away. Most of the nails stay together. Each steel nail has become a little magnet by being near the big magnet. They have become magnetised by **induction**.

If you try the same experiment with soft iron pieces, you will find that these pieces are magnetised by induction more easily, more pieces are held. But when we take the magnet away, all the soft iron pieces fall down. Soft iron does not keep its magnetism after the magnetising magnet has been taken away.

Being attracted towards a magnet does not prove that a piece of metal is magnetised. The real test is whether one end of the metal being tested is repelled by a pole of a magnet.

The earth as a magnet

In the first experiment the floating magnet points north and south. It acts like a compass needle. In fact a magnetic compass is nothing more than a magnetised needle free to swing. The earth behaves as a giant magnet attracting one end of a compass needle and repelling the other. But since the north pole of the compass needle points to the north, there must be a south pole at the earth's 'north pole'. This is confusing, but we can avoid the difficulty if we speak of the pole of the compass that points northwards as the north-seeking pole.

It is possible to make a magnet by holding a piece of steel in a north-south line and hammering it. Files in a workshop are often found to be magnetised if used in this direction.

Demagnetising

Magnetism can be removed by heating the magnet or hammering it, the magnet in this case is best placed in an east–west direction. Another way of destroying magnetism is by holding the magnet in a coil through which a slowly-decreasing alternating current is passed.

Electrical forces

Take a strip of plastic such as polythene or perspex and rub it against some woollen material. Then bring the strip near some small scraps of tissue paper.

The scraps of paper will be attracted towards the plastic. You can make a kind of 'compass' needle from a strip of paper or light card suspended on a pin. When you bring the rubbed plastic strip near each end, it is found that only attraction is possible.

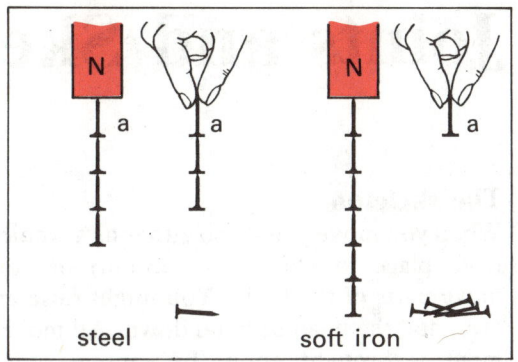

steel   soft iron

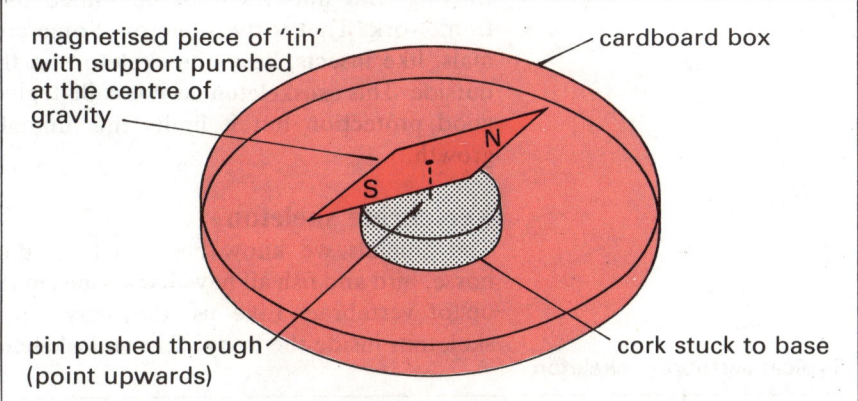

magnetised piece of 'tin' with support punched at the centre of gravity

cardboard box

N

S

pin pushed through (point upwards)

cork stuck to base

Making a compass

There is no sign of repulsion. So although this electrical force looks at first sight as if it is magnetism, it is not.

Here is some magic that you can try! A fine jet of water is coming from a tap but it does not go into a beaker placed below because the beaker is not quite underneath. However, if the rubbed plastic strip is held near the jet, it will attract the water so that it now flows into the beaker.

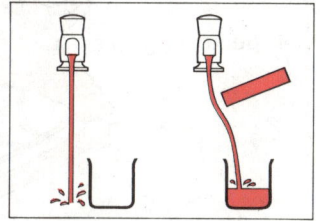

A rubbed plastic strip attracts tissue paper or a jet of water

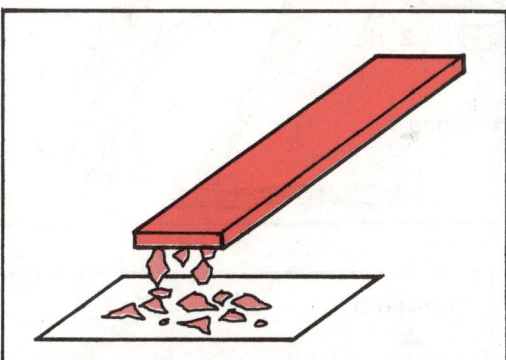

1 A metal rod, A, is not attracted by a magnet. Both ends of rod, B, are attracted by the north-seeking pole of the magnet. One end of rod, C, is attracted, the other end is repelled by the north pole of the magnet. Write down which rod you think is: steel, soft iron, aluminium.

The skeleton

When you move you do so either as a whole, from place to place (locomotion) or you move parts of the body. You might raise an arm, nod the head or bend down. All movement is brought about by muscles contracting. But muscles must be joined to a framework. That is the skeleton. Some animals, like insects, have the skeleton on the outside. This **exoskeleton**, as it is called, gives good protection but it limits the animal's growth.

Vertebrate skeleton

The animals we know best, such as dog, horse, bird and fish all have backbones made up of vertebrae. Like us, they have their skeletons inside the body. This **endoskeleton** supports the body and protects some parts, especially the skull housing the brain. But if all parts of the body were protected as well as the brain is, we should be as heavy as a man in armour!

The vertebrate skeleton is made up of cartilage and bone. Some fish, such as the shark, have skeletons made only of cartilage (often called gristle). **Cartilage** is tough and pliable and is useful in places where some movement is needed, e.g. our external ear (the pinna) and the tip of the nose.

Bone is a firmer material, stronger though less pliable than cartilage. All animals with backbones are built on the same plan. This basic plan, shown simply in the diagram below can be traced in the skeletons of the whale (shown here), the lion and the frog.

1 **Vertebral column** – made up of several small bones (vertebrae) with cushioning discs of cartilage between them.

2 **Skull**

3 **Ribs** – most of which link the backbone to the sternum (breastbone)

4 **Pectoral girdle** (shoulder) – made up of clavicles (collar bones) and scapulas (shoulder blades)

5 **Pelvic girdle** (hip joint) – made up of three pairs of bones:
   ilium – the upper edges are the hips we can feel.
   ischium – these are the bones we sit on.
   pubis – the pubes can be felt at the lower end of the abdomen in front

6 **The limbs** – arms and legs have similar structures.

Typical vertebrate skeleton

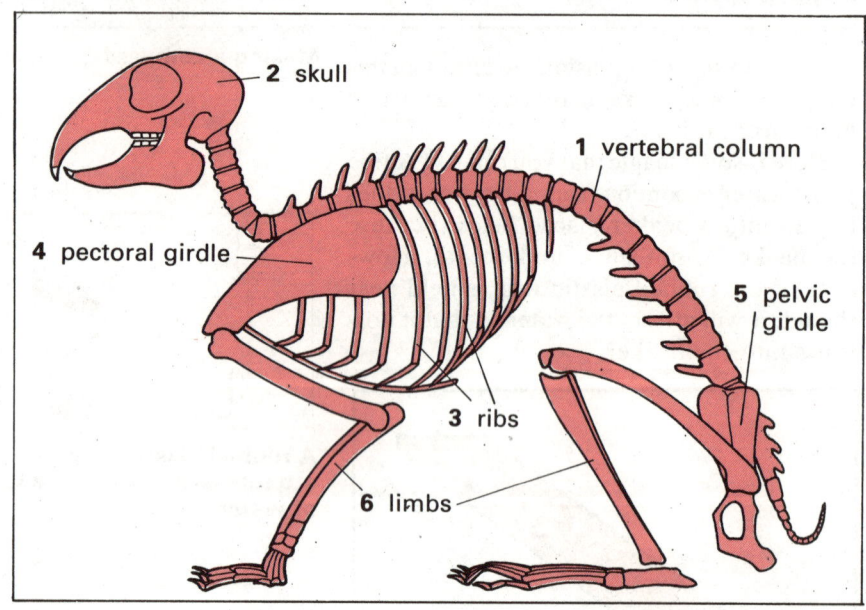

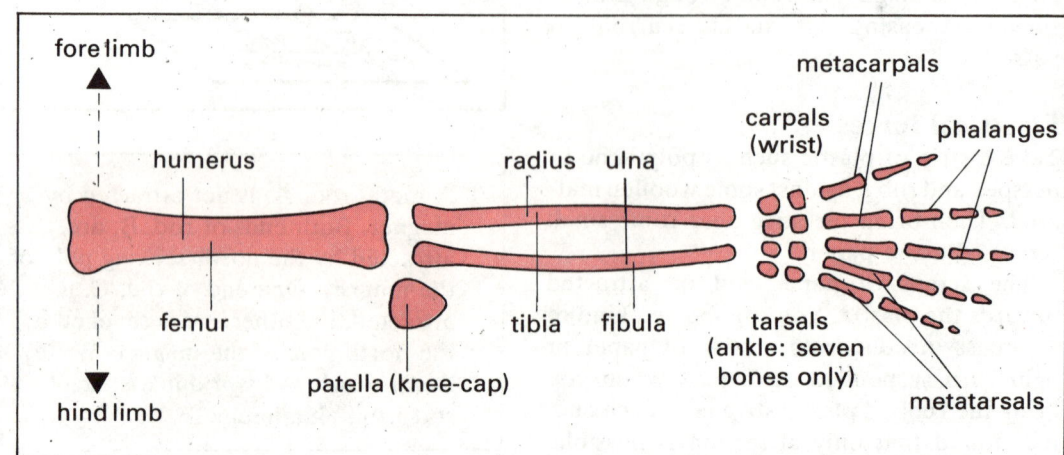

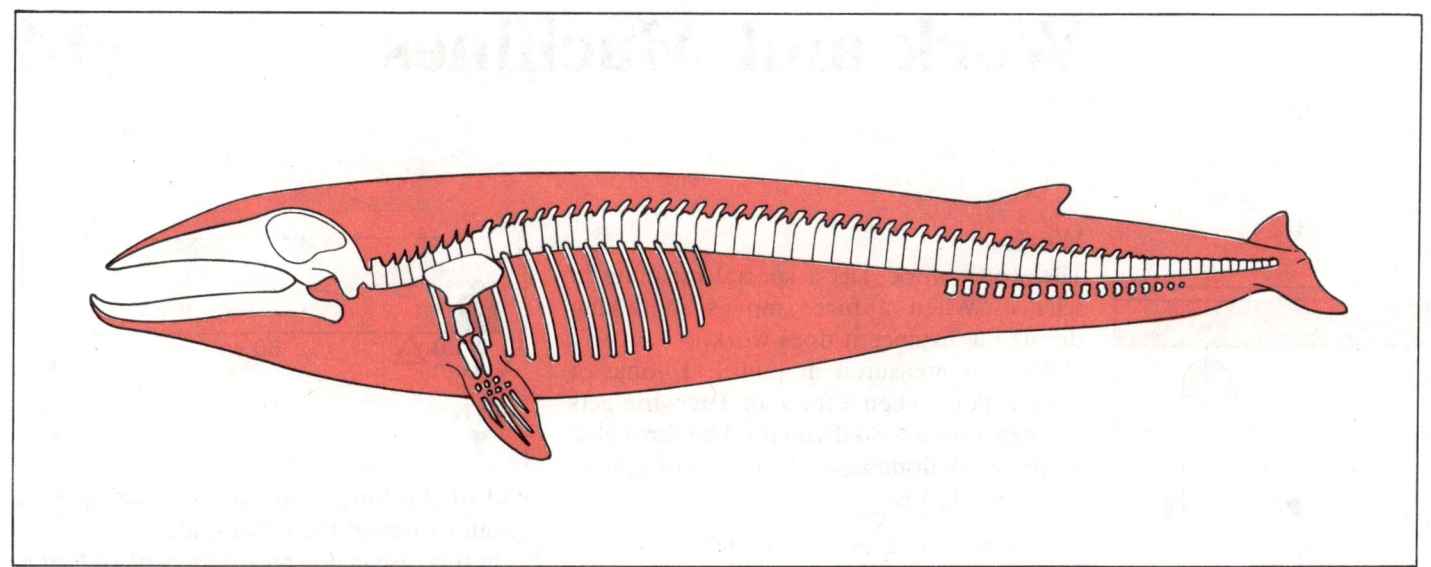

Joints

Bones are joined to each other in such a way that moving is easy. Some joints make limited movement only. The bones of these joints are linked with fibrous tissue or cartilage between them. Other joints where free movement is needed are synovial joints, at which the smooth ends of neighbouring bones glide over each other. These are lubricated by a synovial fluid.

Joints can be classified as:

Hinge joints e.g. knee, elbow, fingers
Double-hinge joints e.g. thumb
Ball-and-socket joints e.g. shoulder, hip
Compound joints e.g. wrist, ankle. These make a wide range of movement possible since there are several joints in one.

Muscles

Bones are moved by muscles attached to the bones.

Most muscles are attached to a bone at each end, the muscle merging into a strong, tough tissue, called a **tendon**. The other end of the tendon merges into the substance of the bone. Some muscles are not attached to bone, e.g. the ring of muscle at the lower end of the stomach, the pyloric sphincter. When this muscle contracts the opening from the stomach into the intestines is closed.

Muscle action

When the biceps muscle in the front of the upper arm contracts the arm bends at the elbow. The muscle fibres shorten, so that the bones of the lower arm are brought out and up. The biceps muscle is now swollen and feels hard. At the same time, the muscle at the back of the upper arm, the triceps, relaxes, so that the biceps contracts smoothly.

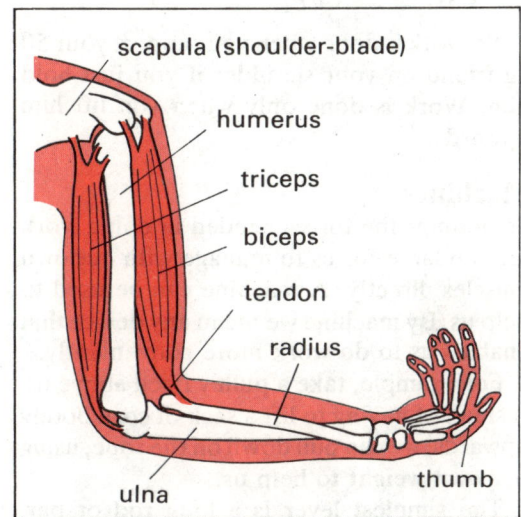

Elbow joint

This balancing of two sets of muscles taking each other's strain is called antagonistic action. It is found all over the body, so that moving is smooth. The damage that sudden contraction could do is avoided. When we straighten and extend the arm, the reverse takes place and one can feel the triceps at the back of the upper arm getting hard and taut.

When the triceps contracts and the biceps relaxes, the arm is straightened.

Summary

Functions of the skeleton are:
1 Support
2 The shape of the body depends on the skeleton.
3 Good protection for the brain. Limited protection (ribs) for heart and lungs
4 Attachment of muscles – making different kinds of movement possible.

Skeleton of whale
Compare this skeleton with the labelled skeleton on the opposite page: it has evolved differently because the whale lives in the sea

—C

Work

The word 'work' has a special meaning in science. When a force moves something through a distance it does work.

Work is measured in joules: 1 joule of work is done when a force of 1 newton acts through 1 metre. So if you lift 3 kg through 2 m, the work done against the pull of gravity (10 N per kg) is:

30 N × 2 m	= 60 J	
Force × distance	= work	
newtons × metres	= joules	
(N) (m)	(J)	

No work is done when you support your 50 kg friend on your shoulder if you just hold him. Work is done only when you lift him upwards.

Machines

Sometimes the forces needed in doing work are too large for us to manage with our own muscles directly. A machine can be used to help us. By machine we mean any device that enables us to do work more conveniently.

For example, take a pulley fixed above us. Instead of having to lift a sack of corn bodily upwards, we can pull down on the rope, using our own weight to help us.

The simplest lever is a long rod or bar, pivoted so that by using a small force at the

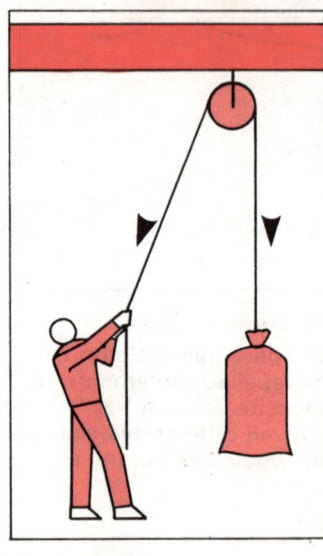

Single pulley

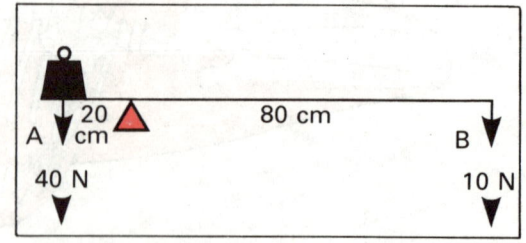

end of the longer portion, we can apply a greater force at the other end.

In this case, a downward force of 10 N at B causes an upward force of 40 N at A, so that 4 kg can be lifted. Without the lever, a force of 40 N would be required.

This might at first sight seem to be magic, but the effort has to be moved through 4 cm for every 1 cm the load is lifted.

Here are a few examples of things in everyday use, in which levers are used. There are three different classes of lever, depending on the relative positions of load (W), effort (E) and fulcrum (pivot) (F).

Look at the diagram of the arm lifting a weight. If a mass of 10 kg is lifted, what effort (T) is applied at E? (Remember – gravitational pull on 10 kg is 100 N.)

Suppose HB (distance from hand to elbow) = 35 cm and EB (distance from elbow to point of attachment of muscle) = 5 cm. Then $5T = 100 \times 35$, so $T = 700$ N. This is equivalent to about 1½ cwt!

Simple levers

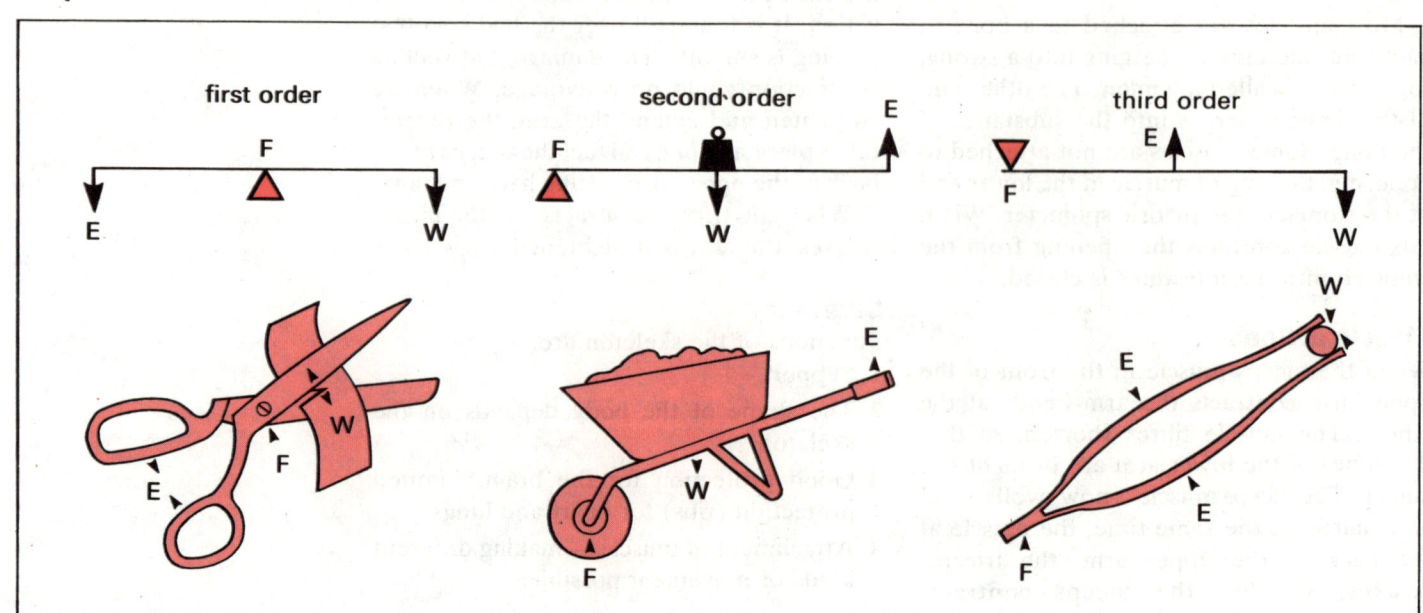

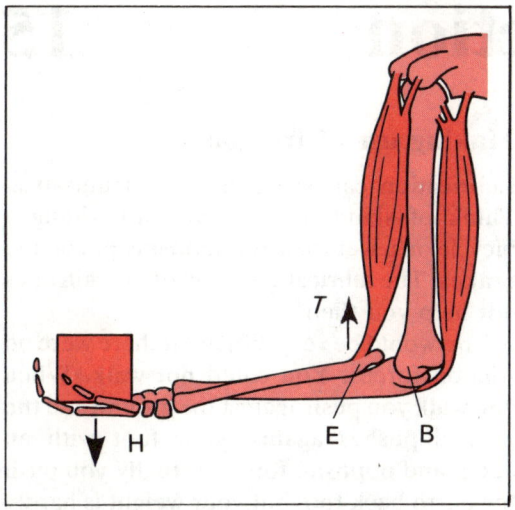

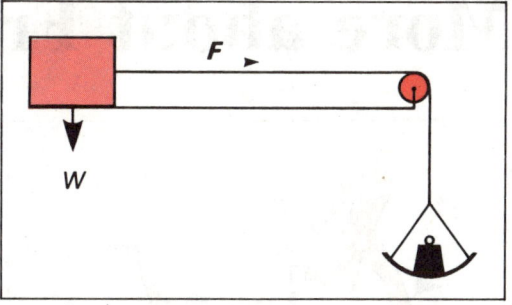

Limiting friction
When a block is just about to move the force (*F*) is the limiting friction force

Inclined plane

Sometimes it is easier to lift a load by moving it along a slope. Running cars on to a car transporter, for instance, needs a smaller force that would be needed to lift the cars up vertically. Perhaps you can think of some examples of this **inclined plane**. You may have seen empty barrels being lifted out of a cellar by drawing them up with a loop of rope.

Roads up the sides of mountains do not usually go straight up, but zig-zag to avoid too great a gradient. Walking up a slope to a mountain top takes much less effort than climbing straight up the rock face.

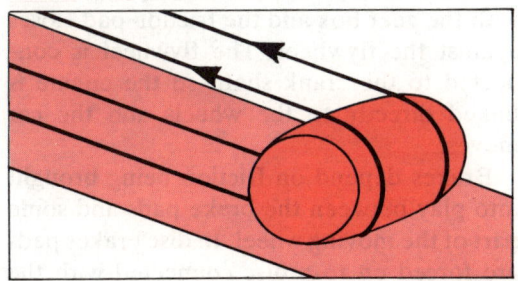

Friction

In practice, some of the work you put into a machine is wasted because of friction.

You can study friction by finding what force is just enough to start a block of wood moving. This force (*F*) acting on the block, weight (*W*), is called **limiting friction** force. Once the block is moving, a smaller force is enough to keep the block moving. This is called the **sliding friction** force.

The heavier the block (*W*) the larger the force (*F*). In fact for any two surfaces *F* is proportional to *W*.

The ratio *F/W* is called the coefficient of friction. It is about 0.4 for wood on wood, about 0.2 for metal on metal and as low as 0.1 for metal on ice.

Overcoming friction

About one half of the work done in running an engine is wasted through friction. This can be reduced by lubricating the moving parts with oil or graphite.

You can see why this is if you think of a 'smooth' surface highly magnified. The rough parts make it difficult for another surface to glide over it. But with oil in between, the surfaces now move over liquid and so friction is less.

Another way of reducing friction is by using 'rolling' things. A heavy load is more easily moved when on rollers. Rolling castors on furniture make moving easier. Ball bearings are commonly used to reduce friction in moving parts, e.g. in the bearing of a cycle wheel.

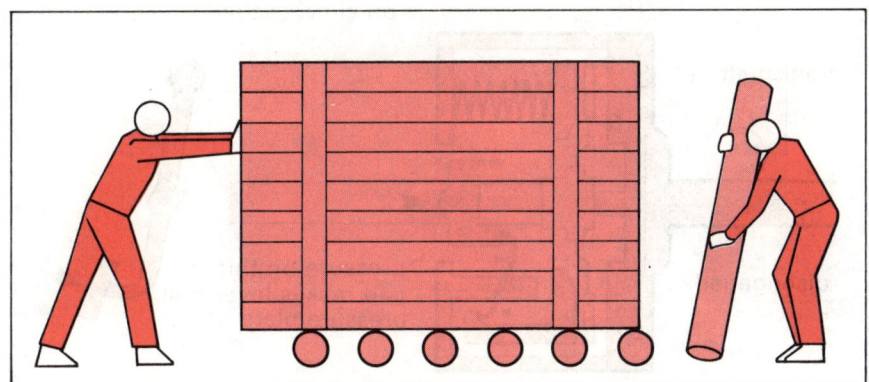

1 Draw sketches to show how you would (a) use a pair of pliers, (b) hold a fishing rod when fishing, (c) hit a ball with either a cricket bat or a golf club. In each case, mark the positions of fulcrum F, load W and effort E.
2 A claw hammer is used to pull a nail out of a piece of wood. The nail is 2 cm from the pivot end of the hammer and a force of 100 N is applied at the end of the handle which is 20 cm long. What force is exerted on the nail?

More about Friction

No friction?
As there is very little friction between boat and water, will he land on the shore?

Making use of friction

Lubrication can sometimes be a nuisance. Think of skidding when you are riding a bicycle on a wet road and suddenly put on the brakes! The lubricating effect of the rain does not help you then.

Life would be very difficult if there were no friction forces. You could not walk. When you walk you push against the ground and the ground pushes against your foot with an equal and opposite force. Actually you push the earth back too, but your weight is hardly likely to move the earth out of its orbit (see Topic 10). Cars would slide to the bottom of the hill and stay there.

To make full use of friction forces we sometimes increase friction by making surfaces rougher. When walking on a slippery surface it is easier if your shoe soles are rough rubber rather then smooth leather. The tracks of bulldozers and tanks are heavily studded for the same reason.

Two interesting uses of friction in a motor car are the clutch and the brakes.

When the clutch pedal is pushed down it causes the springs to be lifted away from the friction pad. But when the clutch pedal is released, the springs push the plate linked with the gear box and the friction pad closes against the flywheel. The flywheel is connected to the crank shaft, so the engine is linked directly to the wheels and the car moves.

Brakes depend on friction being brought into play between the brake pads and some part of the moving wheel. In disc brakes pads are forced on to a disc connected with the wheel. These pads are on both sides of the

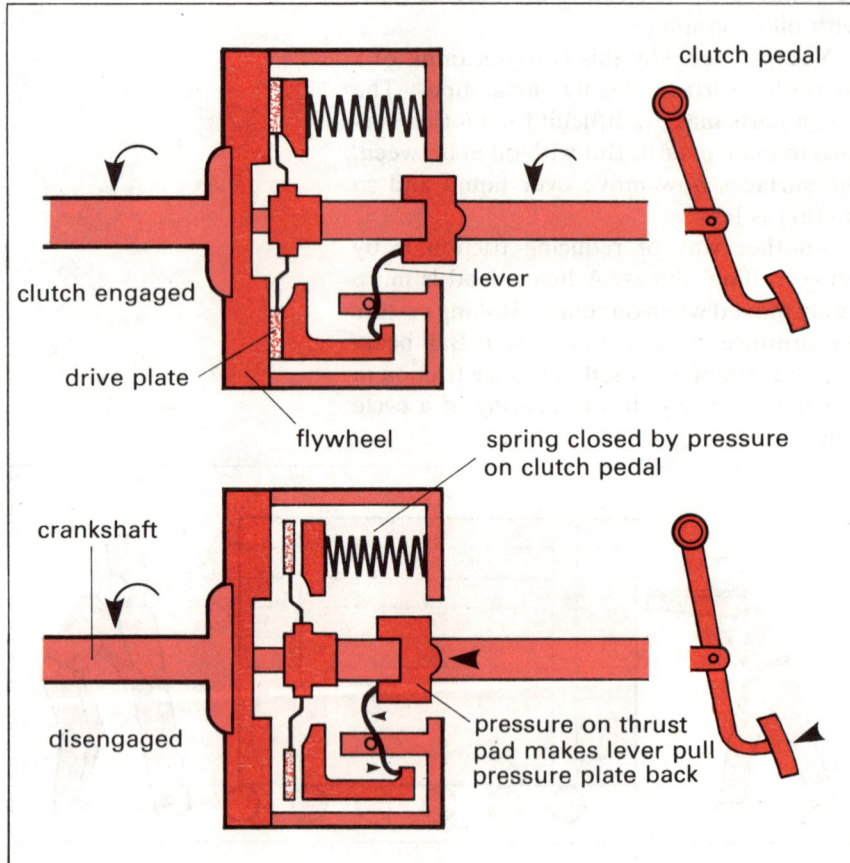

clutch pedal

clutch engaged

drive plate

flywheel

lever

spring closed by pressure on clutch pedal

crankshaft

disengaged

pressure on thrust pad makes lever pull pressure plate back

Clutch
The wheels of a car only turn round because of friction between the clutch plates

Brakes
A car is stopped by the friction between the brake pads and the disc

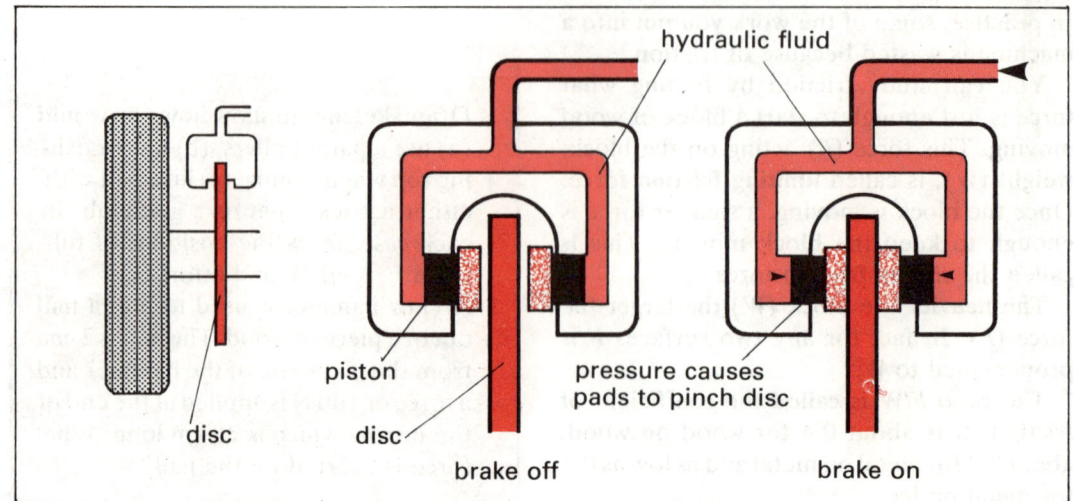

hydraulic fluid

piston

disc

disc

pressure causes pads to pinch disc

brake off

brake on

revolving disc. When the brake pedal is pushed down, hydraulic pressure causes the pads to be pushed against the disc like pincers. This slows down the disc and hence the car.

The hovercraft

Friction makes moving through water very difficult even if the animal or ship is streamlined. Greatest speeds are reached by those craft that skim the surface. In the case of the hovercraft it is as if the vessel is lubricated with a layer of air between itself and the surface of the water. A fan keeps a layer of air at a steady pressure, between the bottom of the craft and the water surface. Much of this air is trapped to a certain extent by the 'skirts' around the edge of the hovercraft.

An interesting simple model to show how the hovercraft works can be made from a piece of balsa wood with a cork glued on. This carries a glass tube, through which a balloon sends a stream of air down to the space between the balsa wood and the surface it is on.

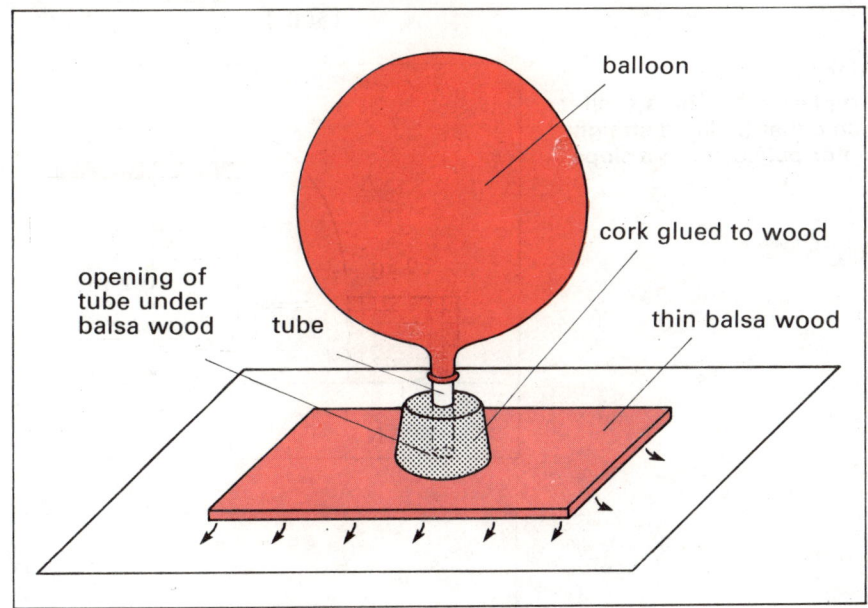

Friction and heat

You will have noticed that if you rub your finger along the table rapidly it gets hot. After sawing or drilling, the saw or drill can be very hot indeed. This heating effect of friction cannot be avoided and it is a loss of energy. In a motor car, it makes a cooling system necessary – usually water cooling. Sometimes air cooling is used but in either case a cooling fan is needed.

Even the friction between a moving object and air can produce tremendous heat. Because of this a space craft returning to earth must have a shield of heat-resisting material.

On a clear night one often sees 'shooting-stars'. These are meteors – lumps of material from outer space that have fallen into the earth's atmosphere and have become white-hot through friction. They usually burn up completely although some pieces may reach the earth as meteorites.

Efficiency of Machines

Efficiency of machines

You can learn more about machines if you look at typical values of the forces involved.

Suppose you have to lift a crate of mass 50 kg from ground level to a shelf, 2 m above ground level. The force needed would be equal to the pull of gravity (10 N per kg) on the crate. This would mean a force of 500 N moving through 2 m upwards.

Useful work done = force × distance through which it acts = 1000 J. But it would be easier to pull the crate up to the shelf along a sloping (inclined) plane. If the slope is 6 m long and the force needed to pull the crate up is a steady 250 N, then:

$$\text{work done} = \text{force} \times \text{distance through which it acts}$$
$$= 250 \text{ N} \times 6 \text{ m}$$
$$= 1500 \text{ J}$$

You have the advantage of not needing such a large force.

$$\text{Mechanical advantage} = \frac{\text{load}}{\text{effort}}$$
$$= \frac{500}{250} = 2$$

but you have to move your effort further than the load is moved.

$$\text{Velocity ratio} = \frac{\text{distance moved by effort}}{\text{distance moved by load}}$$
$$= \frac{6}{2} = 3$$

Of the 1500 J work you do in your effort, only 1000 J is used usefully. So the efficiency of the inclined plane in this case is $\frac{1000}{1500}$

Efficiency is usually stated as a percentage:

$$\frac{\text{useful work done}}{\text{work put into machine}} \times 100\%$$
$$= \frac{1000 \times 100}{1500} \%$$
$$= 66.7\%$$

Thus only two-thirds of the work put into the machine went towards lifting the load. The rest was involved in overcoming friction and some energy was changed into heat and perhaps sound.

Machines

To place a box on a shelf it can either be lifted straight up or pulled along a slope

50 kg

2 m

500 N

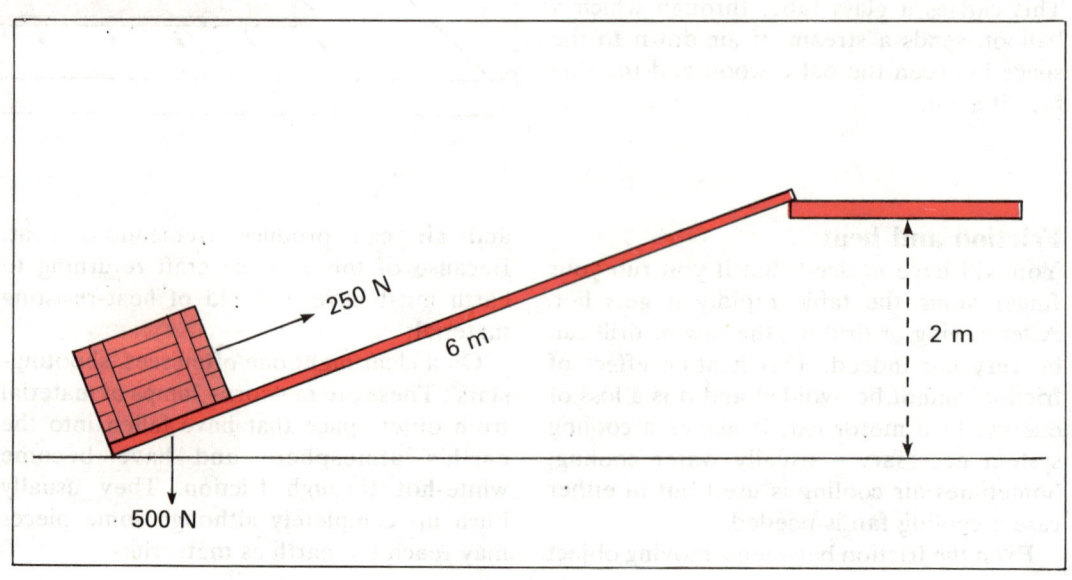

250 N

6 m

2 m

500 N

Pulley system

In the simple system shown below, pulley A is free to move, being held in the loop of string. Pulley A has the load attached to it. Pulley B is fixed.

When the effort (E) pulls up the load (W) 1 cm higher, both parts of the string round pulley A must shorten by 1 cm. So 2 cm pass over pulley B.

$$\text{Velocity ratio} = \frac{\text{distance moved by effort}}{\text{distance moved by load}}$$
$$= 2$$

If a load of 100 N can just be lifted by an effort of 70 N:

Useful work done lifting load 1 m	$= 100\,\text{N} \times 1\,\text{m}$
	$= 100\,\text{J}$
Work done by effort	$= 70\,\text{N} \times 2\,\text{m}$
	$= 140\,\text{J}$
Mechanical advantage	$= \dfrac{100}{70}$
	$= 1.4$
Efficiency	$= \dfrac{100 \times 100}{140}\%$
	$= 71\%$

It might seem possible that a force of only 50 N could lift the 100 N weight, but the effort has to overcome friction and also lift the pulley (A).

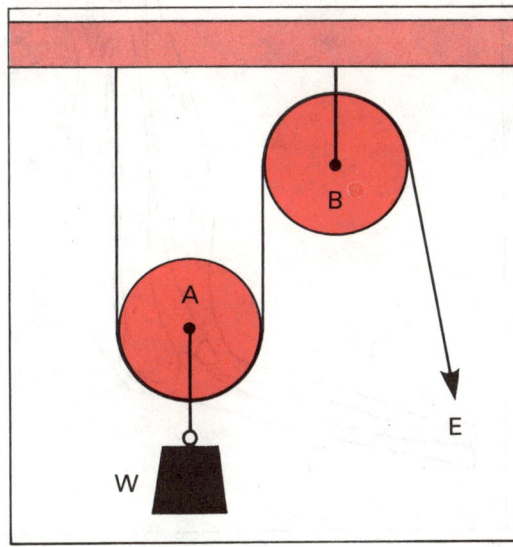

Gear wheels

With different numbers of teeth on two wheels, there is a mechanical advantage whether they engage directly or are linked by a chain, as in the bicycle.

In the gear box of a motor car, different ratios of teeth can be selected by the gear lever.

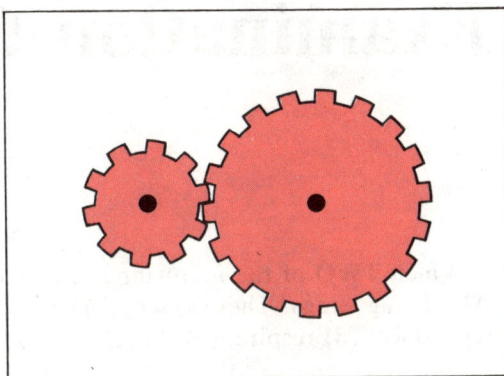

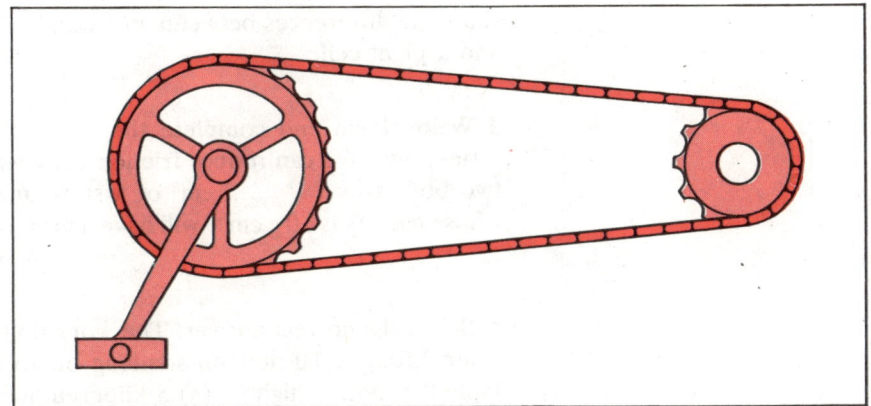

Rate of working

In all the discussion of work so far, we have not thought about speed of working. If 1 joule of work is done in 1 second, the rate of working is 1 watt:

$$\text{so watts (W)} = \frac{\text{joules}}{\text{seconds}} = \frac{\text{J}}{\text{s}}$$

Here is an experiment you can do if you have a stopwatch. A boy of mass 50 kg (weighing 500 N) runs up a flight of ten steps, each 20 cm high, and reaches the top in 2.5 seconds. What is his rate of working?

Vertical distance	$= 20 \times 10\,\text{cm}$
	$= 200\,\text{cm} = 2\,\text{m}$
Work done	$= 500\,\text{N} \times 2\,\text{m}$
	$= 1000\,\text{J}$
Rate of working	$= \dfrac{1000\,\text{J}}{2.5\,\text{s}} = 400\,\text{W}$

(Since 746 W are the same as 1 horse-power, this works out to just over ½ horse-power.)

1 A crane raised 25 kg through 10 m. How much work was done?
2 In the block and tackle shown on the right, there are two pulleys in each block and the velocity ratio is 4. Find the mechanical advantage and the efficiency of the example shown.

Gears

Gear wheels can work when the teeth touch each other (left), or when the teeth are connected by the chain (as in a bicycle; below)

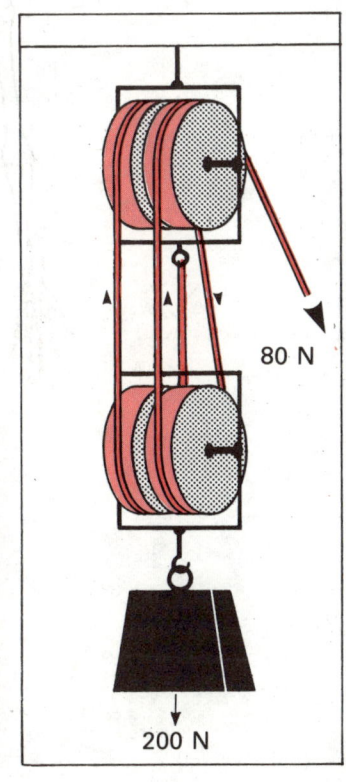

80 N

200 N

Examination Questions: I

1 Which TWO of the following are true of ALL living things? They (a) see, (b) hear, (c) reproduce, (d) respire, (e) think.   . ALS

2 Draw TWO fully labelled diagrams to show the differences between an animal cell and a plant cell.   WM

3 Write down and complete the following statements: We can reduce friction between two bodies by   *   . 5 cm³ of a substance whose density is 10 g cm⁻³ will have a mass of   *   WYL

4 Select the correct answer: The work done when lifting a 10 newton sandbag on to a table 0.5 metres high is: (a) 5 kilogrammes, (b) 50 joules, (c) 5 joules, (d) 2 joules, (e) 5 grams.   Y

5 Give the names of the parts labelled A, B, C, D, E, F and G on diagram 1. Draw and label a diagram of a simple hinged joint and state two places where such a joint may be found on the diagram.   SE

6 Select the correct answer: The mass of an object would be measured in units called the: (a) kilogram, (b) watt, (c) metre, (d) joule, (e) newton.   EM

7 Which two plants never flower? (a) Horse-chestnut tree, (b) mushroom, (c) grass, (d) groundsel, (e) fern, (f) tomato.   EM

8 Look at diagram 2 of the human arm shown:
(a) name the muscles labelled A and B
(b) name the bones C, D and E
(c) name the structure labelled F.   Y

Diagram 1

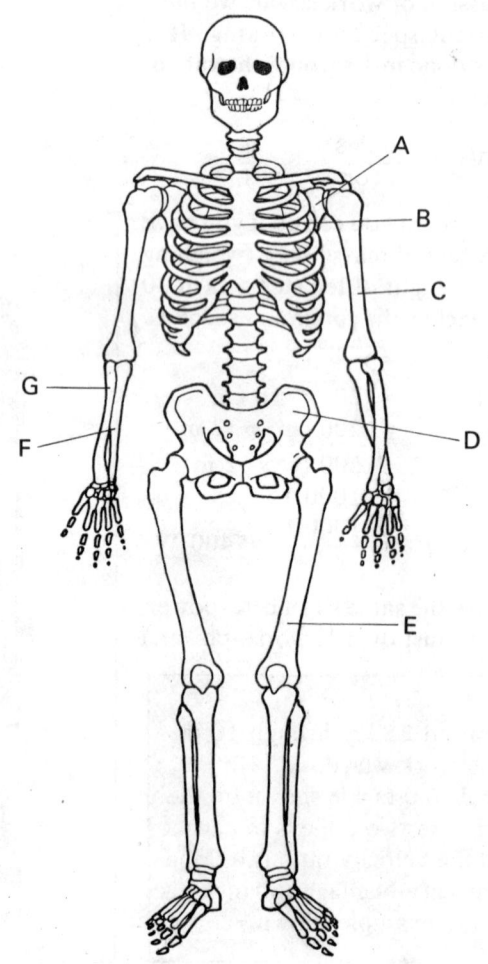

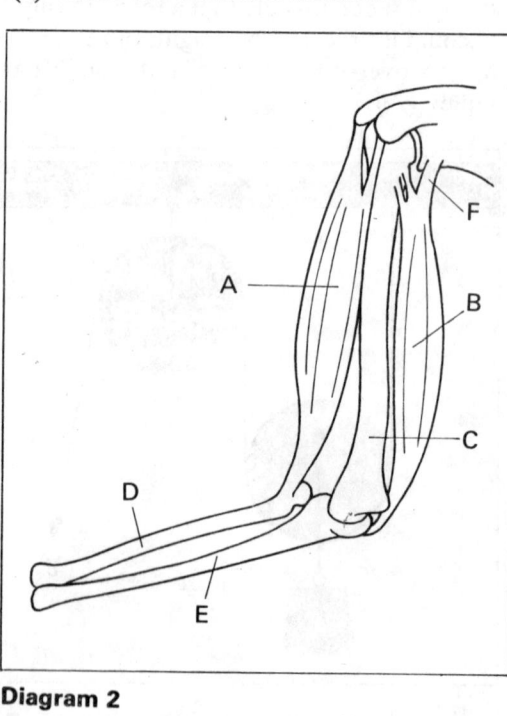

Diagram 2

9 Look at diagram 3 which represents six objects (i) to (vi). Mark with a X the centre of gravity of each of the objects. Objects (i), (ii), (iii) and (iv) are flat; objects (v) and (vi) are like a bunsen burner.
 Explain why object (v) is stable, but object (vi) is unstable.   M

Diagram 3

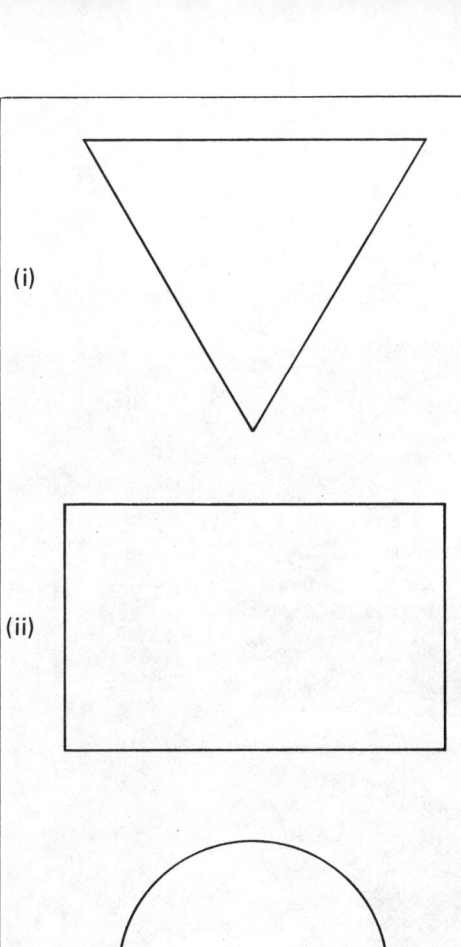

(i)

(ii)

(iii)

(iv)

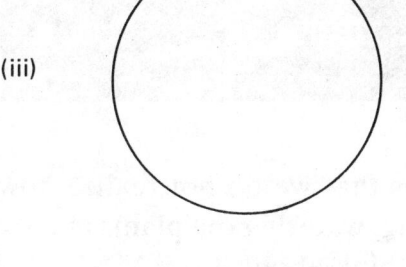

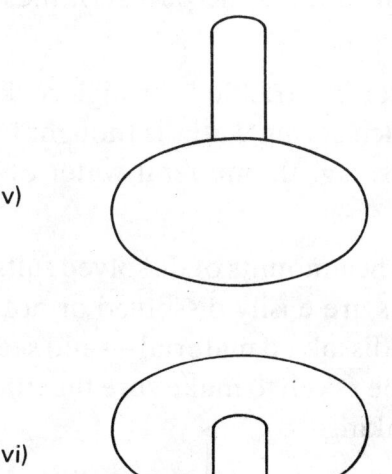

(v)

(vi)

10 Select the correct answer. A piece of metal (density 2 400 kg/m³) has a mass of 600 kg. Its volume is: (a) 250 cm³; (b) 400 cm³; (c) 0.25 m³; (d) 4 m³.   NW

11 (a) Give one example of each of the following types of rock: (i) sedimentary; (ii) metamorphic; (iii) igneous.
(b) Which type of rock is associated with volcanoes?   SW (part)

12 Select the correct answer: The natural organic matter found in the soil is called: clay; loam; chalk; humus; ammonium sulphate.   ALS

13 Select the correct answer: Forces (a) tend to produce motion, (b) always act downwards, (c) are due to gravity, (d) are measured in joules, (e) are measured in newtons.   ALS

14 Select the correct answer. When water is added to calcium oxide (quicklime) (a) carbon dioxide is given off, (b) no reaction occurs, (c) neutralization occurs, (d) a black precipitate forms, (e) heat is given out.   EM

15 Diagram 4 shows part of the body wall of a hydra. How does cell A get the food it needs from inside the gut?   WYL (part)

Diagram 4

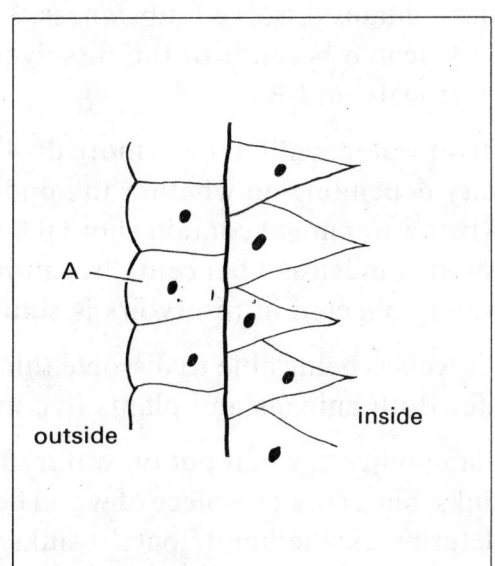

outside   inside

WATER

Buoyancy
Heavy ships float in water
because of the upthrust

Water is taken for granted so much in our lives that we do not realize how important it is. We use it to make tea, for washing, watering the plants and so on. It is only during a severe drought that we think about it.

About three-quarters of the earth's surface is water. Natural water is found as rain, spring water, lakes, rivers and oceans. Rain water is the purest of these, but it contains dust and gases dissolved from the air.

Water is able to dissolve substances. Spring water has trickled through rocks and contains dissolved substances. Sometimes such spring water is thought to be valuable because of the dissolved substances, e.g. the mineral water of Harrogate and Bath.

River water contains even more dissolved salts. The amounts of dissolved salts vary depending on whether the underlying rocks are easily dissolved or not. River water might contain about 0.04 per cent of dissolved material — and sea water as much as 4 per cent. Precautions have to be taken to make sure that the water collected in reservoirs is suitable for drinking.

As well as being able to dissolve things, water is also an important medium for life. Both animals and plants live in it.

Some objects, when put on water, float. Others sink. A lump of iron or lead sinks, but a cork or a piece of wood floats. It seems that the density of the object determines whether it floats or sinks. But a very heavy ship made mostly of iron floats.

Water and Buoyancy

Objects in water

A dense object such as a lump of metal does not seem to weigh as much when suspended in water as it does in air. To investigate this the lump must be completely under water, but not touching the sides of the container.

It seems that the water is pushing upwards on the metal; in other words, there is an **upthrust** on the metal. Does the metal push down on the water? Suppose you compare the weight of a beaker of water with its weight when the metal is suspended in the water. You will find that the metal does push down on the water with the same force.

Lower the lump of metal into a spouted can. You will find that the volume of water displaced (which equals the volume of the metal) weighs the same (1 N) as the loss in weight.

Archimedes' principle

The Greek, Archimedes (in about 250 B.C.) found by experiment that the upthrust or loss of weight is equal to the weight of liquid that the object displaces.

On a body partly or completely immersed in a fluid there is an upthrust equal to the weight of fluid displaced by the body.

The term 'fluid' means that the principle is just as true for other liquids and for gases as it is for water. A balloon filled with hydrogen, for example, is pushed upward by a force equal to the weight of the air displaced by the balloon. The balloon is floating in air.

Relative density

You now have a good way of finding the relative density of a solid.

Weight of a lump of metal = 8.1 N   **(a)**
Weight of metal completely immersed in water = 5.1 N   **(b)**
So upthrust = 3.0 N
But upthrust = weight of water displaced
and volume of water has same volume as metal.

$$\text{Relative density of metal} = \frac{\text{weight of metal}}{\text{weight of same volume of water}}$$
$$= \frac{8.1\ N}{3.0\ N} = 2.7$$

(See Topic 3 to find out if this metal could be aluminium or iron.)

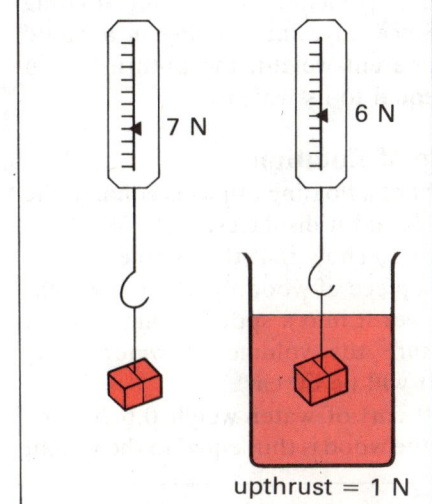

Upthrust
Compare the weight of an object suspended in air and water

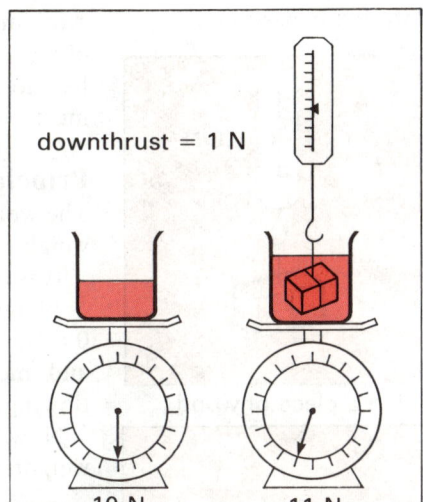

Downthrust (above)
Compare the weight of a water-filled beaker with and without an object suspended in it

Displaced water (left)
The weight of water displaced is the same as the loss in weight

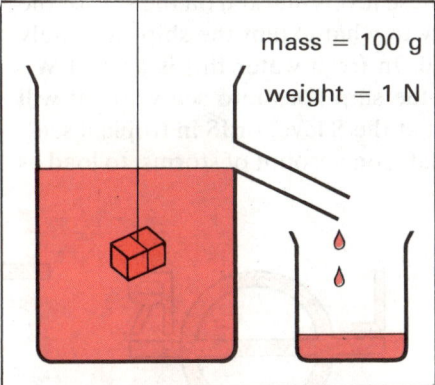

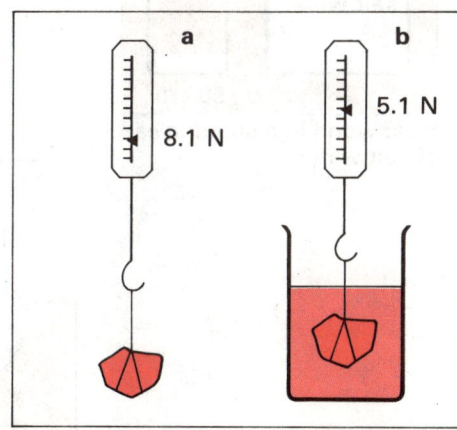

Finding the relative density of a solid

---

1 State whether the substances in the following list would float or sink when put into water: iron, cork, lead, wood, gold. Which of the substances have densities greater than that of water?

2 If you had a sheet of aluminium foil, how could you arrange that it floated when put on water?

3 What would be the mass of 1000 cm³ of gold (relative density 19)?

Flotation experiment

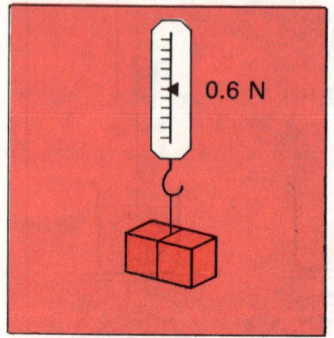

Weigh a piece of wood

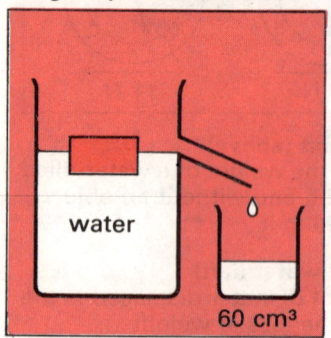

Float wood in a spouted can of water

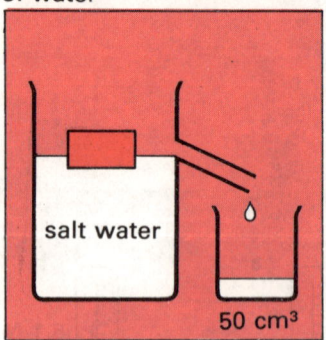

Float wood in a spouted can of salt water

Archimedes' principle is true for floating objects. Since anything floating in a liquid has no apparent weight, the upthrust on it must be equal to its weight.

Principle of flotation

The weight of a floating object is equal to the weight of liquid it displaces.

It is easy to show that this is true.

Weigh a piece of wood. Suppose it weighs 0.6 N. Lower it into a spouted can of water and measure the volume of water overflowing. It will be 60 cm³.

Now 60 cm³ of water weigh 0.6 N. The weight of the wood is thus equal to the weight of water it displaces when floating.

Suppose now, the wood is floated on salt water of relative density 1.2. The wood does not float quite so far down in the salt water. It does not displace quite so much of the liquid. It displaces 50 cm³.

50 cm³ of salt water weighs 0.5 N × 1.2 = 0.6 N. The principle is still true. The weight of salt water displaced is equal to the weight of the floating block.

If we are told that the **displacement** of a ship is 80 000 tons, this means that it displaces 80 000 tons of water. It follows that the ship itself weighs 80 000 tons.

Plimsoll line

Ships have a series of lines painted on their sides. The depth to which an object like a ship floats depends on the relative density of the liquid. These levels marked on the side of the ship show to what extent the ship can safely be loaded. In fresh water this is up to FW.

But if the ship goes into sea water it will now float at the S level or IS in tropical seas. It is not safe, on account of storms, to load as high in winter as in summer. W is for winter, and WNA for winter in the North Atlantic.

A ship made of steel can float because its total weight is not greater than the weight of water it displaces. But suppose water flows into it, in a storm, or after a collision. The total weight of the ship and the water is now greater than the greatest amount of water it can displace, so the ship sinks.

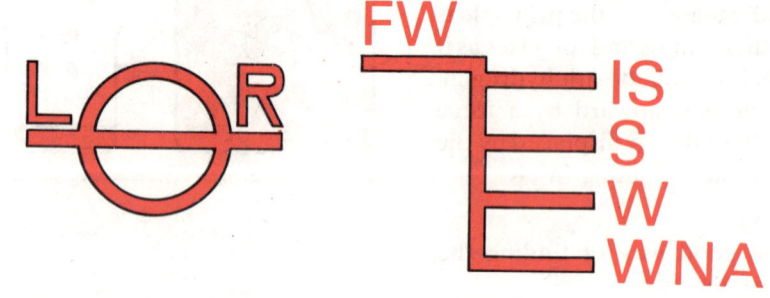

Cross section of a submarine

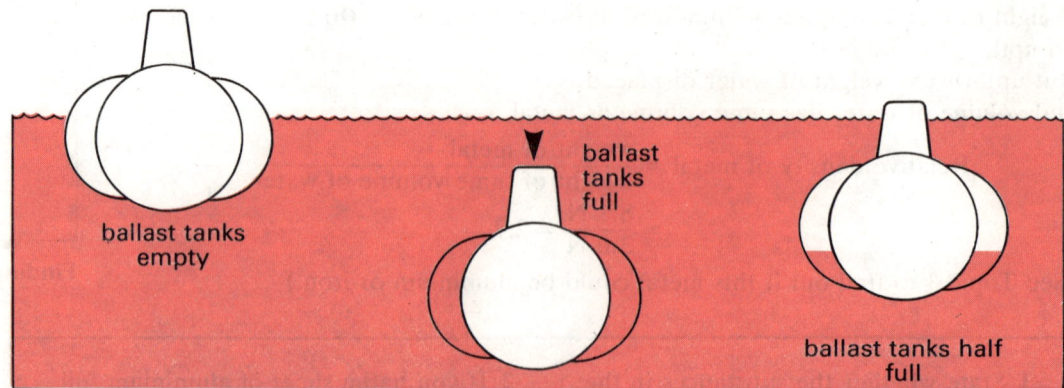

Submarines have ballast tanks on either side of the hull. When these are full of air, the submarine floats on the surface. If the tanks are opened to let water in, the submarine goes down. But it can surface again by forcing the water out of the tanks with compressed air. Once again the total weight of the submarine is less than the water displaced when it is under the water.

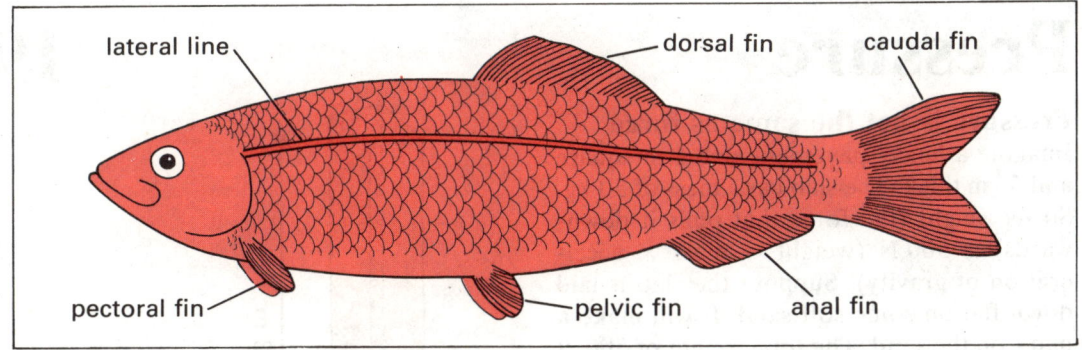

lateral line · dorsal fin · caudal fin · pectoral fin · pelvic fin · anal fin

Fish show some of the features of submarines. The herring, for example, is stream-lined, having a wedged-shaped front end tapering off towards the tail.

There are no large limbs sticking out, only delicate fins. The body is covered with scales, each one overlapping the one behind. The fish thus can move through water with little friction.

It swims with flicking movements of its body and tail. The fins are not used for powerful swimming, only for changing direction.

The fish also has a 'ballast' tank; it is called the **swim-bladder**. It can adjust the size of this so as to make its density the same as that of the water it is swimming in.

Hydrometers

A tube weighted at one end can be floated upright in a liquid. The depth at which it floats depends on the density of the liquid. So the tube can be marked to show the density of the liquid directly. Such an instrument is called a hydrometer.

The diagram below shows how you can make a hydrometer using an ordinary drinking straw and a few lead shot. It can be graduated by noting the level at which it floats in water (1.0), methylated spirit (0.8) and saturated salt solution (1.1).

It is convenient to have one hydrometer marked for relative densities less than 1.0, and another for densities above 1.0.

A special kind of hydrometer is used to check the relative density (or specific gravity) of the acid in a car battery. This hydrometer is a large pipette with a rubber bulb at the top. The lower end is put into the acid in the battery, the rubber bulb is squeezed and released so that some liquid enters the glass

Battery tester
A floating glass bulb shows when the battery needs charging

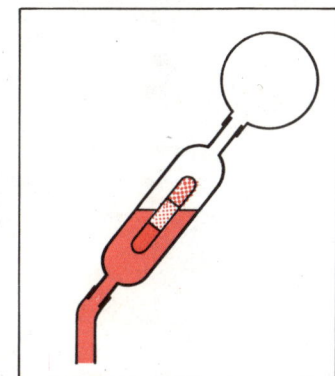

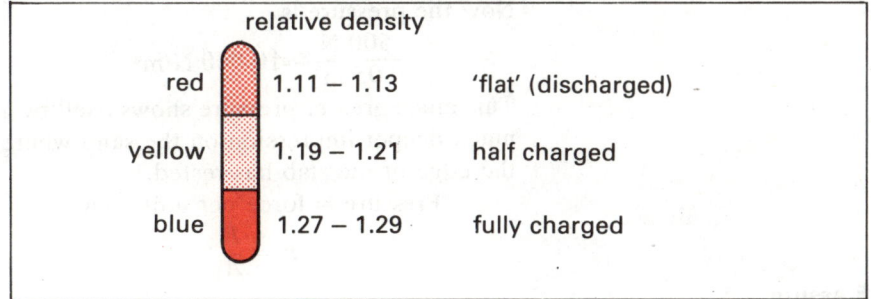

relative density		
red	1.11 – 1.13	'flat' (discharged)
yellow	1.19 – 1.21	half charged
blue	1.27 – 1.29	fully charged

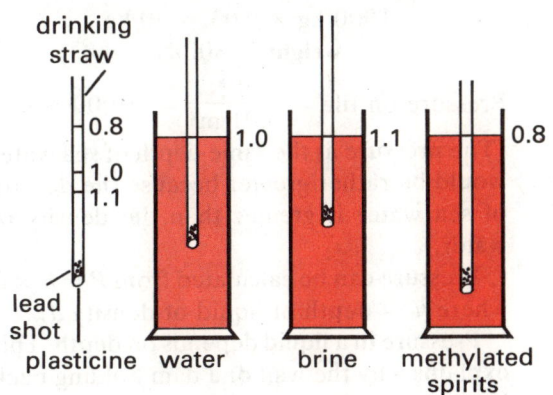

drinking straw

0.8
1.0
1.1

lead shot
plasticine · water · brine · methylated spirits

1.0 · 1.1 · 0.8

How to make a hydrometer

tube. Inside this tube is a glass bulb. Its position, when floating in the sample of acid, shows whether the battery needs charging.

The **carburettor** in a car or motor cycle controls the flow of petrol to the cylinder. The buoyancy of a float is used to cut off the petrol inflow. A needle is pushed up into the inlet when enough petrol has entered the float chamber to raise the float sufficiently.

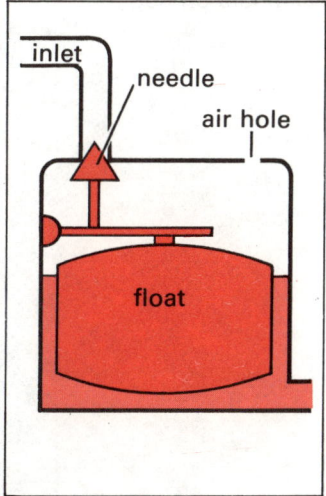

inlet · needle · air hole · float

Carburettor
This controls the flow of petrol into the cylinder

1 A lump of paraffin wax has a volume of 100 cm³ and a mass of 88 g. Find its (a) density and (b) relative density.
2 What will happen when the lump of wax in question 1 is put (a) into water and (b) into methylated spirit of relative density 0.83?
3 What difference would you be likely to notice between swimming in sea water and swimming in fresh water (apart from the taste!)?
4 What might be the cause of petrol not reaching the cylinder of a motor car from the carburettor?

Pressure

Pressure is not the same as force

Imagine a heavy pavement slab 1 m square and 5 cm thick. The slab has a mass of 50 kg. So its weight (the force that pulls it downwards) is 500 N (weight = mass × acceleration of gravity). Suppose the slab is laid down flat on some soft sand. It will make a mark on the sand. The total weight of 500 N is spread over one square metre:

$$\text{pressure} = \frac{500 \text{ N}}{1 \text{ m}^2} = 500 \text{ N/m}^2$$

Pressure
Pavement slab laid on sand

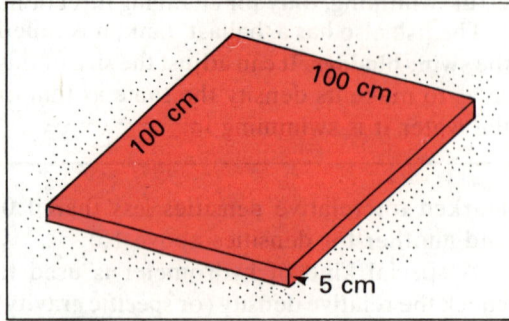

But let us suppose that the slab is now lifted and set down on its edge. The same weight is now exerted downward, but only over a smaller area, 1 m × 5 cm, which is 1 m × 0.05 m = 0.05 m². Now the pressure is

$$\frac{500 \text{ N}}{0.05 \text{ m}^2} = 10\ 000 \text{ N/m}^2$$

This much greater pressure shows itself by a much deeper impression on the sand where the edge of the slab has rested.

$$\text{Pressure} = \text{force per unit area}$$

$$P = \frac{F}{A}$$

Pressure
Slab laid on edge

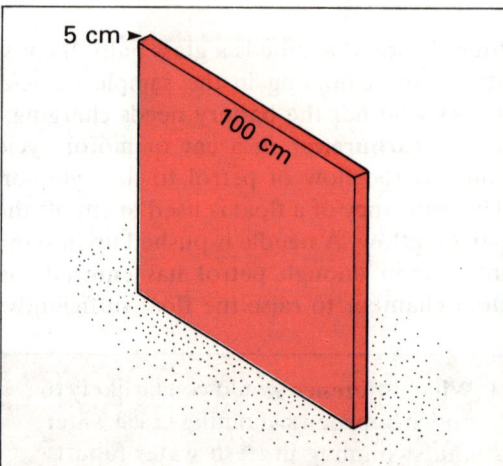

Pressure of water on tile

Pressure in liquids

The water in a swimming bath exerts quite a pressure on the tiles at the bottom.

Think of one of the tiles at the deep end. Suppose it is 10 cm square and there is a depth of 3 m water above it. The tile must support a column of water 3 m high and 10 × 10 cm across. The volume of this water is:

$$0.1 \text{ m} \times 0.1 \text{ m} \times 3 \text{ m} = 0.03 \text{ m}^3$$

Since 1 m³ of water has a mass of 1000 kg, the **mass** of this water column is:

$$1000 \text{ kg} \times 0.03 = 30 \text{ kg}$$
$$\textbf{weight} = 300 \text{ N}$$

$$\text{Pressure on tile} = \frac{300 \text{ N}}{0.01 \text{ m}^2} = 30\ 000 \text{ N/m}^2$$

(The pressure at the same depth of sea water would be rather greater because the density of sea water is greater than the density of water.)

Pressure can be calculated from $P = h \times d$ where h = depth of liquid of density d.

Pressure in a liquid depends on depth. This explains why the wall of a dam holding back large amounts of water is made much thicker at the base. Other important points about pressure are:

(a) pressure at a point in a liquid is the same in all directions;

(b) pressures at points on the same horizontal level are equal.

Hydraulic pressure

Pressure in a liquid is transmitted in all directions. Also water cannot be compressed, so the pressure applied at one place is transferred to another place in the liquid.

Imagine a platform carrying a load (W) sitting on a large cylinder of water. This water supply is connected by a narrow tube to a piston (P). The area of the piston is 1 cm², whereas the area of the platform is 4 m².

A force of 0.1 N is applied at P. So a pressure of 0.1 N/cm² is passed through the water to the lower side of the platform. 0.1 N on every cm means a total upward force of:

$$4 \times 100 \times 100 \times 0.1 = 4000 \text{ N}$$

which would raise 400 kg.

If the platform was the floor of a lift, it means that a small force of 0.1 N could lift 400 kg. This could be the lift itself and several people weighing 70–80 kg each.

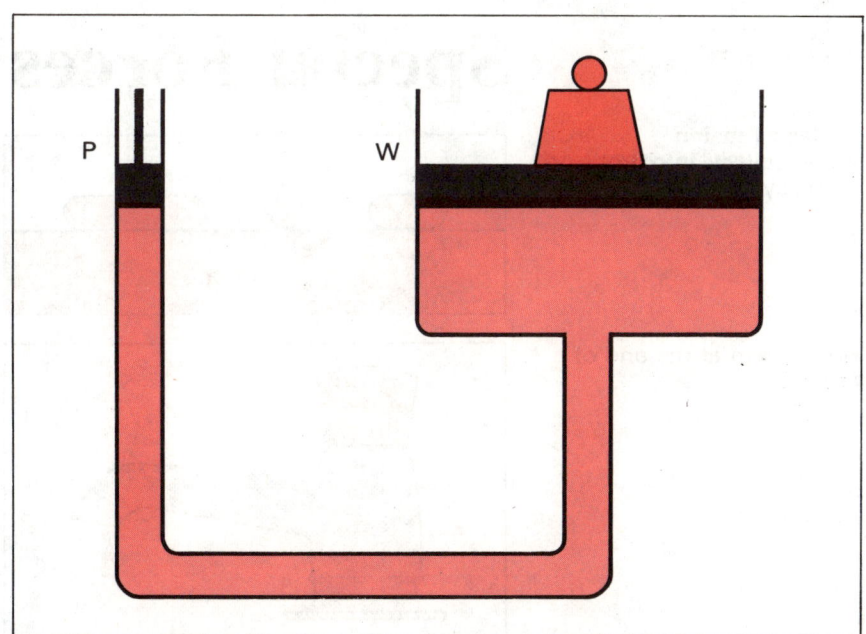

Hydraulic lift
W (large load) on cylinder of water (P = piston)

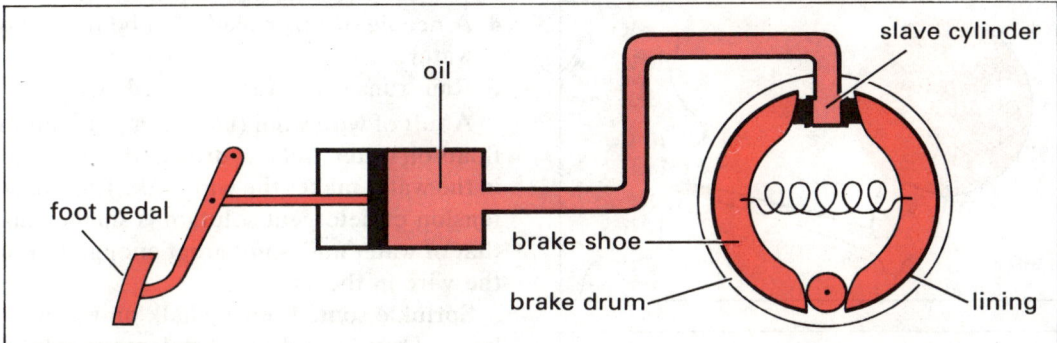

Car hydraulic brakes

Another example of hydraulic pressure can be found in the hydraulic brakes of a car. Oil is used as the liquid, and the pressure of the driver's foot on the brake pedal causes a large force to be applied on the brake shoes in the wheel.

Cartesian diver

A model diver can be made by floating a small bulb upside down in a bottle almost filled with water. There is a bubble of air inside the 'diver' that just keeps it afloat. When this amount of air is correctly adjusted, a little pressure on the cork will make the 'diver' sink. The pressure causes a small amount of water to enter the bulb. This makes the 'diver' heavier and it sinks. Releasing the pressure brings the diver to the surface again.

Cartesian diver

1 Half the weight of a girl of mass 60 kg is carried on one blade of her ice skates. The area of this blade is 10 cm². What pressure is exerted on the ice?

2 What is the mechanical advantage of the hydraulic lift machine described above? (See also Topic 16.)

Special Forces in Water

Surface Tension
Water collects into drops on a greasy surface

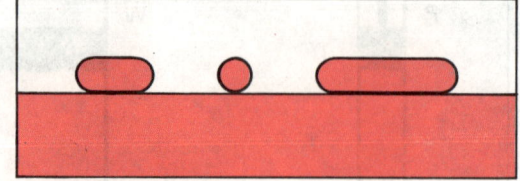

Drops form at the end of a tap

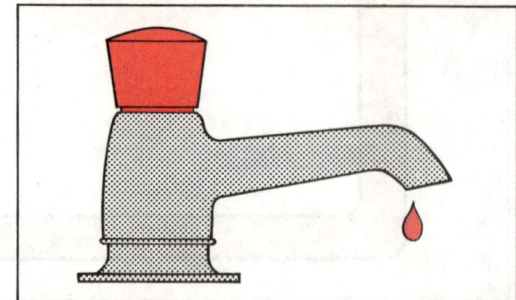

A soap bubble

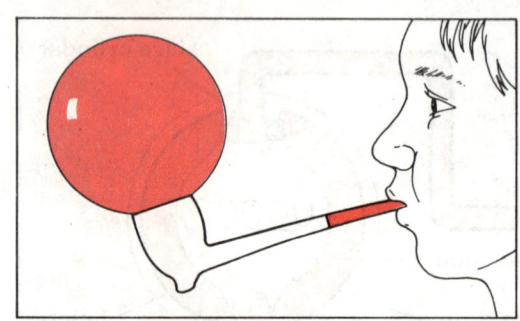

Soap film forming a wire ring

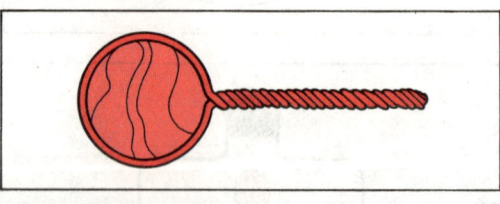

Rain
Forms drops on water-proofed wool

Surface tension

A liquid flows until it 'finds its own level' so the surface of any liquid in a container is horizontal. But this is not exactly true. Put some water on a greasy surface. It does not spread all over the surface, but collects itself into drops.

There seems to be a kind of 'skin' on the surface of the liquid trying to keep it from flowing away. This force in the surface is called **surface tension**.

Here are more examples of surface tension.

1 Water forms drops before it falls from a tap.
2 A soap bubble.
3 A soap film forms on a wire ring dipped into soap (or detergent) solution.
4 A needle or razor blade can be floated on water.
5 Rain runs off a waterproofed surface.

A tuft of wire wool (used as a pot scourer) floats on water, but one drop of detergent put in the water makes the wire sink. The surface tension of detergent solution is smaller than that of water and is not great enough to hold the wire in the surface.

Sprinkle some French chalk on water in a bowl. Then let a drop of detergent solution fall into the middle of the surface. What would you expect to happen? Why do the hairs on a paint brush cling together when the brush is taken out of water?

Hairs on paint brush cling together after being immersed in water

Capillarity

When water is in a glass vessel, it clings to the glass at the edge of the surface, forming a meniscus. It is as if the liquid is trying to climb up the glass. This is shown well if we dip tubes of very fine bore (capillary tubing) into coloured water. The narrower the tube the higher up the liquid goes. This effect is called **capillarity**. Here are some further examples.

A cloth is put into water. The water soaks up into the cloth because there are fine spaces between the fibres of the cloth.

Sandy soil will allow water to drain through more quickly than a clay soil. But the spaces between the particles of clay are very small. This means that the clay soil is kept moist longer because the water lower down is drawn upwards by capillarity.

Molecular explanation

A liquid is a large collection of molecules. These molecules are moving, but all molecules attract others, so they stay together.

The molecules in the surface layer are pulled inwards by the other molecules inside, but there is no force pulling them outwards. So there is a force of **cohesion** keeping the molecules bunched up together. Surface tension is due to cohesion of molecules.

But what about the contact between water and glass molecules? The glass molecules pull up the nearest liquid molecules. This forms the meniscus. There is a force of **adhesion** and this can account for capillarity.

Mercury

Mercury is the only metal that is liquid at ordinary temperature. It is different from most other liquids. The mercury is lower inside a capillary tube than outside. It is as though the mercury is being pushed away by the glass. Its meniscus is downwards. Mercury does not 'wet' the glass. The force of cohesion between mercury molecules is greater than the force of adhesion between mercury molecules and the glass molecules.

> 1 Why is a 'damp course' included in the building of walls of houses?
> 2 How does blotting paper dry ink?
> 3 How can some insects walk on the surface of a pond?

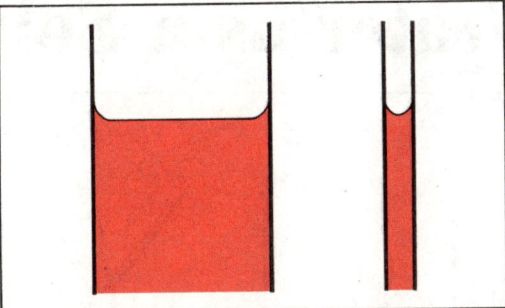

Capillarity
Water clings to the glass forming a meniscus

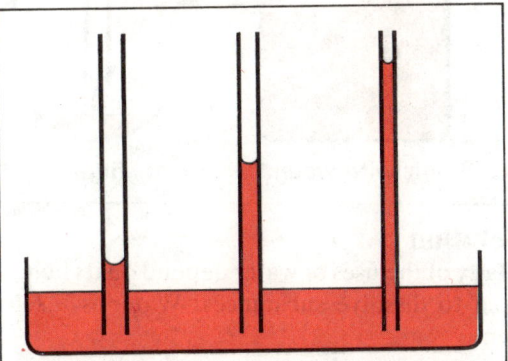

Fine capillary tubes in coloured water
The narrower the tube the higher the liquid climbs

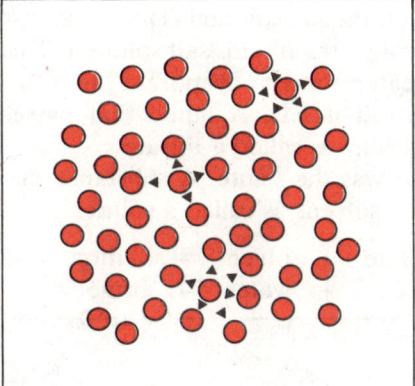

A liquid is a large collection of molecules
There is a force of attraction between the molecules

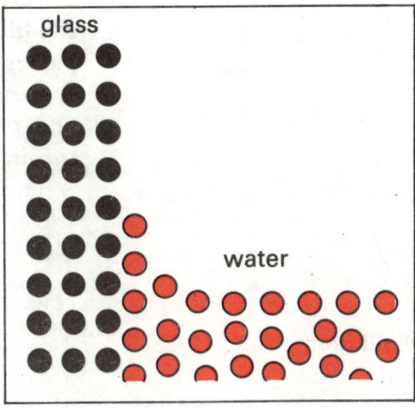

Surface tension
There is also a force of attraction between water and glass molecules

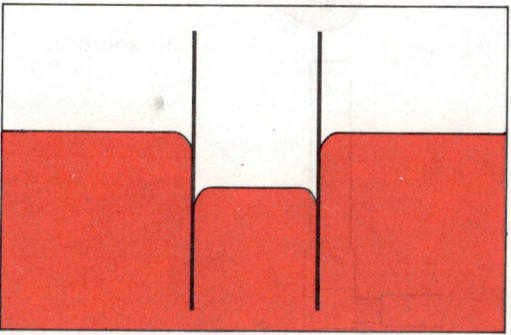

Mercury
The meniscus is downwards

Making pure salt from rock salt

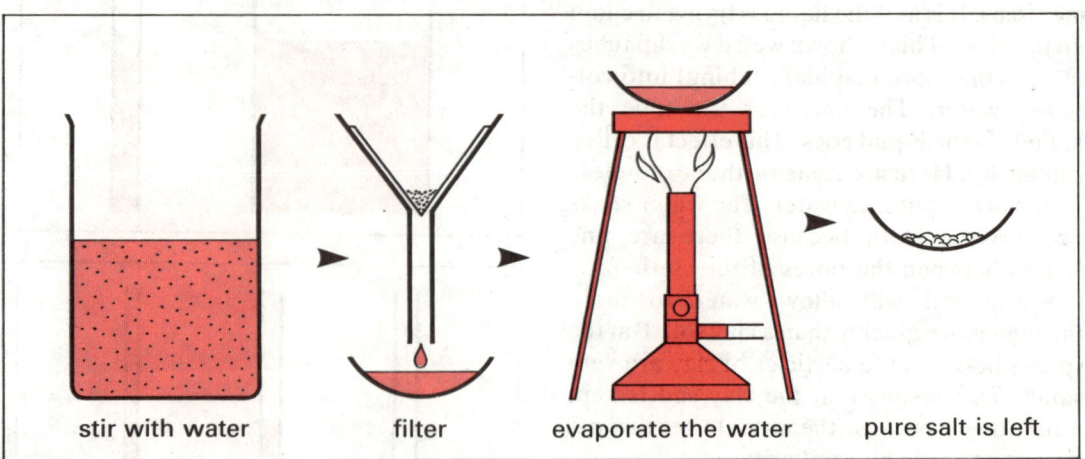

stir with water   filter   evaporate the water   pure salt is left

Solution

Many of the uses of water depend on its being able to dissolve substances. Water is a **solvent**.

Suppose you have some rock salt and you wish to make pure salt from it. You can dissolve the salt by adding water. The sand in the crude rock salt will not dissolve, so you can filter off the salt solution. The liquid that passes through the filter is salt solution. This is then heated so as to evaporate the water and pure salt is left. A liquid that passes through a filter is called a filtrate.

The salt was the **solute**. A substance dissolved in a solvent, is called a solute.

$$\text{solute} + \text{solvent} \rightarrow \text{solution}$$
$$\text{e.g.} \quad \text{salt} + \text{water} \rightarrow \text{brine}$$

If you wanted to get back the water you need to heat the solution, evaporate the water and then condense the steam by cooling it.

$$\text{evaporating} + \text{condensing} = \text{distilling}$$

The distilled water obtained by this process of distillation is pure water and has no substances dissolved in it.

Tests for pure water are:

   it freezes at 0°C

   it boils at 100°C

   it is tasteless

   it does not change the colour of litmus

   it turns colourless anhydrous copper(II) sulphate blue when added to it.

Distillation

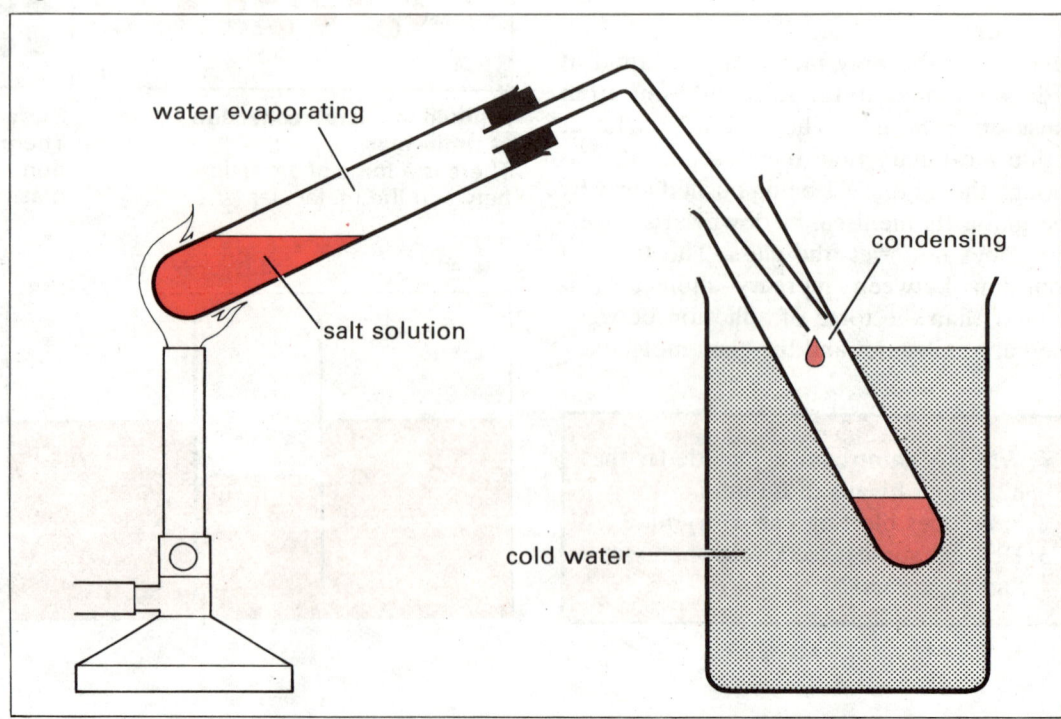

water evaporating

salt solution

condensing

cold water

Saturated solution

How much solute can dissolve in a solvent? When more and more solute is added there is a point at which no more can dissolve. If there is some undissolved solid present in contact with this solution, we say that the solution is saturated.

It is sometimes possible, by cooling or evaporation, to obtain a solution that contains more solute in it than in a saturated solution at the same temperature. Such a solution is called a **super-saturated** solution. But the extra solute will not stay in equilibrium with the saturated solution. Dropping in a small crystal of solute or even just shaking the solution will cause the excess to settle out.

Solubility

Different solutes dissolve to different extents in a solvent. So we need a measure to say how soluble a substance is. The solubility of a substance is the number of grams of solute that dissolve in 100 g of solvent, to make a saturated solution, at a stated temperature.

It is important to state the temperature, because solubility changes with temperature. Most substances are more soluble in hot water than in cold.

Only a few substances are less soluble in hot water. Examples are calcium hydroxide and calcium sulphate.

All gases are less soluble in hot than in cold water. You can drive off ammonia from a solution of this gas in water by warming the solution.

Insoluble substances

Not all substances dissolve in water. It is useful to have a few simple rules to help us.

All nitrates are soluble;
all sodium, potassium and ammonium compounds are soluble.

Some insoluble substances are:

most carbonates;
oxides;
hydroxides;
sulphides;

but not those of sodium, potassium and ammonium.

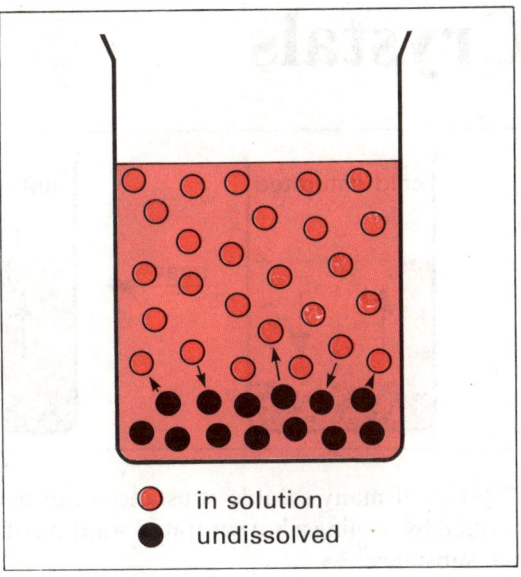

- in solution
- undissolved

Saturated solution
Both dissolved and undissolved solute can be present in a saturated solution

1 What will happen if a solution of potassium nitrate, that is saturated at 50°C, is allowed to cool?
2 If some chalk had become mixed accidentally with some ordinary common salt, how would you get the salt back in a pure state?
3 Which of these substances are insoluble: sodium carbonate; zinc carbonate; ammonium chloride; lead nitrate; copper hydroxide; lead sulphide.

Crystals

Hot saturated solutions form crystals when cooled

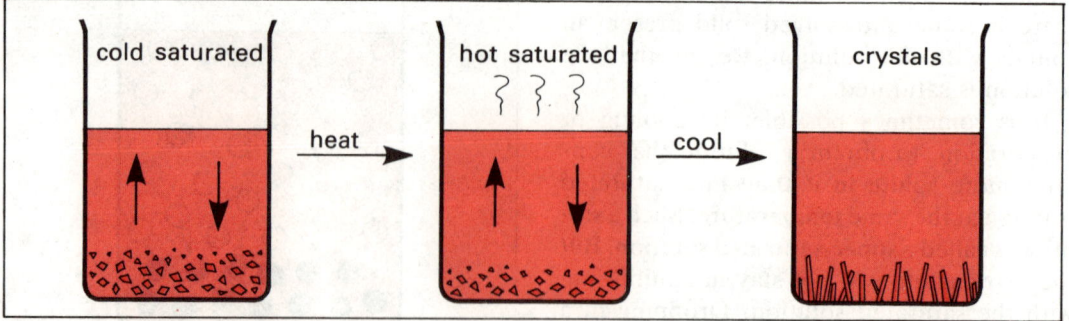

Crystals of many soluble substances can be formed by cooling a hot saturated solution of the substance.

Most substances are more soluble in hot water than in cold. If as much solute as possible is put into solution in hot water, and then the solution cooled the solute will form crystals. If the hot solution is cooled quickly the crystals will be a fine powder, but if it is cooled slowly, large crystals will form.

Making crystals

Crystals of potassium nitrate can be made quite easily as shown in the diagram.

Using a lens the crystals are seen to have regular shapes with smooth shiny faces. If crystals are allowed to form by slowly cooling the solution, the molecules from the solution line up like soldiers on parade. They form regular shapes with sharp edges.

Making crystals of potassium nitrate

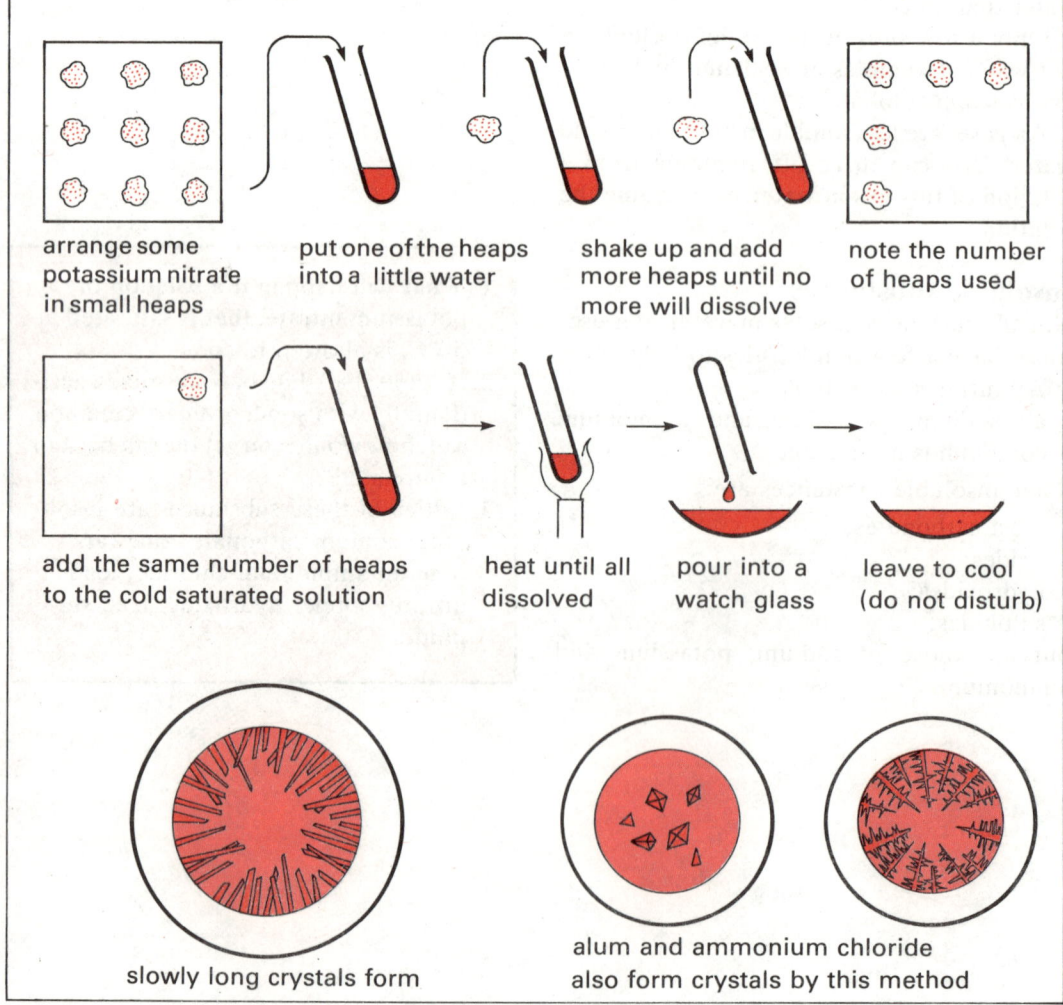

arrange some potassium nitrate in small heaps

put one of the heaps into a little water

shake up and add more heaps until no more will dissolve

note the number of heaps used

add the same number of heaps to the cold saturated solution

heat until all dissolved

pour into a watch glass

leave to cool (do not disturb)

slowly long crystals form

alum and ammonium chloride also form crystals by this method

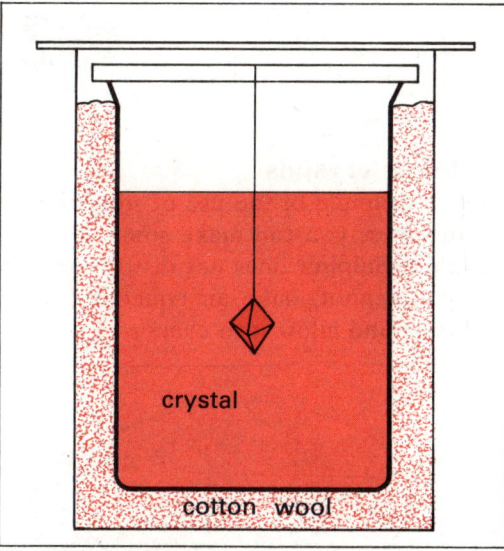
'Growing' a crystal

Lead tintelluride
These crystals have been highly magnified in an electron microscope

Growing crystals

If a small crystal is put in a saturated solution it will slowly become larger. A small crystal, e.g. of alum or of copper sulphate, is suspended in the saturated solution by a thread tied to a glass rod. If the solution were to become heated, the crystal would dissolve, so the beaker must be insulated from heat changes.

Water of crystallisation

When blue crystals of copper(II) sulphate are heated in a test tube, steam is driven off and white powder is left. The dry blue crystals must have contained water. They were **hydrated** crystals.

But this water must be only loosely joined to the rest of the molecules. If water is added to the white (anhydrous) powder, the blue colour is seen again.

$$CuSO_4 \ + \ 5H_2O \ \rightarrow \ CuSO_4.5H_2O$$

anhydrous   hydrated
                     copper sulphate

(white)   (blue)

Many crystalline substances contain water of crystallisation, for example:

Barium chloride	$BaCl_2.2H_2O$
Sodium carbonate	$Na_2CO_3.10H_2O$
Calcium sulphate	$CaSO_4.2H_2O$
Iron(II) sulphate	$FeSO_4.7H_2O$

Some crystalline substances do not contain water of crystallisation, for example:

Carbon (e.g. diamond)	C
Iodine	I_2
Sodium chloride	NaCl
Sulphur	S
Potassium iodide	KI
Silver nitrate	$AgNO_3$

Crystals have regular shapes

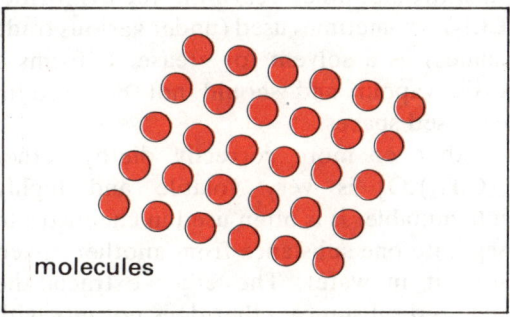
Molecules are lined up in crystals
They form shapes with sharp edges

1 In a solution of potassium nitrate in water, what is (a) the solute and (b) the solvent?
2 Draw a diagram to show the shape of any crystal you have looked at with a lens. Name the substance.
3 Name three substances that contain water of crystallisation. Name three crystalline substances that are not hydrated.

Other Solvents

23

When a substance will not dissolve in water, it is often possible to find another solvent in which it is soluble. An ink stain that cannot be removed with water can usually be removed with methylated spirit, in which the dye is soluble. Also a greasy spot can be removed with petrol or tetrachloromethane because these liquids dissolve fat and oil.

Some important solvents

Acetone $[(CH_3)_2CO]$ is widely used in making plastics. For example, cellulose is treated with acetic acid, then dissolved in acetone. This solution is squeezed through fine holes and when the acetone evaporates, acetate rayon threads are left.

Cylinders of acetylene gas (ethyne) contain this gas under pressure dissolved in acetone.

Nail varnish will not wash off with water, but will dissolve in acetone and amyl acetate. These liquids are used as nail varnish removers.

Alcohol is a name often used for ethanol (C_2H_5OH). Methylated spirit is also mainly ethanol, but it has a blue colour added to it.

Ethanol is used as a solvent for lacquers and varnishes. Coloured shellac dissolved in ethanol forms nail varnish. Tetrachloromethane (carbon tetrachloride, CCl_4) is sometimes used (under various trade names) as a solvent for grease. It forms a toxic vapour and should not be used in enclosed spaces.

Ether or more correctly diethyl ether $[(C_2H_5)_2O]$ is very volatile and highly inflammable. It is often used in chemistry to separate one substance from another mixed with it in water. The ether extracts the required substance; ether does not mix with water, so the substance dissolved in ether can be separated.

Petrol and similar solvents like toluene are good solvents for grease. This kind of solvent is used in 'dry-cleaning' clothes.

Turpentine and the substitute 'White Spirit' are used to dilute paint. Paint is a mixture of driers, pigment and thinners. Turpentine acts as the thinner to dilute the paint, but is not a true solvent in this case.

Warning
Carbon disulphide must only be used in a fume cupboard

Sulphur crystals

As an example of the use of solvents other than water, you can make some crystals of sulphur. Sulphur does not dissolve in water so you cannot make an aqueous (watery) solution and allow it to evaporate.

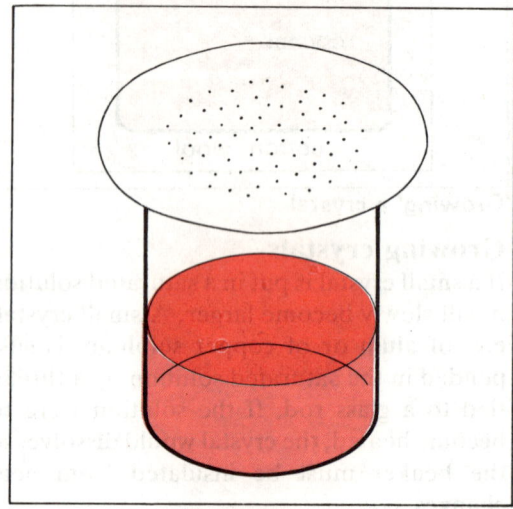

'Growing' sulphur crystals

However, if some roll sulphur is powdered and added to carbon disulphide (NB: as this solvent has a most unpleasant smell and a **very poisonous** vapour, a fume cupboard must be used) then the sulphur dissolves. The carbon disulphide evaporates very quickly, so to slow down the process and obtain larger crystals, a filter paper with a few pin-holes in it is placed over the dish or beaker. After a

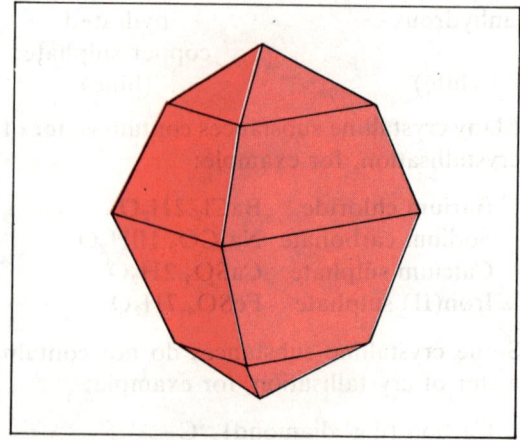

Crystals of rhombic sulphur

few hours, glossy yellow crystals in the shape of rhombs are left. These are crystals of rhombic sulphur.

Next you set up a flask containing toluene (or industrial methylated spirit can be used instead). Put some roll sulphur in the toluene and warm the mixture carefully (**not** over a naked flame). The sulphur dissolves and when the solution is cooled needle shaped crystals form. These are monoclinic crystals of sulphur.

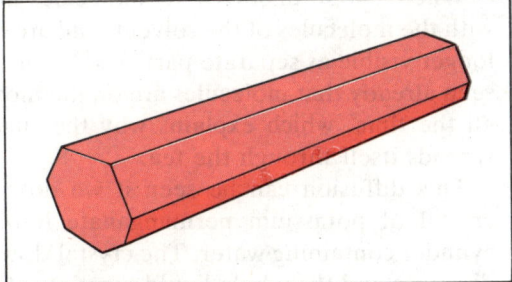

Monoclinic crystals Like needles

Sulphur exists in two different crystalline forms. It is said to be **allotropic**. Rhombic and monoclinic crystals are different allotropic forms of the same element. Carbon is also allotropic and its allotropes are diamond and graphite.

You can make monoclinic crystals of sulphur by a different method. Melt some sulphur slowly in a test tube until it is quite runny. Then pour the liquid into a filter paper folded in the usual way. When the liquid is cool enough to form a crust on the surface,

open up the filter paper. There will be needle-like crystals of monoclinic sulphur inside.

In this case we have prepared crystals by cooling a molten liquid. No solvent has been used.

Crystals can also form when a vapour cools. If iodine is heated in a test tube it all forms a purple vapour. This settles as tiny grey crystals on the cooler part of the test tube.

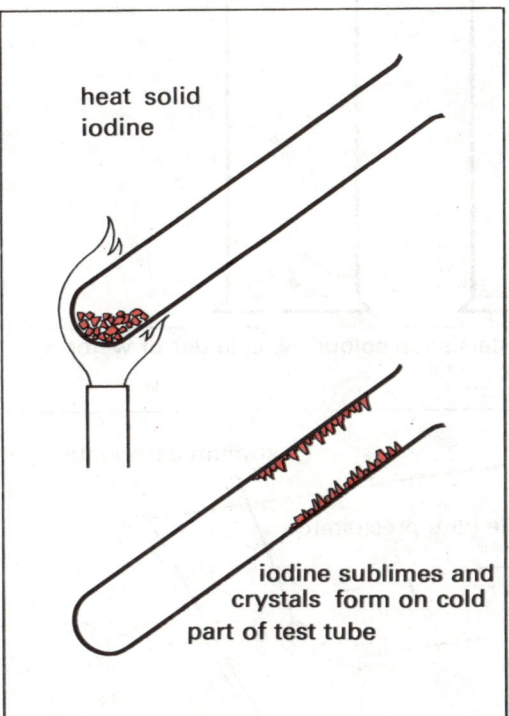

heat solid iodine

iodine sublimes and crystals form on cold part of test tube

Sublimation of iodine

Warning
Toluene must not be warmed with a naked flame

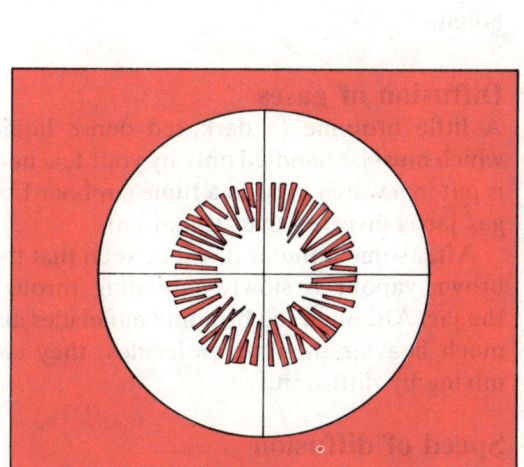

Crystals from the filter paper method

1 Why do you think it is necessary to take great care when working with non-aqueous solvents (i.e. solvents other than water)?
2 For each of the substances mentioned, choose a suitable solvent from those listed. 1 grease; 2 ball-point ink; 3 nail varnish; 4 oil paint. Solvents: A turpentine; B ethanol; C acetone; D tetrachloromethane.

Diffusion

24

Diffusion

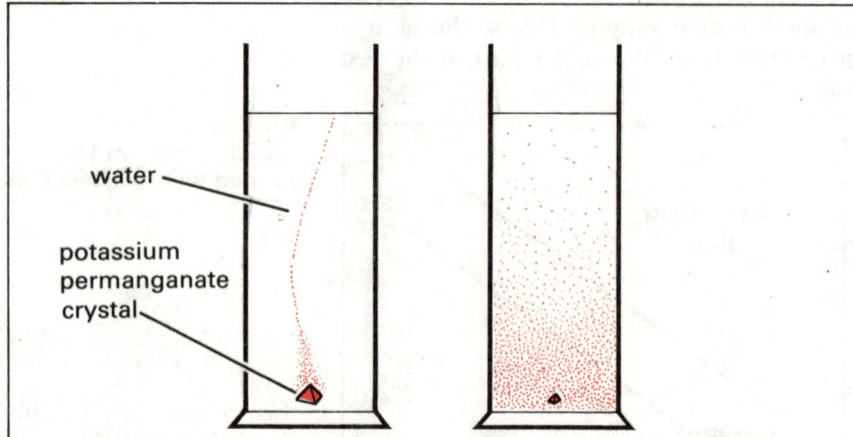

Potassium permanganate crystals soon colour the cylinder of water

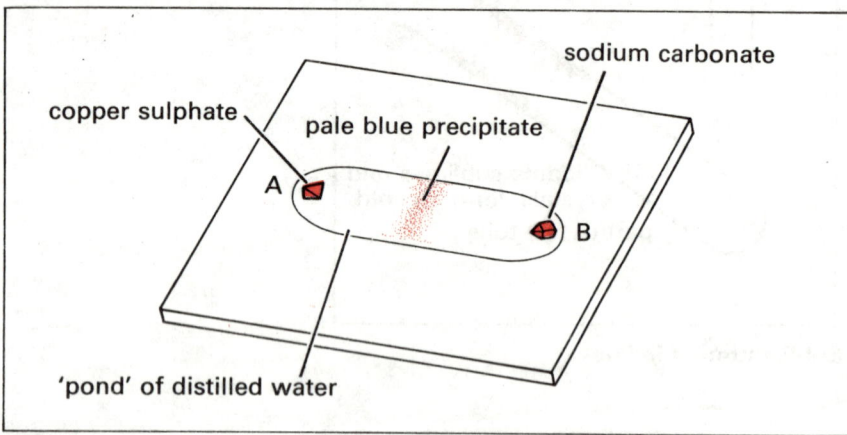

Substances diffuse through the water until they meet and react

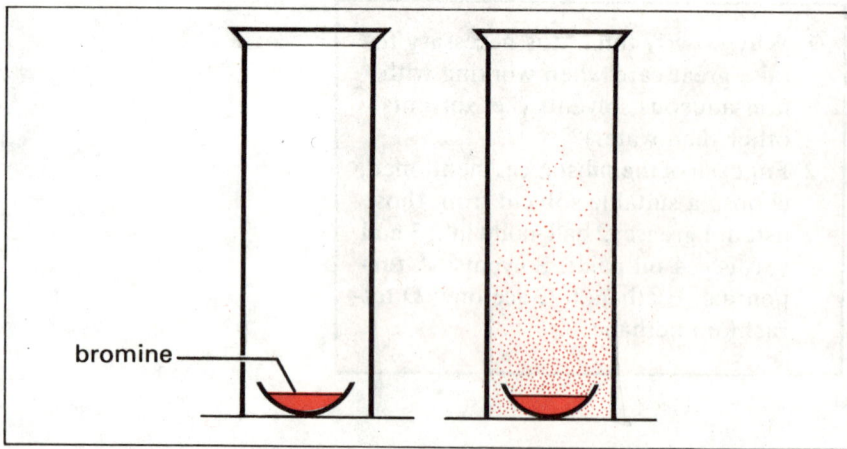

Diffusion of bromine vapour

When a spoonful of sugar is put into a cup of tea, the whole of the tea is sweet after a short time, even without stirring. The sugar has diffused through the liquid.

When a solid dissolves, its molecules mix with the molecules of the solvent, and are no longer visible as separate particles. We have seen already that molecules are on the move all the time, which explains why the sugar spreads itself through the tea.

This diffusion can be seen if we drop a crystal of potassium permanganate into a cylinder containing water. The crystal slowly dissolves and the whole liquid turns purple.

If a solution of sodium carbonate is added to a solution of copper sulphate, a pale blue precipitate is formed.

$$CuSO_4 + Na_2CO_3 \rightarrow CuCO_3 + Na_2SO_4$$

copper   sodium   copper   sodium
sulphate carbonate carbonate sulphate

Now, take a white glazed tile and put a few drops of distilled water on it to make a small 'pond' about 5 cm by 1 cm. Put a small crystal of copper sulphate at A and a small crystal of sodium carbonate at B. Then wait. After a short time a blue precipitate will be seen between the crystals. The substances dissolve and diffuse through the water until they meet and react to form copper carbonate.

Diffusion of gases
A little bromine (a dark red dense liquid which must be handled only by your teacher) is put in a watch glass in a fume cupboard. A gas jar is inverted over it and left.

After some minutes it will be seen that the brown vapour is slowly spreading through the jar. Although the bromine molecules are much heavier than air molecules, they are mixing by diffusion.

Speed of diffusion
Some molecules diffuse faster than others, as you might have noticed in the precipitation experiment.

The fact that hydrogen gas diffuses more quickly than air, can be shown by an experiment. (Hydrogen is the lightest gas known.)

A porous pot (P) is inverted and connected by a rubber stopper to a U-tube with some water in the bend. The other end of the tube is drawn out to a fine jet.

A jar of hydrogen (H) is inverted over the porous pot. Very quickly water is forced out of the jet giving a miniature fountain effect. The hydrogen molecules move into the pot faster than the air molecules diffuse out. So the pressure inside the pot forces the water out. What do you expect will happen if the jar (H) is now taken away?

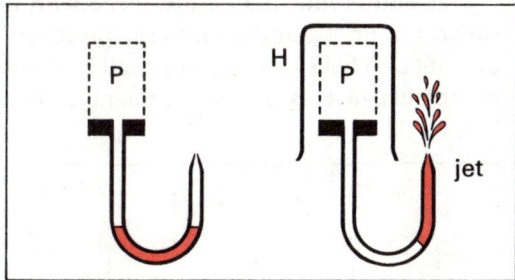

Speed of diffusion

Chromatography

This is a very useful method of separating substances in a mixture. It depends partly on different solubilities and different rates of diffusion.

Experiment 1. Black ink is a mixture of several dyes. They can be separated. A strip is cut towards the centre of a filter paper and bent down so that the end will dip into a mixture of ethanol and water in a glass dish.

A tiny drop of the black ink is put at the centre of the filter paper, allowed to dry. Then another tiny drop is put in the same spot and also left to dry. The filter paper is then put on the dish with the strip drawing up the solvent in the dish (by capillarity). After some time, different coloured circles will be seen, because the different coloured dyes move outwards at different rates.

Experiment 2. The green colouring matter, chlorophyll, in green leaves is really a mixture of pigments. The method above can be used to separate them, but two other methods are given here.

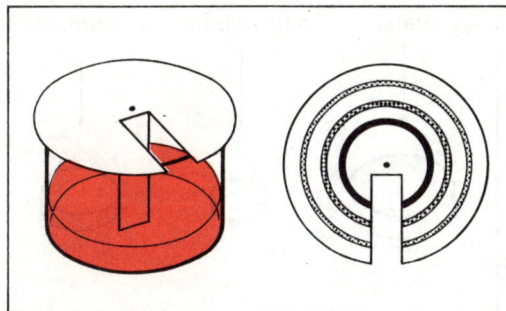

Chromatography
Experiment 1

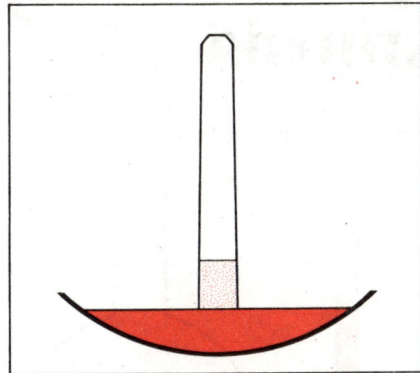
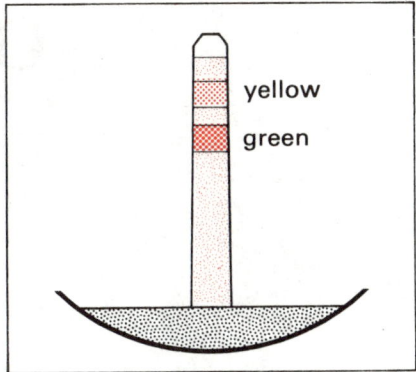

A good way of extracting the colouring matter is to collect some stinging nettle leaves, dry them, then grind them in a mortar with a few cm³ of acetone. A little of this solution is put in a watch glass and a stick of white blackboard-chalk is stood on end in the liquid.

When the green liquid has passed a little way up the chalk, the stick of chalk is taken away and stood in a little alcohol–water mixture. It is left until the liquid has reached almost the top of the stick. Green and yellow bands will be seen at different heights on the chalk.

Experiment 3. Another way is to spot the green chlorophyll solution about 2 cm from the end of a strip of filter paper, which is then suspended from a split cork in a tube. The spotted end dips into the alcohol–water mixture.

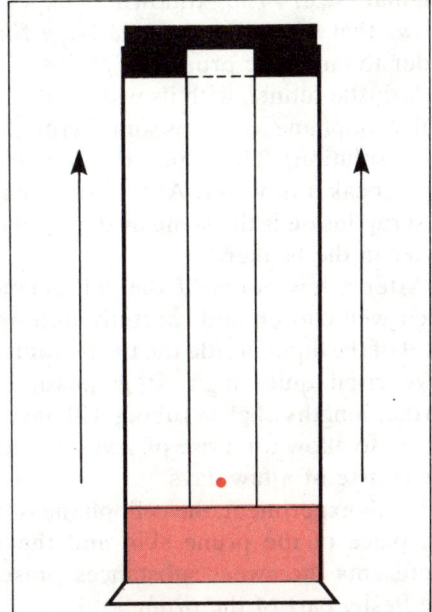

1 Can you explain why acetone is used instead of water to take the green pigment out of leaves?

Chromatography
Experiment 2

Green and yellow bands form on the chalk after insertion in alcohol/water mixture

Chromatography
Experiment 3

Osmosis

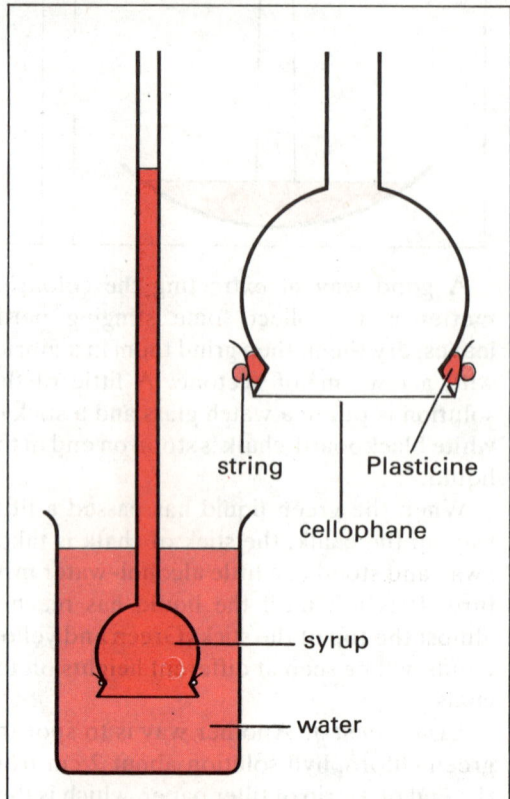

25

Osmosis
Water rises into the funnel

Dry wrinkled prunes, left in water for some time, become smooth and swollen. Rice grains left in water swell to many times their original size. The following experiment shows that the water exerts a large force in order to enter the prunes or grains.

A thistle funnel, with its wide end covered with cellophane, contains some syrup (strong sugar solution). This is put, thistle end down into a beaker of water. At the start, the level of syrup inside is the same as the level of the water in the beaker.

After a few hours, if the cellophane has been well chosen and carefully tied on, the level of the liquid inside the thistle funnel will have risen quite high. It is possible that further lengths of glass tubing will have to be added to allow for a rise of several metres in the course of a few days.

In this experiment, the cellophane is taking the place of the prune skin and the syrup represents the sweet substances present in the fleshy part of the prune.

The cellophane or the prune skin acts as a semi-permeable membrane. It allows water, but not sugar solution to pass through it. This passing of water across a semi-permeable membrane into a solution on the other side is called **osmosis**.

Control experiment 1

Cut a large potato in half. Carve out from each half a tube reaching almost to the cut end. (A cork borer is useful for this.) These halves are then stood in a dish of water. A few grains of sugar are put into the tube of one. There is enough moisture to dissolve the sugar. The other half is left as a control. Water enters into the potato piece with the sugar, but no water goes into the piece without sugar. This shows that the cell walls of a potato can act as a semi-permeable membrane.

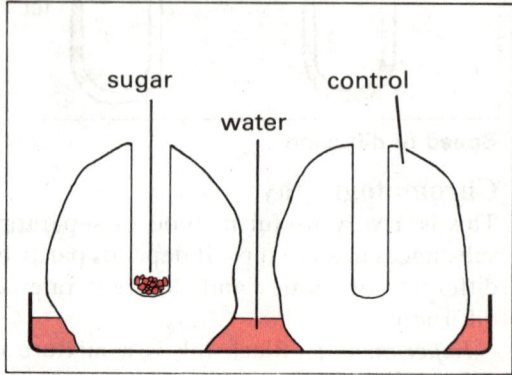

Control experiment 2

Three thin slices of potato or carrot, A, B and C are cut. They are dried with filter paper and weighed. A is put in water, B in strong salt solution and C is left as a control.

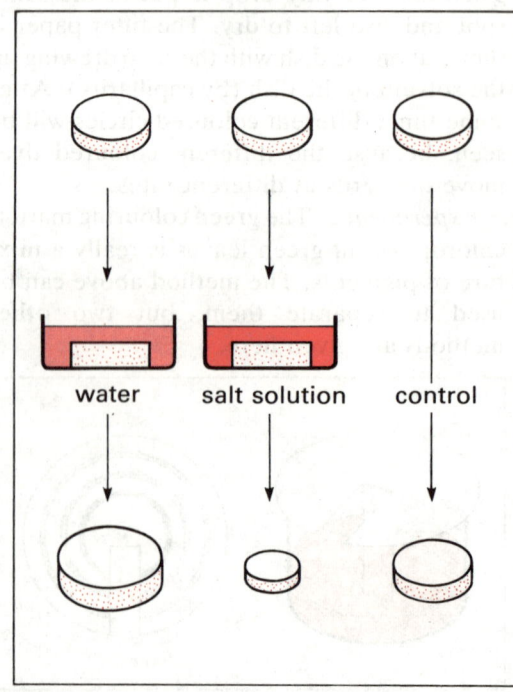

Change in weight of potato pieces
A gains weight; B loses weight; C control

1 Can you explain why A gains weight, B loses weight, as compared with the control C?
2 Why are control experiments needed?

Root hairs

The root of a plant, when studied closely, is found to have very fine hairs near the tip. These are best seen when a plant's roots are in water, because the water supports the slender hairs. These are called root hairs. They are not present at the very tip. This is the region of the root where growth takes place and this would rub off any root hairs. Also there are no root hairs further back. This (older) part of the root grows a corky layer which does not allow water to pass through.

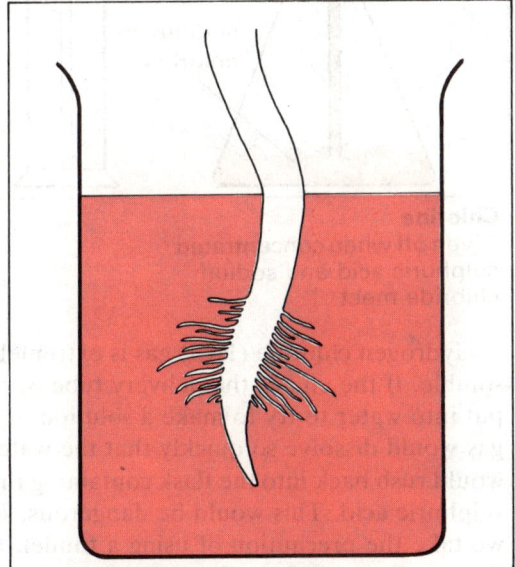

Root hairs
These are best seen in water

Though these root hairs are very fine and only found in this limited region, there is an enormous number of them on the complete root system. Together they are responsible for taking from the soil all the water the plant needs.

Each root hair grows between soil particles. So it is bathed with the film of moisture known as the **soil solution**.

The root hair is an outgrowth of one of the cells in the outermost layer called the epidermis of the root. The cell wall is lined closely by a layer of protoplasm, with a vacuole inside filled with **cell sap**.

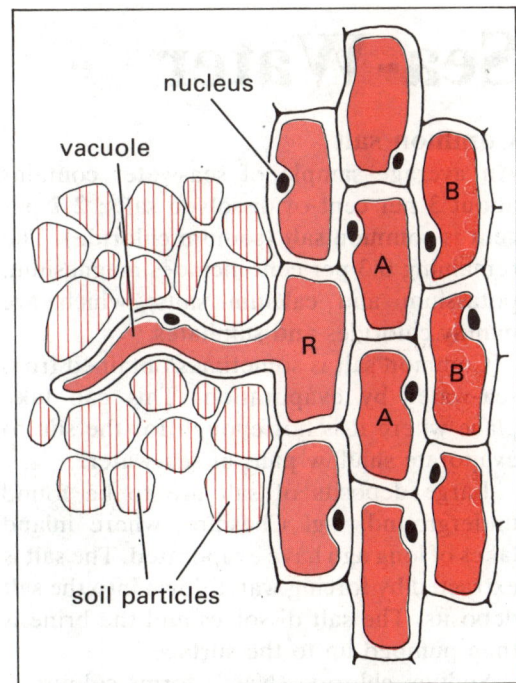

Uptake of water by osmosis
R, Root hair
A, Cells next to R
B, Cells next to A

nucleus
vacuole
soil particles

This cell sap contains several substances in solution. Water passes by osmosis, from the soil solution outside into the stronger cell sap inside, across the semi-permeable membrane of cell wall and protoplasm.

This makes the solution in the root hair R more dilute than the solution in the next cells A. Further osmosis of water occurs from the root hair into the cells labelled A.

The solution in cell A is now more dilute than the solution in the next cell B and water passes from cell A into cell B. This goes on until the water reaches the main vessels that carry water up the root into the stem and then into the leaves.

It is not a good idea to 'feed' your plants with too strong a solution of fertiliser. If the solution you pour on the roots is more concentrated than the solution inside the cells, you will draw water out of the roots, so that the plant will wilt and perhaps die.

Absorption of inorganic salts

A plant needs some inorganic salts, as mentioned in Topic 7. These cannot be taken in by osmosis, so their uptake must occur by diffusion in the stream of liquid going into the plant. Both processes go on at the same time. Both osmosis and diffusion are due to the movement of molecules.

Sea-Water

26

Common salt

An average sample of sea-water contains about 3 per cent of dissolved salts: 2.7 per cent is common salt (sodium chloride). The remaining 0.3 per cent includes magnesium, potassium and calcium salts, which are mainly chlorides and sulphates.

Common salt is sometimes obtained from sea-water by evaporation. This can take place where it is hot enough for the sun to evaporate shallow pans of sea-water.

Large deposits of salt are to be found underground, e.g. Cheshire, where inland lakes of long ago have evaporated. The salt is extracted by forcing water down into the salt deposits. The salt dissolves and the brine is then pumped up to the surface.

Sodium chloride ($NaCl$) forms colourless cubic crystals. It is a very stable substance (does not decompose easily). It melts at about 800°C and boils at 1465°C.

Warning
The preparation of hydrogen chloride must only be demonstrated in a fume cupboard by a teacher

Cubic crystals of salt

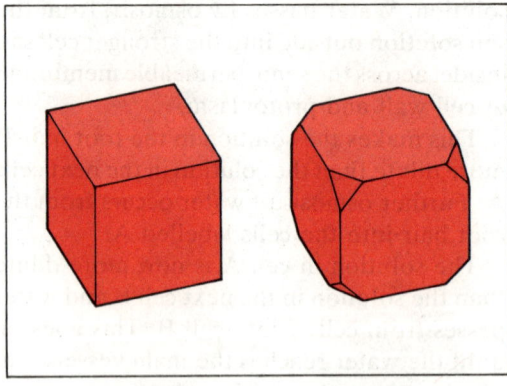

It is soluble in water. It is not much more soluble in hot water than in cold so to form crystals a solution must be evaporated. It is no good cooling a hot saturated solution. Sodium chloride is an important chemical. It is the source from which many sodium and chlorine compounds are made.

Our blood contains sodium chloride and we must take some into the body with our food, about 10 g per day. This is one of the reasons why we like salt with our food. Salt is also used for preserving food (e.g. pork and fish).

Crude salt becomes damp after a time. This is because some magnesium chloride is present as an impurity and this salt absorbs moisture from the air. But a fine dry salt is obtained by adding a trace of phosphate which reacts with the magnesium chloride. The magnesium phosphate formed does not absorb moisture.

Hydrogen chloride

When concentrated sulphuric acid is added to sodium chloride (with some warming if necessary) a colourless gas if given off which may form steamy fumes if the air is moist. The gas is denser than air so it can be collected downwards in a gas jar.

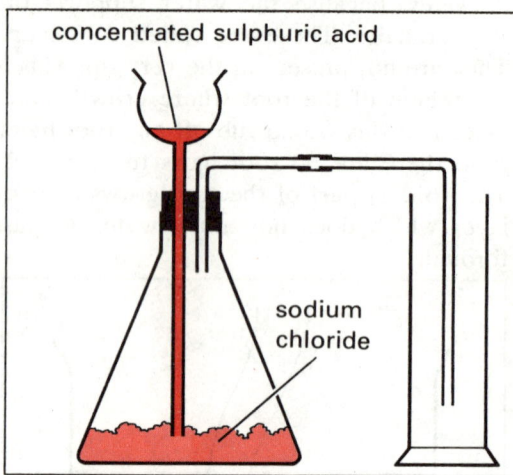

concentrated sulphuric acid

sodium chloride

Chlorine
Given off when concentrated sulphuric acid and sodium chloride meet

Hydrogen chloride (HCl) gas is extremely soluble. If the end of the delivery tube were put into water to try to make a solution, the gas would dissolve so quickly that the water would rush back into the flask containing the sulphuric acid. This would be dangerous, so we take the precaution of using a funnel. If the gas does 'suck back', the water in the beaker drops below the lower end of the funnel and the solution in the funnel just falls back again.

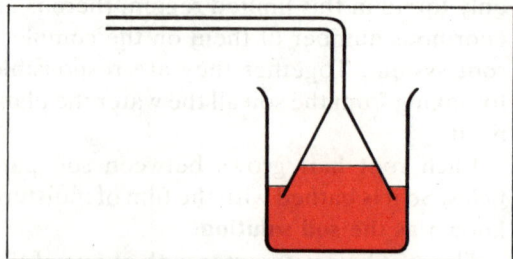

Funnel precaution in case of 'suck back'

Hydrochloric acid

A solution formed in this way is called hydrochloric acid. The concentrated acid is a colourless liquid that fumes in moist air. The properties of this and other acids are dealt with later (see Topic 41).

Hydrochloric acid is used in making plastics, dyes, drugs and other chemicals. It is used in 'pickling' (cleaning) metals before they are galvanized or electroplated.

Chlorine

If we remove the hydrogen from molecules of hydrogen chloride we make the poisonous greenish-yellow gas chlorine (Cl_2).

This is done by making the hydrogen come away from the chlorine and join with oxygen:

$$O + \begin{array}{c} H-Cl \\ \\ H-Cl \end{array} \rightarrow O\begin{array}{c} H \\ \\ H \end{array} + Cl_2$$

water (H_2O)

The oxygen is provided either by heating hydrochloric acid with manganese(IV) oxide or by dropping the acid on to potassium permanganate.

Chlorine has many uses. It is a very reactive element and is used in the manufacture of many chemicals. Examples are D.D.T., T.C.P., weed-killers, chloroform and carbon tetrachloride. Domestic bleaches like 'Domestos' and 'Parazone' work because they easily set free chlorine and the chlorine kills harmful bacteria.

Because chlorine is a disinfectant, a very small amount of chlorine is dissolved in water supplies (not more than 1 part per million) and also in the water of swimming baths.

Salt pans
Common salt can be obtained by the evaporation of sea-water

1 Which of the following processes could be used to make sea-water fit for drinking: distillation; evaporation; filtration; chromatography; solution?

2 Why is hydrogen chloride sometimes called 'salt gas'?

3 Chlorine gas may be collected either by downward delivery into a gas jar or by displacement of water. What do these facts tell you about the density and solubility of chlorine?

Stalactites
These are formed in limestone caves where the water that drips through the roof leaves calcium carbonate when it evaporates

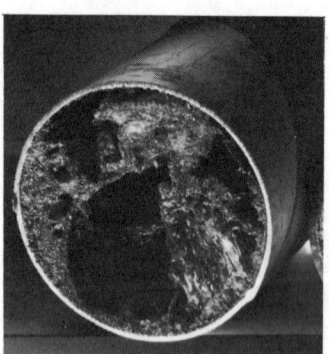

Furred Pipes
Hard water can cause deposits to form in pipes

If a little distilled water is evaporated in a glass basin there will be nothing left. Distilled water is pure water and it has all evaporated away.

Some samples of tap water will leave a whitish powder in the basin. This means that there must have been some solid dissolved in the tap water. If the water has come through chalk or limestone rocks or gypsum, the residue left will be quite noticeable. If you live in Scotland or the north of England, you may find little or no residue.

Tap water that has calcium or magnesium salts dissolved in it is said to be hard. It is difficult to get a lather with soap.

Temporary hardness

Rain may pass through chalk or limestone on its way to rivers and reservoirs where drinking water is collected. Although chalk (calcium carbonate) is not soluble, it will react with water containing carbon dioxide to form soluble calcium hydrogencarbonate.

$$CaCO_3 \ + CO_2 + H_2O \rightarrow Ca(HCO_3)_2$$

calcium   calcium
carbonate   hydrogen-
                              carbonate
(insoluble)   (soluble)

Rain dissolves some carbon dioxide as it falls through the air. It may collect more when it trickles through soil in which there is decaying vegetation. So the water reaching the reservoir contains calcium hydrogen-carbonate.

Soap is a sodium or potassium salt of certain organic acids, stearic, oleic or palmitic. Perhaps this will help you to remember the names:

    Stearate   **O**leate   **A**nd   **P**almitate

When one of these substances meets calcium compounds it forms a scum, e.g.

potassium + calcium → calcium + sodium
  oleate   salt   oleate   salt
 (soap)   (scum)

So the soap is used up and cannot form a frothy lather until all the calcium salts are removed. The sodium or potassium salts left in the water do not interfere with the action of soap.

Removing temporary hardness. When water containing calcium hydrogen-carbonate is boiled for some minutes, calcium carbonate is formed. This is insoluble:

$$Ca(HCO_3)_2 \rightarrow \ CaCO_3 \ + CO_2 + H_2O$$

calcium   calcium
hydrogen-   carbonate
carbonate

This reaction happens inside a kettle over and over again when temporarily hard water is boiled. The calcium carbonate deposited is called 'fur'. But the boiling has made the water soft and a lather can now be formed with soap.

Permanent hardness

Water that passes through rocks containing calcium sulphate or magnesium sulphate will dissolve some of the salt. Such water will be permanently hard. It cannot be made soft by boiling.

To sum up: temporary hardness is caused by calcium hydrogencarbonate or magnesium hydrogencarbonate.

Permanent hardness is caused by any of the salts — calcium sulphate, magnesium sulphate, calcium chloride, or magnesium chloride.

Softening hard water

The soluble calcium or magnesium salts must be changed to insoluble calcium or magnesium compounds which will not react with soap.

Method 1: By adding sodium carbonate.

calcium + sodium → calcium ↓ + sodium
salt   carbonate   carbonate↓   salt

This is why washing soda crystals (hydrated sodium carbonate, $Na_2CO_3.10H_2O$) were used before the days of detergents.

Bath salts are crystals of sodium carbonate, usually coloured and with scent added.

Method 2: Base exchange. Hard water can be made soft with a substance called a **zeolite**. As water flows over it, the zeolite takes the calcium out of the water in exchange for sodium:

zeolite + calcium → calcium + sodium
      salt   salt of   salt
             zeolite
    (in hard   (in-   (soluble)
    water)   soluble)

When the zeolite is all used up it can be recovered by running a solution of common salt (sodium chloride) through the softener:

spent + sodium → restored + calcium
zeolite   chloride   zeolite   chloride

The calcium chloride formed is washed away. Both kinds of hardness can be removed in this way.

Hard water can also be softened by distilling it, but this is an expensive method.

Water treatment beds

Treatment of public water supplies

Rain water falling over a large area is collected in reservoirs. The water is allowed to settle then pumped into filter beds made with sand and gravel with the finer sand at the top.

These filter beds remove any suspended material. But to make sure no bacteria get into our drinking water, a very small amount of chlorine (less than 1 part per million) is often added (see Topic 26).

The water sent out from the water works is tested from time to time to make sure that no harmful bacteria or solids are present. The presence of ammonia means that there may be serious contamination by sewage.

Model filter bed

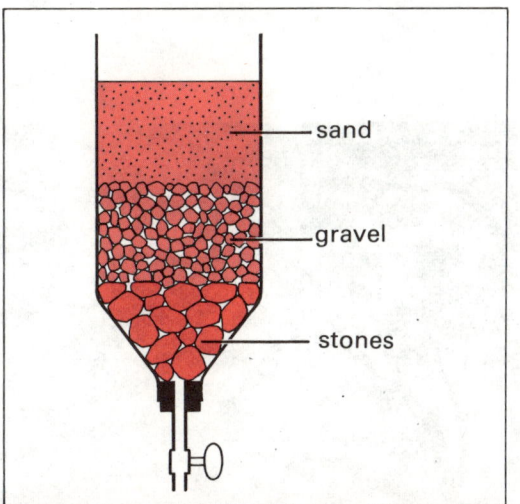

sand

gravel

stones

1 What might be the cause of hardness in a sample of tap water that can be softened by boiling for a few minutes?
2 What is the white deposit called 'fur' sometimes found inside a kettle?
3 How would you soften water that could not be softened by boiling?

AIR

There is a layer of air round the earth about 300 miles thick. It is held there by gravitational force. We are most interested in the first 7 miles of the atmosphere. This region is called the **troposphere** and it is where clouds are formed.

This atmosphere exerts a pressure on everything. We can measure this pressure with a barometer.

You can make use of this air pressure in many ways, such as when you use a drinking straw or a cycle pump.

But the air is not a single substance, it is a mixture of gases principally nitrogen and oxygen. Oxygen is essential for animals to breathe and enables them to live in air.

Carbon dioxide also, although only present in very small amounts, is essential for plant life.

You will not be surprised to learn that there is water vapour in the air. It condenses as rain (too often, some think!).

The Atmosphere

How great is air pressure?

1 A little water is put into a can (a large syrup tin is suitable).

2 The water is boiled for a few minutes to drive out the air.

3 Heating is stopped and the can quickly closed with an airtight rubber stopper.

4 As the can cools it begins to collapse. Only steam was inside above the water. As the steam condenses there is no pressure inside to balance the atmosphere pressing on the outside.

5 The collapse is very rapid if cold water is poured over the tin.

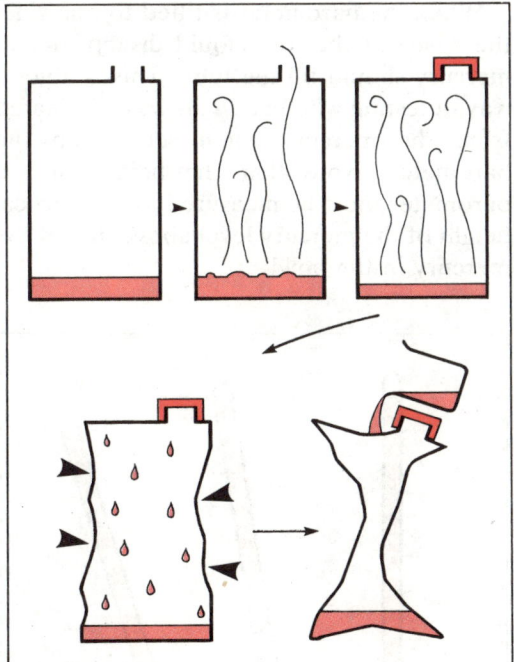

Air pressure
A tin can is collapsed by the pressure of the air

Magdeburg hemispheres

A demonstration of this great atmospheric pressure was given on the large scale in 1654 in the presence of Emperor Ferdinand III. Otto von Guericke made two large hemispheres of bronze, that fitted snugly together, but could quite easily be separated. But when the air was pumped out of the sphere, the pressure holding the two hemispheres together was so great that two teams of eight horses each, pulling in opposite directions, could not pull them apart. The hemisphere only came apart when air was allowed in.

Before deciding how to measure this atmospheric pressure, let us study a way of balancing pressures.

Balancing liquid columns

A U-tube is set up with some mercury filling the bend. Two different liquids are poured on to the mercury. Water in the left limb and paraffin in the right. The amount of these liquids are adjusted until the levels of mercury in the two limbs are the same. Then the pressure on the mercury in the right-hand limb is equal to the pressure on the mercury in the left-hand limb, since the bottom of each column of water and paraffin is at the same level.

$$\text{So} \qquad H_w d_w g = H_p d_p g$$

where H_w and d_w are the height and density of water, and H_p and d_p are the height of the paraffin column and the density of paraffin. g is the acceleration due to gravity.

$$\text{Hence} \qquad d_p = d_w \times \frac{H_w}{H_p}$$

You can find the density of paraffin by measuring the lengths of the columns.

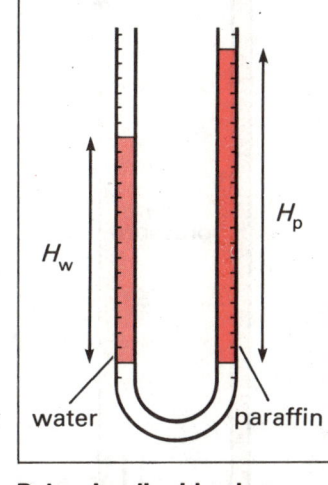

Balancing liquid columns

U-tube manometer

The U-tube balance can be used to measure the pressure of gases. Suppose one limb is connected by rubber tubing to the tap of the gas supply. The U-tube contains water. When you open the tap the gas pressure forces the water to a greater height in one limb.

Suppose the level in this limb is 15 cm above the one in the other. Then the pressure in the gas supply is the pressure of 15 cm column of water plus atmospheric pressure.

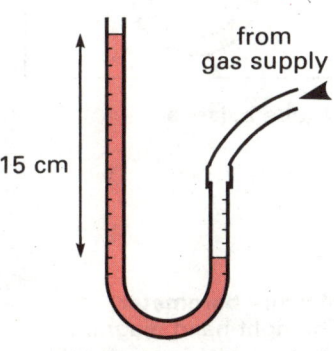

U-tube manometer

1 In the experiment with mercury above, it was found that a water column 17.4 cm high balanced a column of paraffin, 20 cm high. Find the density of paraffin.

2 Why is it dangerous to draw air out of a thin glass flask, using a vacuum pump?

3 Could the Magdeburg hemispheres have been pulled apart more easily if the experiment had been tried at the top of Mont Blanc? (Mont Blanc is about 4800 metres high.) Give the reasons for your answer.

-E

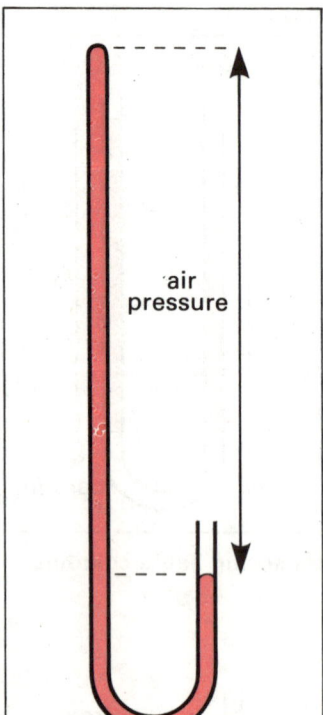

U-tube balance

Can we use the U-tube balance to measure the pressure of the atmosphere? This would need one limb closed so that the atmosphere pressing on the open side would be balanced by a mercury column in the closed tube.

Mercury barometer

A thick-walled glass tube about 5 mm internal diameter, and about 1 m long, is sealed at one end. This tube is carefully filled with mercury. A long wire is used to remove air bubbles from the mercury. With the thumb on the open end, the tube is then inverted in a bowl of mercury. (**Mercury is poisonous** so rubber gloves must be worn and care taken not to spill any.) APPARATUS

When this aparatus is set up, it is found that the atmosphere supports a column of mercury about 76 mm high as measured from the mercury level in the bowl. In 1608, an Italian scientist, suggested that the space above the mercury in a barometer is a vacuum. (Actually a slight amount of mercury vapour is bound to be present.)

✱ TORRICELLI

Mercury barometer
The right-hand diagram shows a test for air bubbles

Warning
Mercury is poisonous: wear rubber gloves

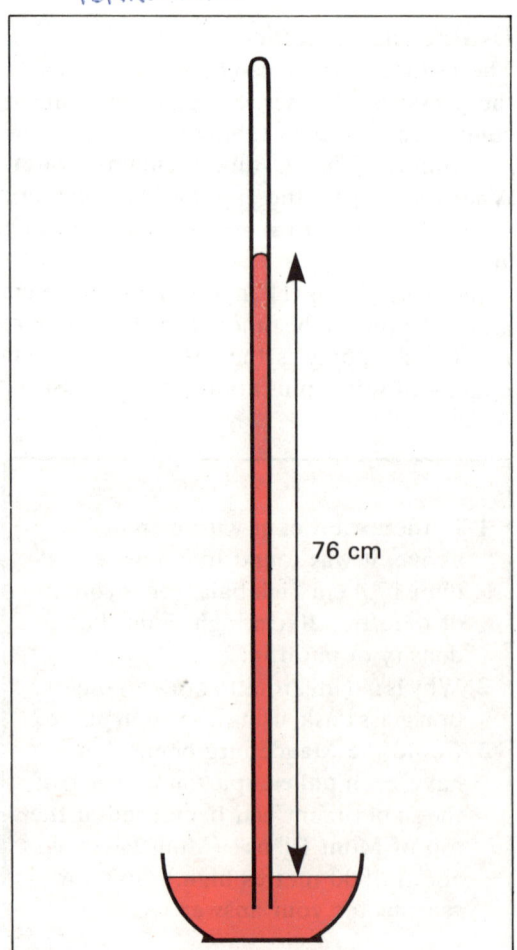

76 cm

1 Would there be any reason why one could not use capillary tubing in order to reduce the amount of the mercury used?
2 Suppose you used water instead of mercury, how high would this water barometer be? (Relative density of mercury = 13.6.)
3 How would the height of the mercury column be affected if there were a little air above the mercury?

When the barometer is tilted to one side, the space at the top should disappear and mercury should fill the tube. This is a good way of testing whether all air was eliminated from the mercury when setting up the barometer. Note that the height of the barometer must be measured as the vertical height of the mercury level above that of the mercury in the bowl.

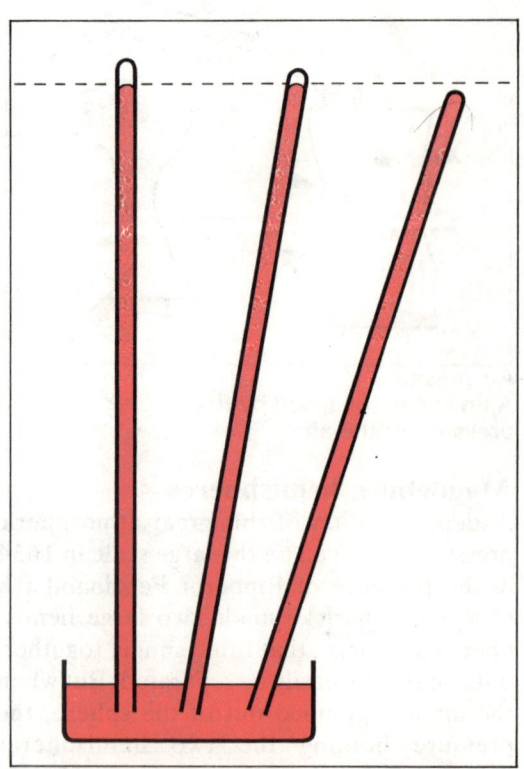

4 What would you expect to happen to the pressure recorded if the barometer were taken to the top of a high mountain?

Aneroid barometer

This barometer does not contain mercury. Instead there is a capsule made of corrugated metal. The atmospheric pressure affects the volume of this thin capsule. The changes in volume are transmitted by a system of levers to a pointer moving over a scale.

In the barograph, the lever system moves a long pointer fitted with an inking point. So a continuous record of the pressure is made on graph paper on a slowly rotating drum.

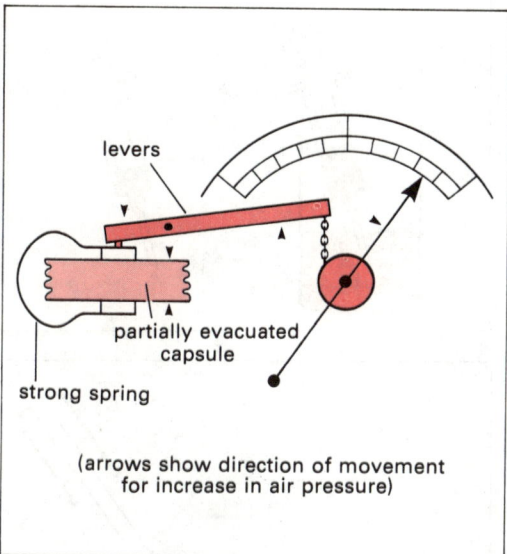

Aneroid barometer

Boyle's Law

The change in volume of air inside the capsule in the aneroid barometer is an example of a general rule.

The volume of a fixed mass of gas is decreased when the pressure on it is increased. Take for instance, the case of holding your finger on the end of a bicycle pump and pressing down on the handle. As the pressure increases, the volume decreases.

Robert Boyle (1627–1691) found that there is an exact connection between volume and pressure. If pressure is p and volume V, then

$$p \times V \text{ is a constant.}$$

In words, Boyle's Law states that the volume of a fixed mass of gas varies inversely as the pressure, provided that its temperature remains constant.

If a glass vessel full of air is pushed down under water until it is half full of water, the pressure of the air inside is twice as great as the atmospheric pressure.

Density of air

Air must have weight. But how can air be weighed when there is air all around? Here is one way:

A round-bottomed glass flask is fitted with rubber stopper. The stopper has a glass tube fitted with rubber tubing. (i) A few cm³ of water is placed in the flask. (ii) The water is boiled. (iii) After steam has been coming off for a few seconds, the flask is removed and the clip fastened. The whole is weighed when cool on an accurate balance. (iv) The clip is then opened and air rushes in. The flask is weighed again, not forgetting the clip. This weight is greater than the first by the weight of the volume of air. (v) This volume can be found by pouring water in to fill the flask to the level of the rubber stopper.

5 In an experiment, it was found that 650 cm³ of air weighed 0.84 g. What is the density of air (a) in g/dm³ (b) in kg/m³?

6 A balloon having this same volume, 650 cm³, was weighed and it was found that the air it contained weighed much more than the 0.84 g. How would you explain this?

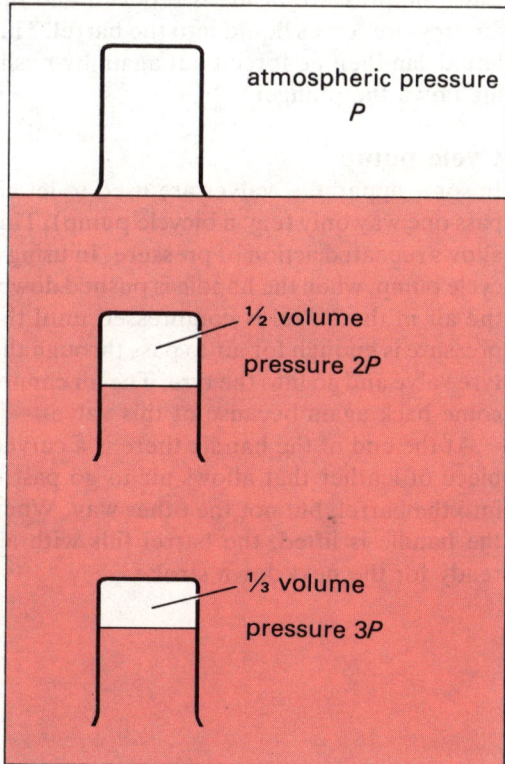

Pressures

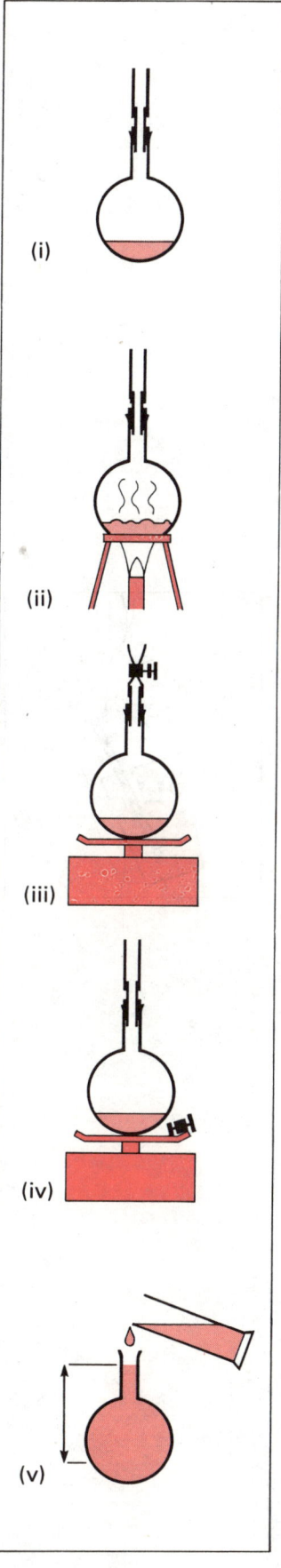

Weighing air

67

A tumbler filled with water with a card placed on top can be turned upside down without the water falling out. This might at first look like magic. But when we remember how great air pressure is (equal to about 10 N/cm²) it is not surprising that it can support a few inches depth of water and a piece of card.

The atmosphere which holds up 76 cm of mercury would hold up 76 × 13.6 cm of water, that is over 10 m.

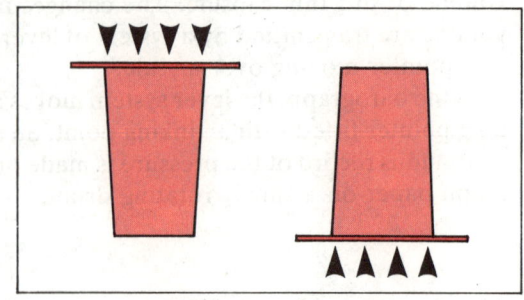

> **1** Why is 76 multiplied by 13.6 in the above calculation?

The following are some everyday examples of the use of air pressure.

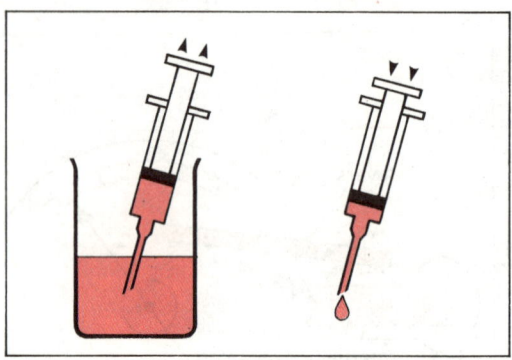

Drinking straw

Air pressure often helps us. Take a simple case. You use a drinking straw to draw up your drink. But you have no magic power in sucking. It is air pressing on the drink in the glass that forces it into your mouth when you reduce the pressure inside the straw.

Syringe

Lifting the piston of a syringe enlarges the volume of air inside, so, according to Boyle's Law, the pressure inside is reduced and the air pressure forces liquid into the barrel. This liquid can then be forced out again by pushing down the plunger.

Cycle pump

In some apparatus, valves are used to let air pass one way only (e.g. a bicycle pump). This allows repeated action of pressure. In using a cycle pump, when the handle is pushed down, the air in the barrel is compressed until the pressure is enough for air to pass through the tyre valve and go into the tyre. The air cannot come back again because of this valve.

At the end of the handle there is a curved piece of leather that allows air to go past it into the barrel, but not the other way. When the handle is lifted, the barrel fills with air ready for the next down stroke.

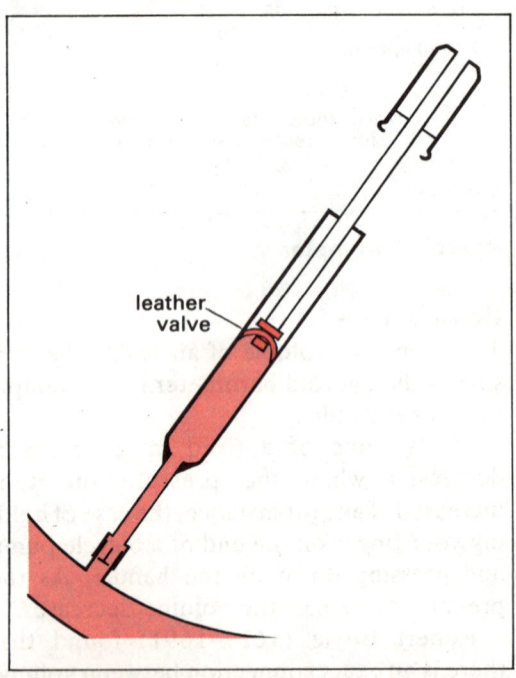

leather valve

> **2** What could the cycle pump be used for if the curved leather valve at the end of the handle were to be turned round the other way, that is with the curved end facing towards the handle?

A lift pump

The part that air pressure plays in the action of this pump can be seen in the diagram. In the upward stroke, the valve (a) is closed and the water above the valve is lifted out. Valve (b) is open and the air pressure forces water into the barrel.

> **3** Write down what happens during the down stroke of the pump.

Remember that the air pressure can support not more than about 10 metres of water, you see that this pump cannot lift water more than this.

> **4** How could you arrange two lift pumps to raise water 15 metres?

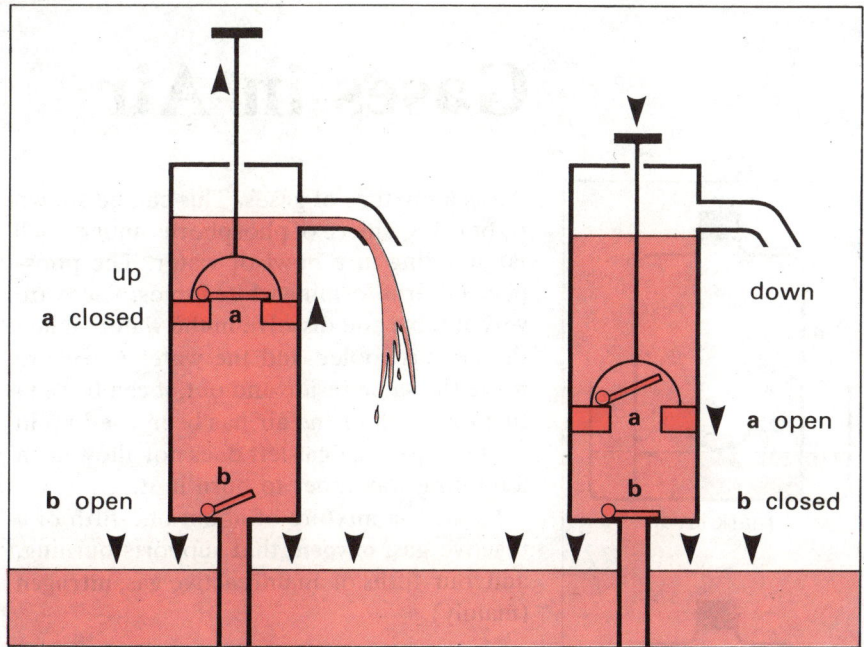

Lift pump

Vacuum cleaner

A fan driven by an electric motor pushes air out from the body of the cleaner, so reducing the pressure inside the dust bag. Air pressure forces air into the cleaner, together with dust particles. Often the movement of the dust is helped by a brush as well.

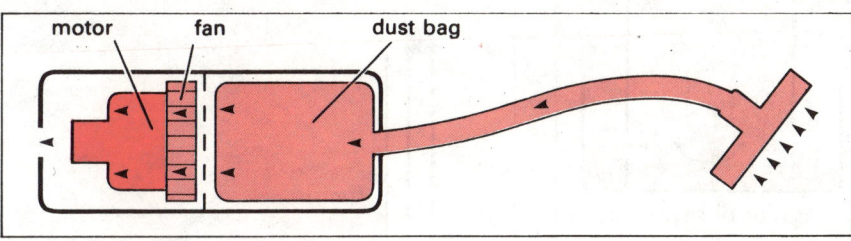

Vacuum cleaner

Siphon

Liquid can be drawn off from a high position to a lower one by a tube so long as the tube is filled with liquid and the lower end of the siphon tube is below the top level of liquid. In this case, the atmospheric pressure acts on the water in both vessels. It used to be thought that the siphon works because the pressure at A is greater than at B by the weight of the column of liquid (h). But since it has been found that a siphon still works in a vacuum, it seems that the cohesion of the liquid in the tube may partly explain the action.

In the flushing cistern, pushing down the handle lifts the plunger. This lifts water into the top bend of the siphon, so siphon action starts and empties the cistern. The floating ball falls down with the water and this opens the inlet for water to come in to fill the tank again. The ball is carried up with the level of water, until it closes the inlet valve and stops more water coming in.

> **5** Try to draw a diagram of how self-flushing cistern empties itself.

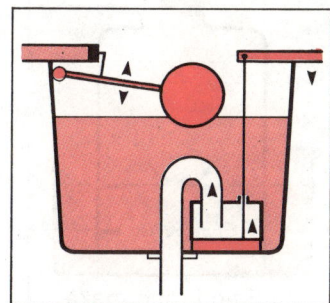

Flushing cistern

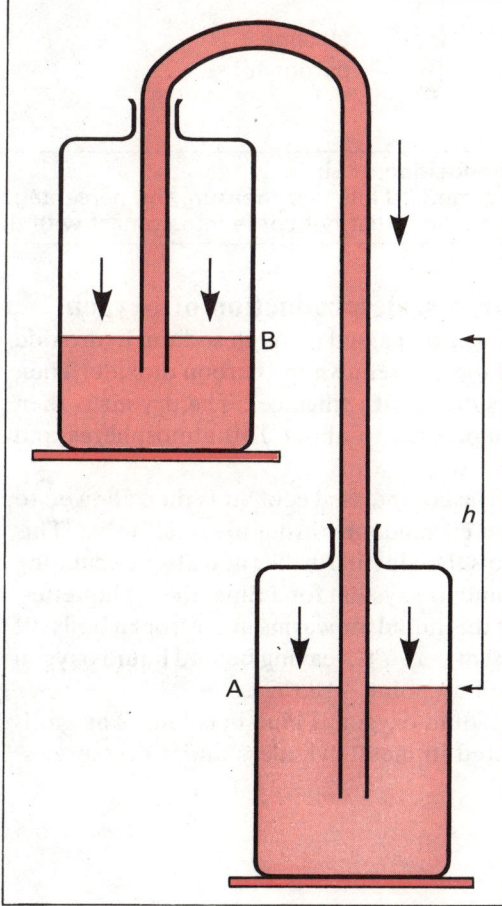

Siphon

Gases in Air

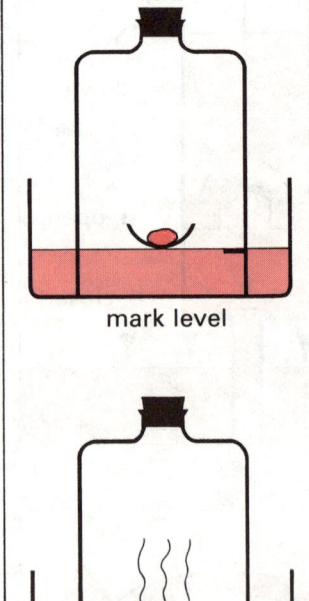

mark level

ignite phosphorus

cool and then make levels equal

gas left does not support burning

Composition of air
Air is one-fifth oxygen and four-fifths nitrogen

Air is a mixture of gases. This can be shown by burning a piece of phosphorus under a bell jar standing in a bowl of water. The phosphorus burns forming white fumes. These are very soluble and dissolve in the water. When the jar has cooled and the water levels are made the same inside and out, it can be seen that one-fifth of the air has been used up in the burning. The gas left does not allow even a burning wax taper to burn in it.

So air is a mixture of about one-fifth of a reactive gas, **oxygen**, that supports burning, and four-fifths of an unreactive gas, **nitrogen** (mainly).

If a burning candle is used instead of phosphorus, in the above experiment, only about one-seventh of the air is used. This is because the candle does not have the chance of using up all the oxygen before the carbon dioxide, which is a dense gas, puts the candle out.

Percentage of oxygen in air
The proportion of oxygen can be found more accurately by shaking up air in a long tube with an alkaline solution of pyrogallol. This solution removes the oxygen and the percentage removed can be found by opening the tube under water.

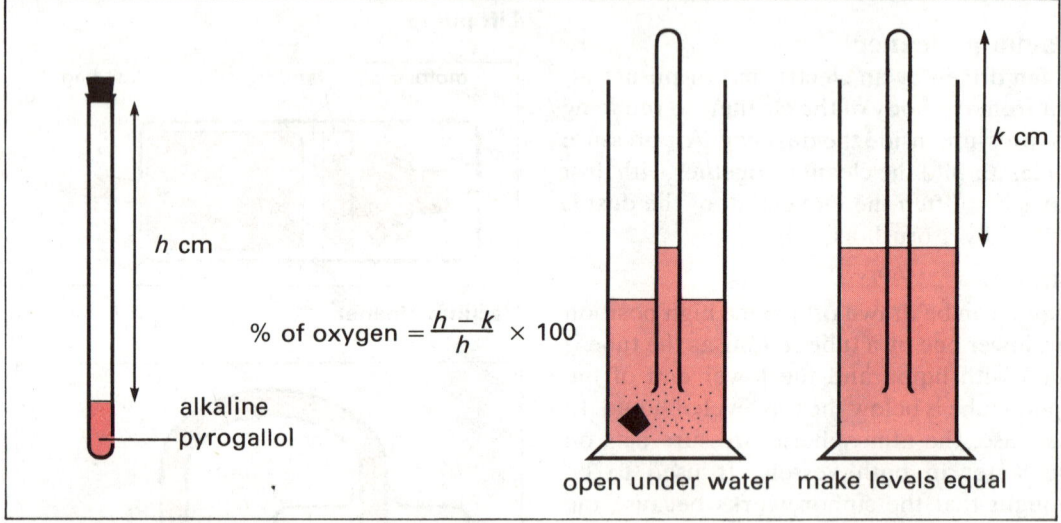

h cm

$$\% \text{ of oxygen} = \frac{h - k}{h} \times 100$$

alkaline pyrogallol

k cm

open under water   make levels equal

Composition of air
This method lets you measure the percentage of oxygen in air more accurately. Alkaline pyrogallol must not come into contact with the skin. Rubber gloves must be worn.

Large scale production of oxygen
Air is first passed through sodium hydroxide solution to remove the carbon dioxide. Then it is dried with silica gel. The dry air is then compressed to about 150 atmospheres and cooled.

The compressed cool air is then allowed to expand suddenly through a small valve. This cools the air still further and after circulating round this system for a time, the air liquefies. As this liquid air warms up, nitrogen boils off first at −196°C, leaving behind liquid oxygen (boiling point −183°C).

Liquid oxygen is blue in colour. The gas is stored in metal cylinders under pressure.

Oxygen in use
This man is inserting a lance supplying oxygen to a steel-making furnace

Laboratory preparation of oxygen

If a cylinder of the gas is not available, oxygen can be prepared by heating a substance that contains a lot of oxygen, such as potassium permanganate, $KMnO_4$.

A convenient method of preparing the gas is to drop hydrogen peroxide on to a little manganese(IV) oxide, (manganese dioxide, MnO_2) in a flask. Oxygen is collected over water.

$$2H_2O_2 \rightarrow 2H_2O + O_2$$
hydrogen   water   oxygen
peroxide

Hydrogen peroxide gives off oxygen slowly on its own, but the action is speeded up by the manganese(IV) oxide. The manganese dioxide acts as a catalyst. Although it increases the speed of the reaction, it can be recovered unchanged at the end of the experiment.

Properties of oxygen

It is a colourless gas and has no smell or taste. It is only slightly soluble in water. Animals and plants living in water depend on oxygen dissolved in the water. Oxygen is a very active gas. It supports burning but does not itself burn. Sulphur, magnesium and carbon all burn more brightly in oxygen than in air. Even iron wire can be made to burn brightly in the gas. In all these cases the elements form oxides.

Acidic and basic oxides

A striking difference is found if these oxides are dissolved in water. When litmus is added to the solutions, some turn the litmus red while others turn it blue.

Element	Oxide	Dissolved in water	Colour litmus turns
Carbon	CO_2	H_2CO_3	red
Sulphur	SO_2	H_2SO_3	red
Phosphorus	P_4O_{10}	H_3PO_4	red
Sodium	Na_2O	$NaOH$	blue
Magnesium	MgO	$Mg(OH)_2$	blue

Acidic oxides are formed by the elements carbon, sulphur and phosphorus; **basic** oxides are formed by the elements sodium and magnesium.

That is: metals combine with oxygen to form basic oxides; non-metals form acidic oxides.

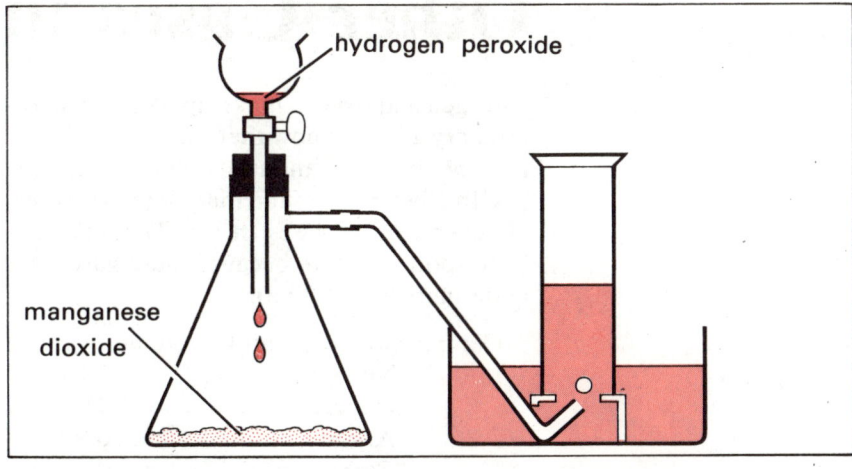

Preparation of oxygen

Uses of oxygen

1 As an oxidising agent in making steel from pig iron.
2 To aid rapid burning of fuels in rockets.
3 As oxy-hydrogen and oxy-acetylene flames in welding.
4 Used to help patients' breathing, e.g. oxygen tent.
5 It is also mixed with anaesthetics for surgical and dental operations.
6 Compressed oxygen may be carried in breathing apparatus by divers, firemen, climbers and pilots.

Oxygen in use
Here a swimmer is being revived with oxygen from a cylinder

1 Which of the following substances would probably give off oxygen when heated: potassium dichromate, $K_2Cr_2O_7$; magnesium oxide, MgO; sodium peroxide, Na_2O_2; ammonium chloride, NH_4Cl?
2 Write down the following oxides in two columns according to whether they are acidic or basic oxides: Sodium oxide, Na_2O; mercury(II) oxide, HgO sulphur trioxide, SO_3; magnesium oxide, MgO; carbon dioxide, CO_2; copper(II) oxide, CuO.

Other Gases in Air

32

Nitrogen and oxygen make up 99 per cent of ordinary air. We know that carbon dioxide is present, because animals breathe out this gas. In 1894 it was found that there are traces of several other gases in air. This had not been spotted before because these gases are quite inert (unreactive).

Composition of dry air by volume

Nitrogen	78%
Oxygen	21%
Argon	0.9%
Carbon dioxide	0.03%
Helium	
Neon	}traces
Krypton	
Xenon	

Argon, helium, neon, krypton and xenon are called 'noble' gases. Because of their inertness, they are used in gas-filled lamps and discharge tubes. Helium is a useful non-flammable substitute for hydrogen in balloons.

Nitrogen, although inactive as an element forms important compounds. Carbon dioxide is only present in very small amounts, but it plays a very important part in the life of plants (see Topic 65). We already know that it is given off when calcium carbonate is strongly heated.

$$CaCO_3(s) \rightarrow CaO(s) + CO_2(g)$$

(s) stands for solid, and (g) for gas.

Action of acids on calcium carbonate

When dilute hydrochloric or nitric acid is added to calcium carbonate, carbon dioxide is given off with much fizzing

$$\underset{\substack{\text{calcium} \\ \text{carbonate}}}{CaCO_3} + \underset{\substack{\text{hydrochloric} \\ \text{acid}}}{2HCl} \longrightarrow \underset{\substack{\text{calcium} \\ \text{chloride}}}{CaCl_2} + \underset{\text{water}}{H_2O} + \underset{\substack{\text{carbon} \\ \text{dioxide}}}{CO_2}$$

Without writing an equation, it would not be easy to see that water is produced in the reaction.

The dilute hydrochloric acid is poured down a thistle funnel on to marble chips in a flask. Because the gas is very dense it can be collected by the displacement of air as in the diagram. The gas can also be collected over water since it is not very soluble.

Preparation of carbon dioxide

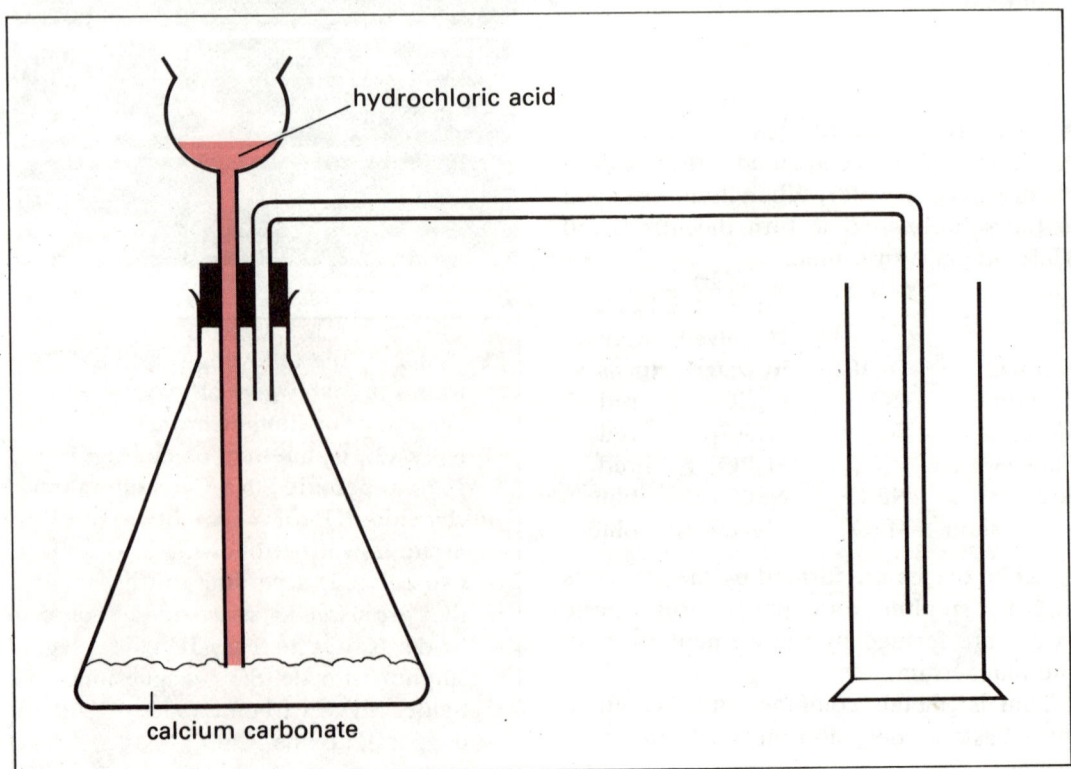

hydrochloric acid

calcium carbonate

Properties of carbon dioxide

It is a colourless gas with a slight not unpleasant smell and a taste of soda water. At low temperature, it forms a white solid known as 'dry ice'.

When passed into limewater (calcium hydroxide solution) a precipitate of calcium carbonate is formed. This dissolves when excess CO_2 is passed (see Topic 8).

Carbon dioxide is an acidic oxide. With water it forms a weak solution of carbonic acid

$$H_2O + CO_2 \rightarrow H_2CO_2$$

It reacts readily with sodium hydroxide forming first the carbonate. Excess will form sodium hydrogencarbonate

$$2NaOH + CO_2 \rightarrow Na_2CO_3 + H_2O$$
sodium   carbon   sodium   water
hydroxide dioxide carbonate

$$NaOH + CO_2 \rightarrow NaHCO_3$$
sodium hydrogen-carbonate

Most substances will not burn in carbon dioxide. An exception is burning magnesium which continues to burn forming white fumes of magnesium oxide and leaving black spots of carbon on the jar.

$$2\,Mg + CO_2 \rightarrow 2MgO + C$$
magnesium carbon   magnesium carbon
dioxide

Uses of carbon dioxide

The various uses depend on its properties.
1 Under pressure the gas dissolves in water forming a solution with a pleasant taste. Fizzy drinks contain this solution of carbon dioxide. Almost all the carbon dioxide manufactured in industry is used in making bottled beer and soft drinks.
2 'Dry ice' is used to keep ice cream cold. It is much more efficient and less messy than ordinary ice for this purpose.

3 Carbon dioxide is denser than air. An interesting experiment is to drop a soap bubble into a jar half-filled with carbon dioxide. The bubble will float at the junction of the carbon dioxide and the air. Since the gas does not support combustion, it is used in some fire extinguishers. The gas is dense and can be poured over a fire so it extinguishes the fire by preventing air from getting to the burning material.

Another type of fire extinguisher contains a solution of sodium hydrogencarbonate and a sealed bottle of strong acid. When the extinguisher is used a plunger breaks the bottle of acid and a jet of water and carbon dioxide can be directed at the fire.

$$NaHCO_3 + HCl \rightarrow NaCl + CO_2 + H_2O$$

All green plants need carbon dioxide (see Topic 65).

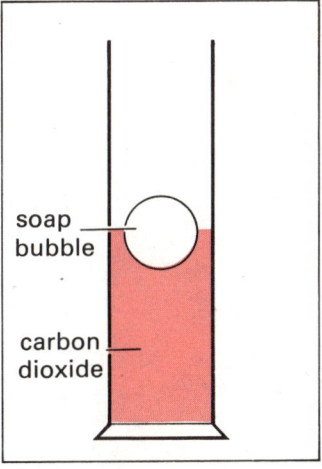

Carbon dioxide is a dense gas
A soap bubble will 'float' on carbon dioxide

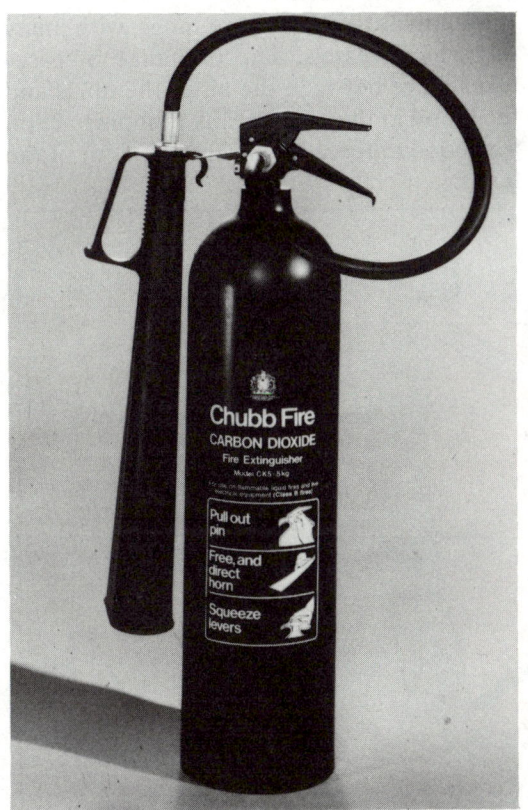

Fire extinguisher
Carbon dioxide 'blankets' a fire and prevents burning

---

1 Why does a can of 'Coke' fizz when the seal is broken?
2 What would you see if you mixed soda-water with limewater?

3 Why would nitrogen be less effective than carbon dioxide as a fire extinguisher?

Living in Air

33

Most animals need to breathe in oxygen and get rid of the carbon dioxide that forms in the tissues of the body. Animals living in water can use the oxygen that is dissolved in water and their waste carbon dioxide will dissolve in the water. But animals that live in air still need a watery surface for this exchange of gases to take place.

Different kinds of organs are involved in different animals. Fishes have gills. Insects have tracheal tubes leading from tiny pores into the body. Frogs can breathe partly through their moist skin.

Mammals like man have lungs, with many tiny blood vessels. The exchange of gases takes place between the air in the lungs and the blood in these vessels. A pumping action is needed to force the air into and out of the lungs.

Mechanism of breathing

When you breathe in, two sets of muscles contract. These are the muscles between one rib and the next and the muscles of the diaphragm. The muscles between the ribs are called the **intercostal** muscles. When these muscles contract the chest cavity is made larger. The lungs are inside the chest cavity or thorax. They are separated from the wall of the thorax by a double layer of thin tissue. The two layers are separated by a little fluid. The inside of the lungs is connected with the atmosphere through the trachea and mouth and nose. When the thoracic wall moves outwards, the lungs are pushed out with it and more air enters the lungs.

When you breathe out, the muscles between the ribs and the diaphragm muscles relax.

The complete cavity of the lungs is 4–5 dm^3, although the **tidal air** (that is the amount of air normally breathed in and out) is only about 500 cm^3 (0.5 dm^3).

Breathing
When you breathe in muscles between the ribs contract: this increases the volume of the lungs

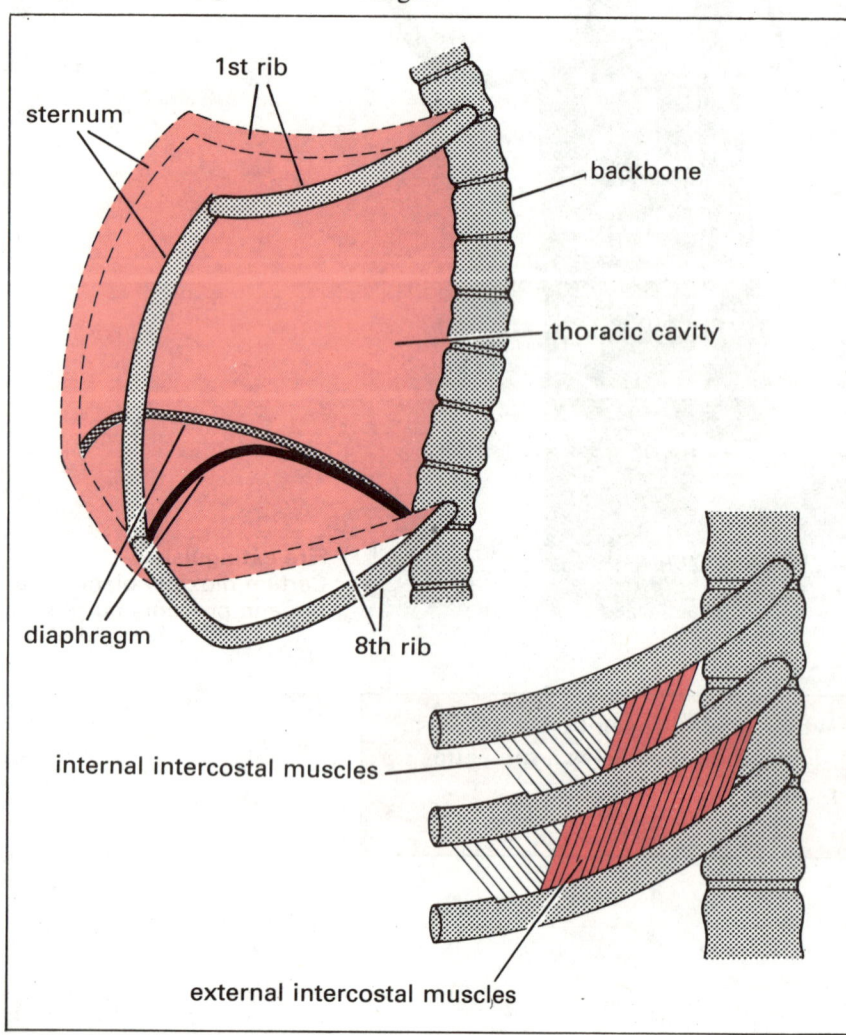

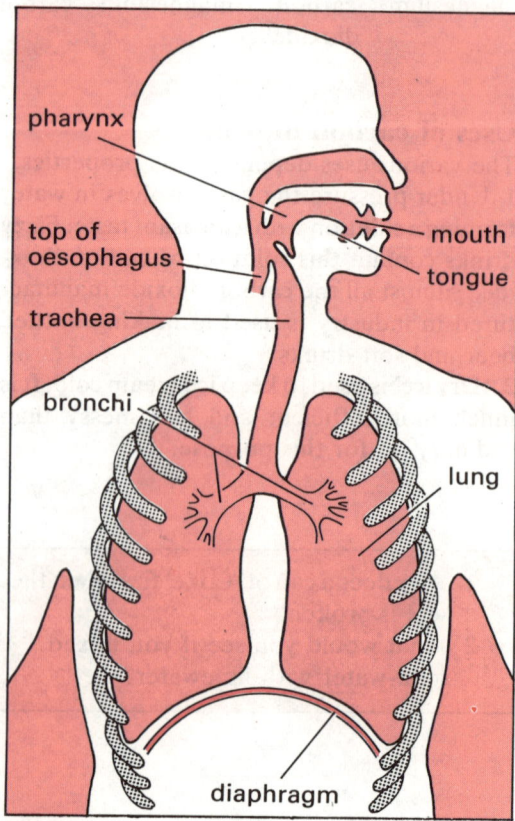

The lungs are connected to the nose and mouth by the trachea

Effect of breathing on air

When we compare the air we breathe out with the air breathed in we find several differences.

Air breathed in	Air breathed out
oxygen 21%	oxygen 16–17%
carbon dioxide 0.03%	carbon dioxide 3–4%
can be cold	warmer
can be dry	moist
can be dusty	free from dust

There is only a slight increase in the carbon dioxide content. But it can be demonstrated rather neatly by using the apparatus shown.

If you breathe in and out in a normal way (preferably with the nose clipped) so that all the air passes in and out through the apparatus, two things will be seen. First, the slight precipitate in the limewater in flask A shows that, there is some carbon dioxide in the air breathed in. Secondly, the precipitate formed in flask B is much greater. This shows that carbon dioxide must be produced in the human body and added to the air breathed out.

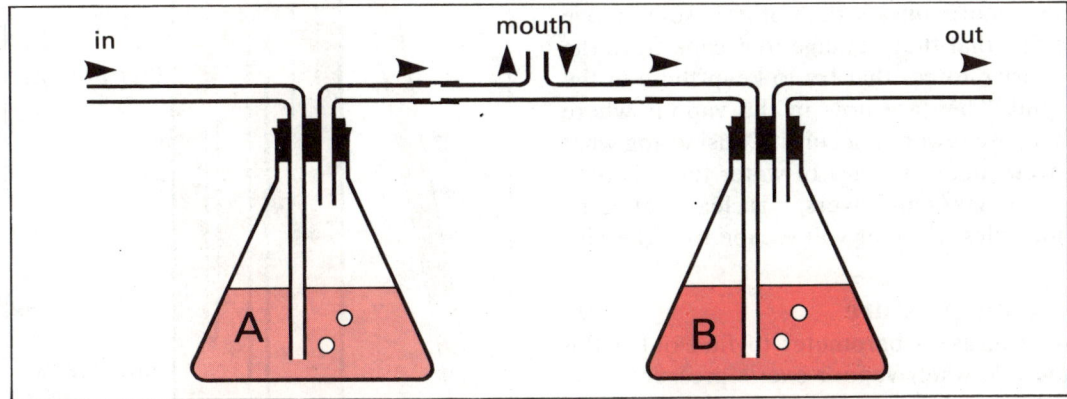

Composition of air
The air you breathe out contains more carbon dioxide than the air you breathe in

Flying

Although man lives in air, he is not able to fly, like birds and insects, without mechanical aid. One mammal, the bat, is able to fly.

Birds are well-designed for flying. Their bodies are very light with hollow bones and they are stream-lined. Also their chest muscles are very powerful.

They fly by flapping their wings and by gliding. A bird can glide without moving its wings because the wings are shaped like aerofoils.

Insects also beat their wings by muscle action. The honey bee makes about 250 wing beats a second. Even then it reaches a speed of only 5–6 miles/hour. But a dragon-fly can travel over 15 miles/hour with only 30–40 wing beats per second.

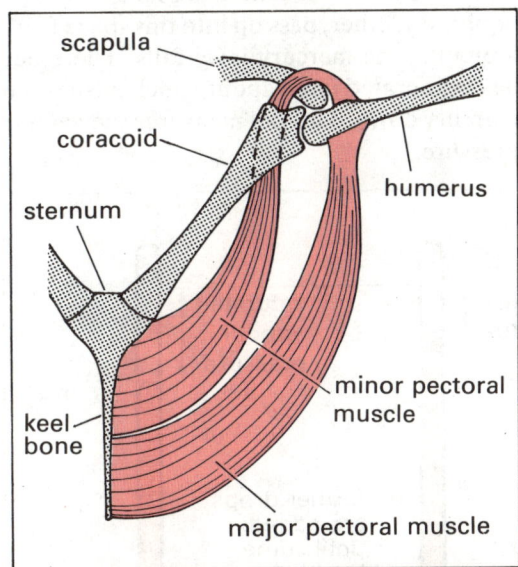

Flying
Birds have very strong chest muscles to help them to fly

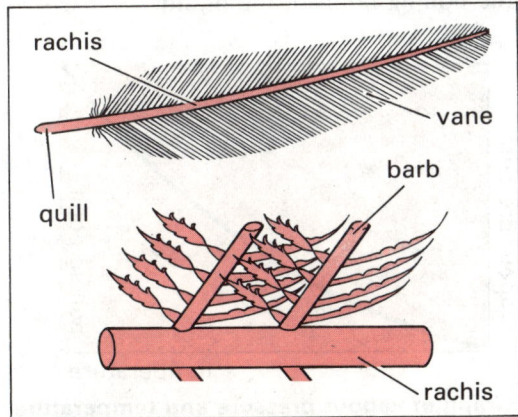

Structure of a bird's feather

1 When we breathe in what part is played by (a) the intercostal muscles, (b) the diaphragm?
2 Write down three ways in which the air we breathe out is different from the air we breathe in.
3 Which of the following animals can fly: wasp, ladybird, cockroach, bat, swallow, swallow-tail butterfly, emu, pheasant?

Air contains water vapour. This varies in amount from day to day, even hour to hour.

During very cold weather the windows of a room sometimes get 'steamed up' on the inside. This is particularly noticeable when there are several people in the room (all breathing, of course). So when air containing water vapour is cooled it will deposit condensed water on the cold surface.

Think of the molecular theory. The molecules of a liquid are in motion. Some move more quickly than others; some move so fast that they manage to escape from the cohesive forces that try to keep them in the liquid. They are now in the vapour where there are fewer molecules. Considering what a tremendous surface of water there is over oceans, lakes and rivers, it is clear that many molecules of water will escape into the air.

Vapour pressure

We can use a barometer to find out if this invisible water vapour exerts pressure.

Above the mercury in a mercury barometer there is a vacuum. If we let a drop of a liquid, say, ether, pass up into this space from a pipette, the mercury level falls. The ether has evaporated into vapour which pushes the mercury down. This fall measures the vapour pressure.

When no more liquid will evaporate the **saturated vapour pressure** has been reached.

Suppose the whole barometer is put in a glass jacket through which we can pass water at different temperatures. We find that the saturated vapour pressure will increase in value as the temperature rises.

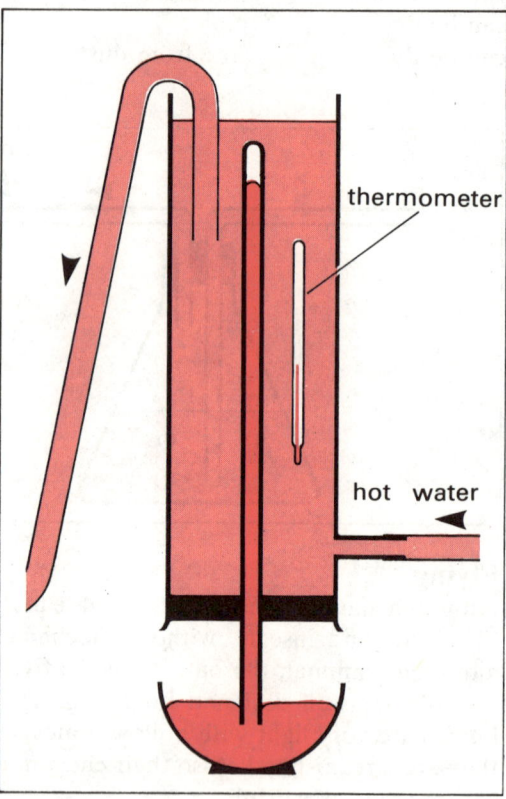

thermometer

hot water

Vapour pressure
Vapour pressure
The temperature of the barometer can be changed with a water jacket

The results can be shown in a graph.

The more water vapour there is in the air, the greater will be the vapour pressure. The higher the temperature the more water vapour the air can hold without condensing.

The saturated vapour pressure of any liquid is equal to the atmospheric pressure at the boiling point of the liquid.

Vapour pressure

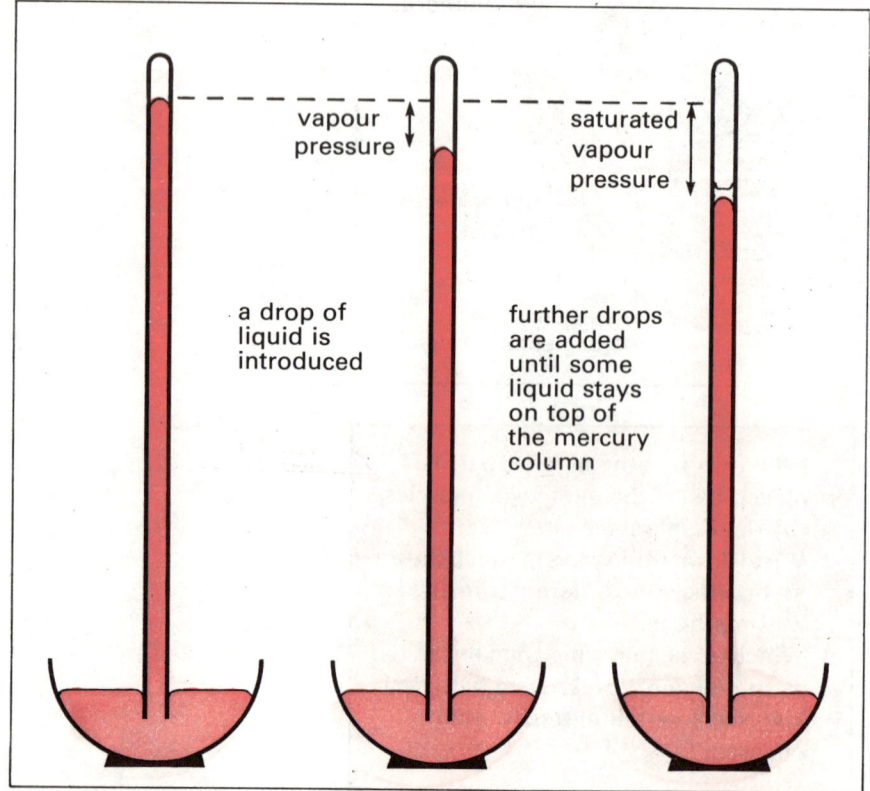

vapour pressure

saturated vapour pressure

a drop of liquid is introduced

further drops are added until some liquid stays on top of the mercury column

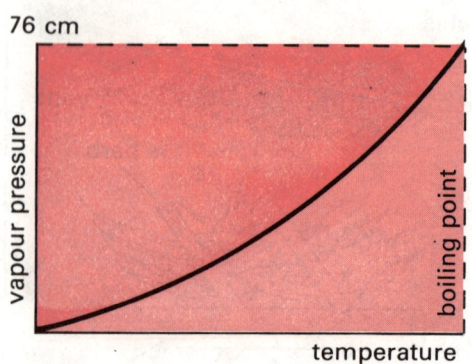

76 cm

vapour pressure

temperature

boiling point

Graph of vapour pressure and temperature

Humidity

This shows how much water vapour is in the air. A mass of air that is not saturated with water vapour will deposit moisture (dew) when cooled. The temperature at which this dew forms first is called the dew point. We can think of relative humidity as:

$$\frac{\text{saturated vapour pressure at dew point}}{\text{saturated vapour pressure at air temperature}}$$

Hygrometers

A hygrometer is an instrument that is used for finding the relative humidity. One kind commonly used is the wet and dry bulb hygrometer.

Two thermometers are set up, one with a muslin cloth round the bulb. The muslin cloth has the other end dipping into water. So the bulb is kept damp and as the water evaporates, a cooling effect is produced (see Topic 51).

The more humid the air is the less the cooling will be, and the less the difference between the readings of the two thermometers. The relative humidity is found by using tables.

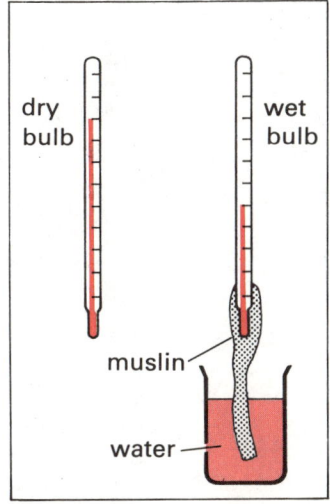

Hygrometer

Transpiration

A potted green plant is left under a bell jar for a few hours. Drops of moisture appear on the inside of the jar. These could not have come from the soil in the pot because a sheet of plastic covered the pot. So the leaves probably gave off water vapour. This loss of water vapour from the leaves of a plant is called transpiration. How much water is lost in this way?

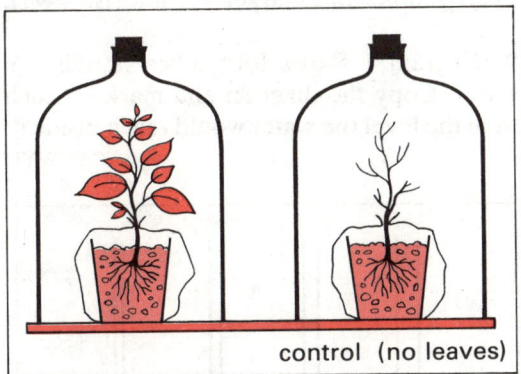

control (no leaves)

A specimen tube with leafy twig dipping into water is used. An oil layer on the water prevents direct evaporation of water. The tube is weighed when set up then again a few hours later. These weights are compared with the corresponding weighings for a similar tube with the leaves removed from the twig. This acts as a control.

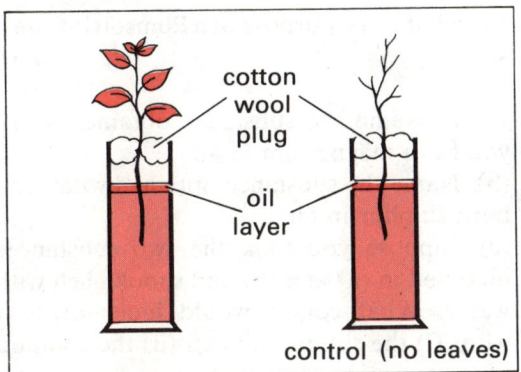

Transpiration experiment

Stomata

If the surface layer on the underside of a green leaf is pulled off and looked at under a microscope it is seen that a number of tiny pores are present between the cells of the epidermis (outer layer of cells).

These pores are mouth-like. They are called **stomata** (singular: stoma). Water vapour escapes through these stomata from the tissues inside the leaf.

The rate of transpiration is greater when the air round the plant is moving and dry rather than moist and still. Transpiration is also faster when it is warm rather than cold.

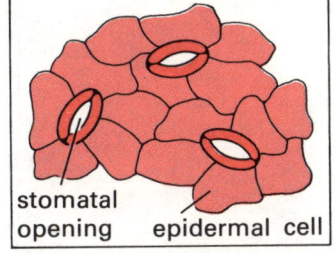

Stomata in underside of leaf

Fog and mist

When the water vapour in the air is saturated, a fall in temperature will cause water to condense. If the air is reasonably clean, these water droplets will form mist.

But the air of some towns carries a lot of soot and dust particles and when the water droplets condense on these, fog is formed. Since smokeless zones have been introduced, there are not so many fogs.

1 What is the difference between fog and mist?
2 Write down the missing words in the following statements. A plant loses water from its * . This water passes out through the * . The process of water loss is called * .
3 Why does the grass verge, but not the pavement, become damp as the air cools in the evening?

Examination Questions: II

1 A freshly cut cylinder of potato exactly 25 mm long was placed in a petri dish containing strong salt solution.
(a) What would you expect to happen to the length of the cylinder after 15 minutes? (b) State briefly a reason for your answer to (a).
ALS

2 Choose words from the list below to complete the following passage: semipermeable; food; transpiration; leaves; root hairs; osmosis; respiration; mineral salts; permeable; stem.
Water passes into a plant through its * by a process called * which can only take place through a * membrane. This water contains the * necessary for healthy growth and development.
EM

3 Write down a complete statement, using one of the phrases (a)–(e)
Substances which are more dense than water . . .
(a) must be solid; (b) float on water; (c) sink in water; (d) cannot be liquids; (e) are always metals.
EM

4 Which two answers are correct?
The pressure in water . . .
(a) is the same at all depths; (b) increases with depth; (c) is zero; (d) only acts downwards; (e) only acts upwards; (f) acts equally in all directions.
EM

5 Pick out the answer you think is correct and write out the complete statement.
'Two elements found in air are . . .'
(A) carbon dioxide. (B) oxygen and carbon dioxide. (C) nitrogen and water vapour. (D) oxygen and nitrogen. (E) carbon dioxide and water vapour.
Y

6 Choose the correct answer and write out the complete statement.
Compared with inspired air, expired air . . .
(A) is cooler. (B) contains less oxygen. (C) contains less nitrogen. (D) contains less carbon dioxide. (E) contains less water vapour.
Y

7 Write out the following statements with the missing spaces correctly filled.
When it rains the water dissolves * gas as it falls through the air. This makes the rain slightly * . When the rain falls on limestone or * , the chemical name for which is * , this dissolves to form a solution of * which makes the water * so that with soap it will not * easily. If this water is * , it becomes softer and for this reason it is often termed * . Sometimes water cannot be completely softened in this way. This is because * might be dissolved in the water. In this case a softener which could be used is * or *
M

8 Explain briefly why a balloon, filled with hydrogen becomes larger when it rises. WYL

9 Diagram 1 shows four tubes standing in water. Copy the diagram and mark on each tube the level the water would reach inside it.
WYL

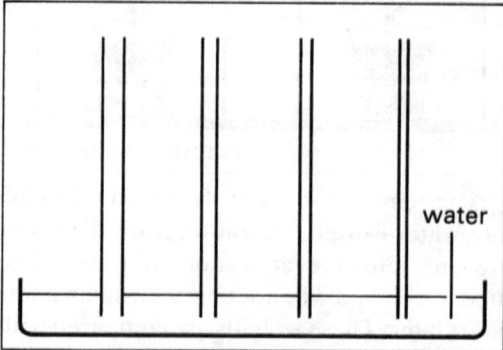

water

Diagram 1

10 What is the purpose of a Plimsoll line on a ship?
WYL

11 (a) Name the substance obtained when you burn magnesium in air.
(b) Name the substance obtained when you burn sulphur in air.
(c) Suppose you took the two substances obtained in (a) and (b) and shook each with water. What colour would indicator turn with: (i) the solution of (a); (ii) the solution of (b)?
WYL

12 Three porous pots (which have walls containing many holes) were set up as in diagram 2.

(a) What causes bubbles to come out of the tube in experiment 1?

(b) What causes the water to rise up the tube in experiment 2?

(c) What do these experiments indicate about carbon dioxide and hydrogen?

(d) Experiment 3 was used as a control. Suggest a reason why this was necessary. WM

13 Use diagrams to describe how air is moved into and out of the human lungs from the atmosphere.   EM (part)

13 Three men, A, B and C each had a barometer. A was at sea level, B was at the top of a mountain and C was down a deep mine. (i) State whose barometer showed the highest pressure. (ii) State whose barometer showed the lowest pressure.   W

14 State two different ways in which the *rate* of dissolving sugar can be increased.   W

15 Choose the correct answer and write out the correct statement. The crystallisation of a salt from a hot solution on cooling depends on (a) the use of an evaporating basin and a bunsen burner, (b) the purity of the salt, (c) the salt's greater solubility in hot water than cold, (d) the salt's greater solubility in cold water than hot.   NW

16

Table of densities (all given in g/cm³)

Alcohol	0.79	Bone	1.90
Ash	0.75	Cork	0.25
Balsa wood	0.22	Ice	0.92
Brine	1.20	Nylon	1.15

Use the table given above to list (i) the solids that will float in water (density 1.00 g/cm³); (ii) the solid that will float in brine, but not in water.   SW (part)

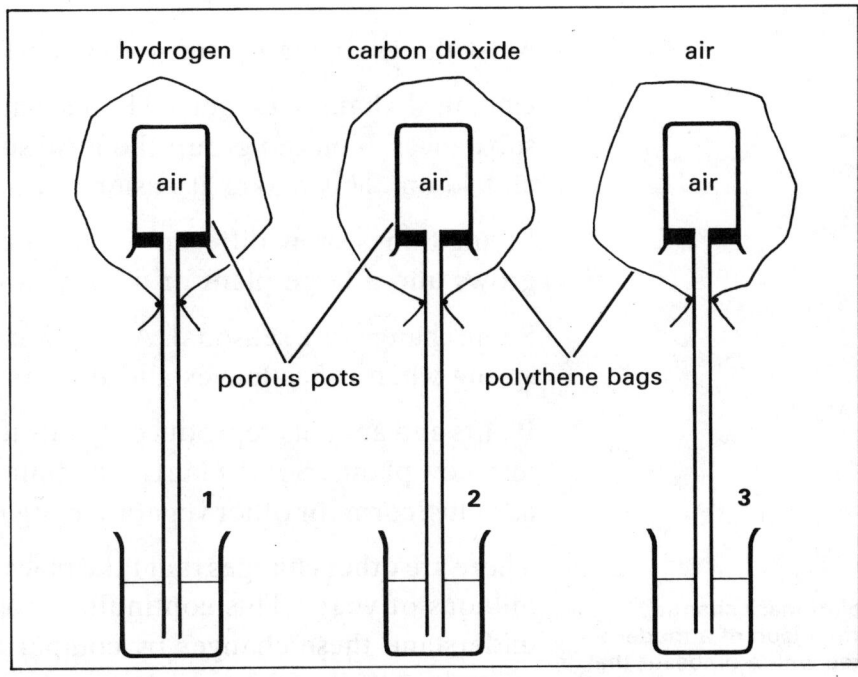

Diagram 2

17 (a) Describe and explain how you would obtain, from muddy sea-water, (i) clean sea-water: (ii) a sample of the salts in sea-water, (iii) pure water.

(b) Why would the constant use of temporary hard water lead to problems in heating systems?

(c) In what way is hard water beneficial to health?   W

18 Here is a list of some chemical processes: chromatography; crystallisation; dissolving; reduction; fractional distillation; oxidation. State which process is best for each of the following purposes: (a) Removing grease spots from a coat. (b) Obtaining a solid from a solution of a solid. (c) Separating the coloured substances present in grass. (d) Separating crude oil into several different liquids.   EAN

CHANGE & GROWTH

Most things, living or non-living, undergo change.

Chemical change happens when substances are heated or react with other substances. You can group chemical substances into classes, like metals, acids and so on. This makes it easier to find rules about chemical changes.

Living things show different kinds of change. They grow, as when a tiny seed grows into a large plant or a baby boy grows into a man.

Some changes are seasonal. A tree that is bare in winter looks quite different in spring when it has leaves and flowers.

Plants and animals reproduce. Plants usually produce seeds and these develop into new plants. Some plants, like tulips and gladioli survive during the winter as bulbs, corms or other vegetative organs. These grow into plants in the spring.

There are other changes that take place in living things very slowly, over many millions of years. This continuing process is called **evolution**. We can partly understand these changes by comparing fossils of prehistoric plants and animals with the plants and animals living today.

Evolutionary change
Comparison of a modern horse with a dinosaur that might have lived about 150 million years ago

Chemical & Physical Change

Physical change

When sulphur melts it forms a liquid which is still sulphur. There has just been a change of **state**, not of chemical nature. Solid sulphur forms when the liquid cools.

Water boils to form steam, but this is not a different substance, it is still water. Steam becomes water when it cools.

Salt can easily be got back from solution by heating to drive off the water.

These are all **physical changes**; no new substances are formed.

Chemical change: iron and sulphur experiment

Iron is attracted by a magnet, but sulphur is not. Iron is attacked by dilute hydrochloric acid and hydrogen is given off. Sulphur is not attacked. You can distinguish between iron and sulphur by these tests.

Now suppose some iron filings and powdered sulphur are mixed thoroughly together. You can still take the iron away from this mixture easily by using a magnet, leaving the sulphur.

If the mixture is heated in a test tube a **chemical** change takes place. One sign of this change is a bright red glow that soon spreads through the mixed powder. Also, the test tube gets very hot even if it is taken away from the flame.

When the test tube is cool the residue is found to be a grey-black solid. This solid is not attracted by a magnet. It does not set free hydrogen when acid is added to it, although it does give off a different gas called hydrogen sulphide, which smells like bad eggs.

This grey-black solid is formed as a result of chemical change. It is iron(II) sulphide. It is not easy to change it back into iron and sulphur.

$$Fe + S \rightarrow FeS$$
$$\text{iron} \quad \text{sulphur} \quad \text{iron(II) sulphide}$$

Heat produced

Many chemical changes can be spotted because a lot of heat is produced when they take place. Heat is given off when metals burn in oxygen, or when water is added to quickline. This type of reaction is called an **exothermic** reaction. (In an **endothermic** reaction heat is absorbed.)

When hydrochloric acid is added to calcium carbonate, there is a brisk reaction and carbon dioxide is given off. The mixture gets quite hot. A chemical reaction has taken place.

If the liquid left in the flask afterwards is evaporated to dryness, a white solid is obtained. This solid is calcium chloride, $CaCl_2$. It is quite different from the calcium carbonate present at the start of the reaction. It is very soluble, and it absorbs water from the air, becomes quite wet and finally dissolves in the water is absorbs. The solid is said to be **deliquescent**.

Sometimes heat is given off during physical changes. One example is dissolving sodium hydroxide in water, the solution gets quite hot. But usually heat-changes in such cases are not as great as in chemical changes.

Physical change and chemical change

Chemical
1 Different substances are formed with different properties.
2 Large heat changes.
3 Reversing the change is not easy.
4 Atoms alter their groupings.

Physical
1 No new substances formed, but there may be a change of state.
2 Heat changes are usually small.
3 Change is easily reversed.
4 Atoms stay in same grouping.

Chemical change
In fireworks (above) a chemical change takes place: much heat is given off. Brewing beer (below) is also a chemical change: sugar is turned to alcohol.

1 For each of the following, state whether it is a physical or chemical change: iron rusting; sulphur burning in air; milk turning sour; dissolving sugar in water; ice melting; making toast; a fire burning.
2 How can calcium carbonate be changed into (a) calcium chloride, (b) calcium oxide and (c) calcium hydrogencarbonate.
3 Does a chemical change take place when blue copper(II) sulphate crystals are heated.

F

Metals

Some interesting chemical changes involve metals. Silver, gold, copper and tin have been known to man for centuries. Other well-known metals are iron, zinc, lead, aluminium and magnesium.

Physical properties

These are all solid elements. Mercury is the only metal that is liquid at ordinary temperatures.

Metals are shiny and can be polished. They are quite different from non-metallic elements like carbon, oxygen and sulphur.

Metals are hard and make a ringing sound when struck.

They are ductile, i.e. they can be pulled out into long wires. They are malleable, they can be hammered out into thin sheets. Gold leaf, for instance can be only a few millionths of a centimetre thick. Sulphur or carbon could not be made into thin sheets like this.

Metals conduct heat and electricity. With the exception of carbon, non-metals are poor conductors.

Generally, metals are denser than non-metals.

Chemical properties

The physical properties are not a safe guide to deciding which elements are metals. We need to turn to chemical properties to be certain.

Metals burn in oxygen to form **basic oxides** (see Topic 31). Non-metallic elements form **acidic oxides**, e.g. sulphur dioxide.

The properties of chlorides also help us to decide. Most metallic chlorides are quite stable, often crystalline salts. Nearly all dissolve in water, e.g. sodium chloride and calcium chloride.

But the chlorides of non-metals are usually decomposed by water, forming hydrochloric acid and another acid. One chloride of phosphorus, for instance, reacts with water to form phosphoric acid and hydrochloric acid. The chloride of carbon (tetrachloromethane; CCl_4) is an exception. It is a colourless liquid not decomposed by water.

Oxidation

If a strip of copper foil is heated in a bunsen flame, the golden-brown colour of the copper disappears. The metal becomes dark brown, then black. A black powder can be scraped off the surface of the copper. This black powder is copper(II) oxide (CuO).

The copper has joined chemically with the oxygen from the air:

$$\text{copper} + \text{oxygen} \rightarrow \text{copper oxide}$$
$$2Cu + O_2 \rightarrow 2CuO$$

This reaction can be used to find the percentage of oxygen in the air. A measured volume (V_1) of dry air in one syringe is passed slowly over heated copper. The air is passed back and forth over the copper and the final volume of nitrogen left, when cool, is measured (V_2). Then the percentage of oxygen in the air is:

$$100 \times \frac{V_1 - V_2}{V_1}$$

Finding the percentage of oxygen in air

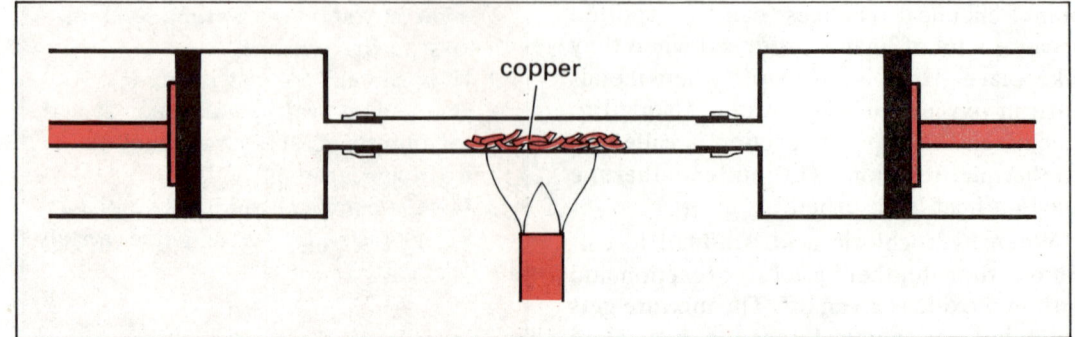

copper

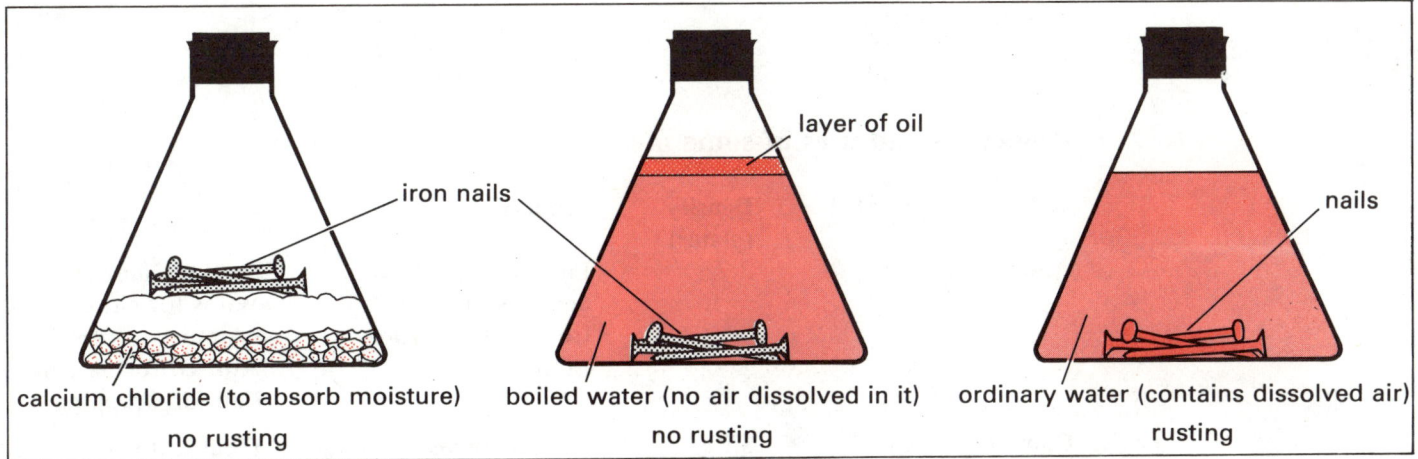

iron nails

layer of oil

nails

calcium chloride (to absorb moisture)

boiled water (no air dissolved in it)

ordinary water (contains dissolved air)

no rusting

no rusting

rusting

Most metals are tarnished or corroded in air. Aluminium, zinc and magnesium all become dull, because a covering of the oxide forms. This coat of oxide prevents further action. An aluminium saucepan, for example, soon loses its shiny look, but can still be used for years.

Copper is not easily corroded, but when exposed on roofs of buildings, it becomes covered with a green powder. In smoky city air, this green powder might be a basic sulphate but near the sea, it could be a basic chloride of copper.

Rusting

Iron rusts on exposure to moist air. Experiments can be carried out to show that both air and water are needed for iron to rust. Iron is such a useful metal that it must be protected from air and water that would cause rusting. The protection can be by:

Galvanizing A thin layer of zinc is put on the iron. The iron is dipped into molten zinc.

Electroplating Usually with nickel followed by chromium.

Tinning The iron sheet is covered with a thin layer of tin by being dipped into molten tin. What we call a 'tin can' is really tinned iron. When the tin covering wears away, the iron underneath will rust.

Painting The paint excludes both air and water.

Uses

A table of the uses of some metals is given on the next page.

---

1 Write down a list of objects that are made of (a) galvanized iron and (b) chromium-plated iron.

2 You suspect that an article is made of painted iron or iron covered with plastic. How can you find out if you are correct without removing the paint or the plastic covering?

Rusting
What causes rust to form?

Forth rail bridge
All the iron girders in this bridge have to be painted to stop rusting

Properties and uses of some metals

Metal	Symbol	Density (g/cm³)	Properties	Uses
Aluminium	Al	2.7	Film of oxide forms and this protects it from further action of air	Cooking utensils Wires for electric grid system Anti-corrosion paint Foil for wrapping
Copper	Cu	8.9	Resists corrosion, but goes green in city air or near sea	Electrical wires Water tanks and pipes Transatlantic cables
Gold	Au	19.3	Not easily corroded. Attacked only by aqua regia (nitric acid and hydrochloric acid)	Standard of currency Jewellery
Iron	Fe	7.7–7.9	Needs to be protected from rusting	Lamp posts Car parts Horseshoes Ornamental work
Lead	Pb	11.4	Soft, dense Corrodes easily	Gutters Roofing Storage batteries
Magnesium	Mg	1.74	Very low density Oxide film protects surface	Fireworks and flares Incendiary bombs Alloys in aircraft
Nickel	Ni	8.9	Not greatly affected by air or water	Nickel plating Alloys to withstand high temperatures and hard wear
Silver	Ag	10.5	Attacked by hydrogen sulphide A soft metal	Table ware Coinage Silver plating
Tin	Sn	7.3	Not attacked by air or water	Tin cans Solder Pewter
Zinc	Zn	7.1	Corrodes Melts in bunsen flame	Galvanizing
Platinum	Pt	21.5	Not easily corroded	Jewellery Dentistry Catalysts

When different metals are melted together, they may form alloys. In some cases, a chemical compound is formed but most alloys are mixtures. They are special kinds of mixtures, in which there are definite parts of each metal.

Alloys have quite different properties from any of the metals they contain: Brass is a mixture of 60–70 per cent copper with zinc. Bronze contains about 90 per cent copper with tin. Brass and bronze are much stronger than copper, zinc or tin. These alloys do not corrode in air as easily as the separate metals.

Alloys of aluminium and magnesium are much harder and do not dent as easily as either of these metals, yet the alloys have the low densities that make them useful in aircraft construction.

If 3 g of lead, 2 g of tin, 7 g of bismuth and 1 g of cadmium are heated together in a test tube, an alloy called 'Wood's metal' is formed. This alloy melts at 71°C, which is much lower than the melting point of any of these metals. It can even be melted in hot water!

Steel alloys are very important. There is a very large number of different steel alloys. Steel is iron with a small amount of carbon dissolved in it.

Traces of other added metals make alloys with special properties. If manganese or tungsten is added the steel is very hard and suitable for tools.

A non-magnetic steel can be made by adding silicon. Cobalt steel makes permanent magnets.

Stainless steel is a name given to several different alloys. One is iron with 18 per cent chromium and 12 per cent nickel.

Mercury forms alloys with most metals; they are called amalgams. It forms an amalgam even with gold. (So do not let mercury get on your gold ring!)

Solder is an alloy of tin and lead. When melted it stays pasty long enough to be worked. Plumbers can 'wipe' joints in lead pipes because of this.

Brass
Brass is an alloy of copper and zinc

1 What three metals would you expect to find occurring naturally in the earth as minerals?
2 Name the two lightest metals. Name the two most dense metals.
3 Name three articles that are made of aluminium, and three articles made of brass.

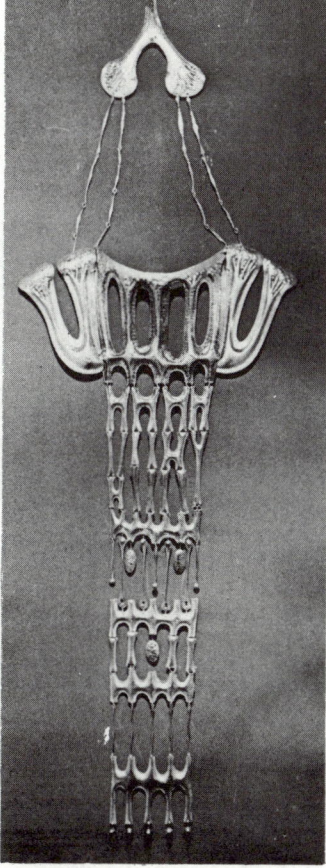

Jewellery
Even 'gold' jewellery is often an alloy of gold and other metals

Attack on Metals

38

Corrosion
Sea-water soon attacks the iron that the anchor is made of

Silver coin
Silver does not rust easily and so it was often used to make coins (this one is over 2500 years old)

Warning
Sodium metal must only be used by the teacher. Potassium metal should not be used in schools

Sodium reacts with water
The hydrogen evolved burns if the sodium is prevented from moving around

Gold, platinum and mercury are not affected by air or water, and silver and copper are stable enough to be used in coinage. But some metals are more easily affected:

Iron rusts slowly in moist air, but some other metals react more violently with air and water.

Sodium is a soft metal that can be cut easily with a knife. The cut surface is bright and silvery, but quickly clouds over because sodium oxide forms.

Potassium is another soft metal which is oxidised even more rapidly after it is cut.

Metals and water

A small piece of sodium is dried with filter paper (it must **not** be handled). It is then dropped into water in a basin. Immediately the metal hisses and darts around on the surface of the water. Bubbles of gas are given off and the sodium becomes so hot that it melts into a silvery ball (globule). The gas is given off unevenly over the surface and this accounts for the random movement of the globule in one direction after another.

Some fumes are seen and, when nearly all the sodium has reacted, it usually becomes stranded at the edge of the basin. It may explode with a sharp crack and a yellow flash.

When a small piece of potassium is dropped onto water, the reaction is even more violent. The gas burns with a pale lilac-coloured flame immediately. (The yellow colour for sodium or the lilac colour for potassium is seen when any compound of these elements is strongly heated.)

The gas given off in the reaction of sodium with water burns with the yellow flame if the sodium is kept in one place. This can be done by putting a piece of filter paper under the sodium before floating it in water.

The gas given off in each case is hydrogen. The liquids left turn litmus blue. They are sodium hydroxide and potassium hydroxide:

$$2Na + 2H_2O \rightarrow 2NaOH + H_2$$
sodium   sodium   hydrogen
                     hydroxide

$$2K + 2H_2O \rightarrow 2KOH + H_2$$
potassium   potassium   hydrogen
                     hydroxide

These soluble hydroxides, NaOH and KOH are examples of **alkalis**.

The metal calcium reacts with water in a similar way, but the action is slow. The calcium hydroxide formed is not very soluble, but it dissolves enough in water to form a weak alkali known as **limewater**.

$$Ca + 2H_2O \rightarrow Ca(OH)_2 + H_2$$
calcium   calcium   hydrogen
                   hydroxide

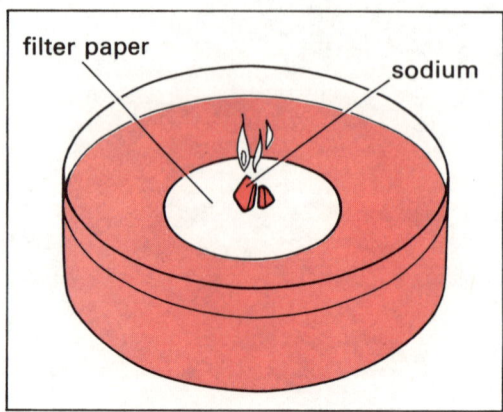

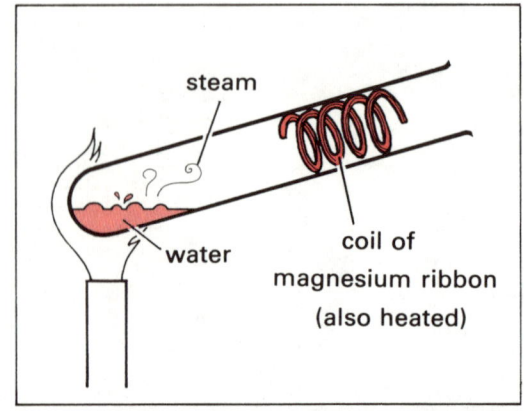

Magnesium reacts violently with steam

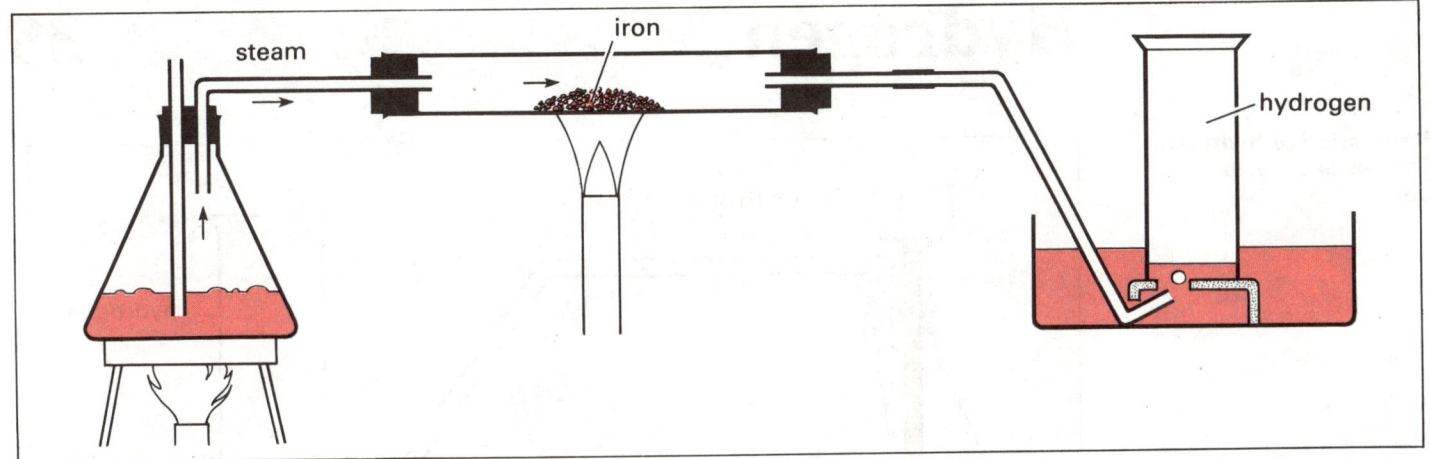

Magnesium reacts only very slowly with water at ordinary temperatures. It reacts violently with steam.

$$Mg + H_2O \rightarrow MgO + H_2$$
magnesium   magnesium hydrogen
     oxide

Iron reacts with steam when it is heated with steam passing over:

$$3Fe + 4H_2O \rightarrow Fe_3O_4 + 4H_2$$
iron   iron oxide   hydrogen

Zinc is another metal attacked by steam. But steam has no action on tin, copper, silver or gold.

Metals and acids

Hydrochloric acid reacts with most metals. Tin and lead, which are not affected by water, react slowly with the acid. Zinc and magnesium, which are only slowly attacked by water, react with dilute hydrochloric acid. The gas hydrogen is given off:

$$Zn + 2HCl \rightarrow ZnCl_2 + H_2$$
zinc   hydro-   zinc   hydrogen
     chloric   chloride
     acid

The hydrogen can be collected by passing it upwards into a test tube. With the thumb over the top, the tube is turned up near a flame. The thumb is taken away. The gas will ignite and burn quietly if it is quite pure, but more often it explodes with a shriek because the hydrogen is mixed with a little air. This is a good test for hydrogen.

Copper, silver, gold and platinum do not react with hydrochoric acid. Copper and silver do react with nitric acid.

Dilute sulphuric acid does not react with copper. It is true that concentrated sulphuric acid does react with copper, but no hydrogen is formed.

Nitric acid, either dilute or concentrated, reacts violently with copper and produces brown fumes.

Magnesium is the only metal which sets free hydrogen from nitric acid, so long as the acid is very dilute.

Sodium and potassium react even more violently with acids than they do with water. You must **not** add even dilute acid to either sodium or potassium, because the reaction is explosive.

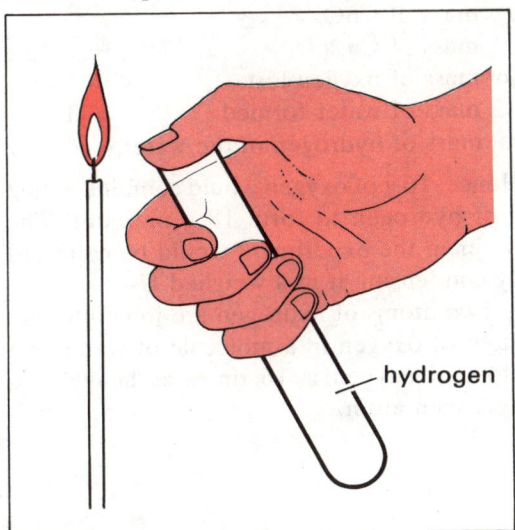

A test tube full of hydrogen often burns with a loud pop

1 Sodium and potassium are soft, like plasticine. So why do you call them metals?
2 Hydrogen can be collected by pouring it upwards. What does this tell you about the gas?
3 What happens when dilute hydrochloric acid is added to magnesium? Write an equation.

Hydrogen

39

Preparation of hydrogen
The gas is collected over water

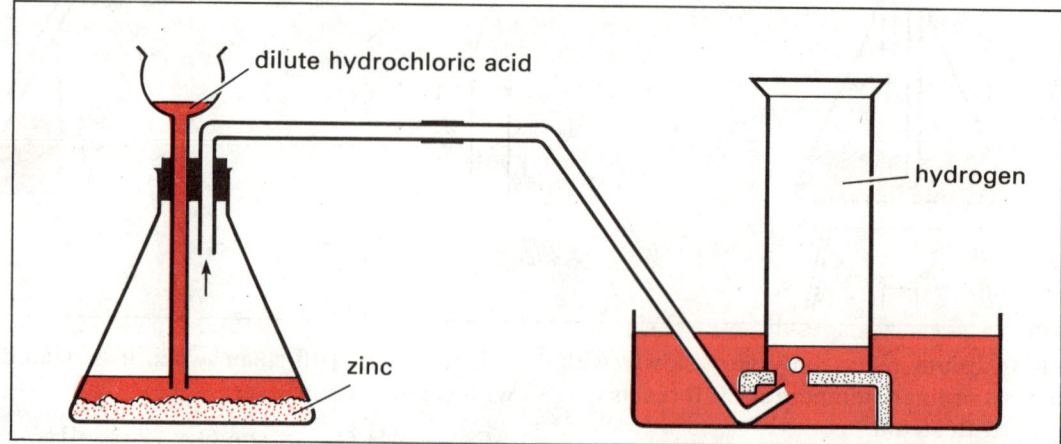

Hydrogen can be made by adding dilute hydrochloric or sulphuric acid to zinc:

$$Zn + H_2SO_4 \rightarrow ZnSO_4 + H_2$$
zinc   sulphuric   zinc   hydrogen
      acid   sulphate

If the zinc is quite pure, this action may be slow. It can be speeded up by adding a little copper sulphate.

Properties

Hydrogen is a colourless gas, with no smell. It is the lightest gas known. It is used in balloons, but this can be dangerous because hydrogen burns in air with a hot flame. It forms an explosive mixture with air.

If a hydrogen flame is brought against a dry flask with cold water inside, drops of colourless liquid are formed on the outside. This liquid is water. So hydrogen has combined with something in the air to form water. But with what?

Hydrogen, carefully dried by bubbling through concentrated sulphuric acid, is passed over copper(II) oxide. The oxide is heated in a combustion tube. Excess hydrogen is burnt at the outlet.

Quite soon a red glow is seen to pass through the black copper oxide powder. Heat is produced (chemical change). Droplets of liquid form on the cooler part of the tube and tests show it to be water. The powder left in the tube is now reddish-brown and can be shown to be copper.

Composition of water

So hydrogen has joined with oxygen from copper oxide to form water:

$$H_2 + CuO \rightarrow Cu + H_2O$$
hydrogen   copper(II)   copper   water
         oxide   (metal)

If the copper(II) oxide and the final copper are weighed, it is possible to work out what masses of hydrogen and oxygen join to form a known mass of water. For example:

mass of CuO	= 7.95 g
mass of Cu left	= 6.35 g
so mass of oxygen lost	= 1.60 g
mass of water formed	= 1.80 g
so mass of hydrogen in the water	= 0.20 g

Hence, 16 g of oxygen would combine with 2 g of hydrogen to form 18 g of water. The water in the experiment would be collected by condensing it in a weighed U-tube.

Two atoms of hydrogen are joined to one atom of oxygen in a molecule of water. An atom of oxygen is 16 times as heavy as a hydrogen atom.

Burning hydrogen
Water is formed and condenses on the cold flask

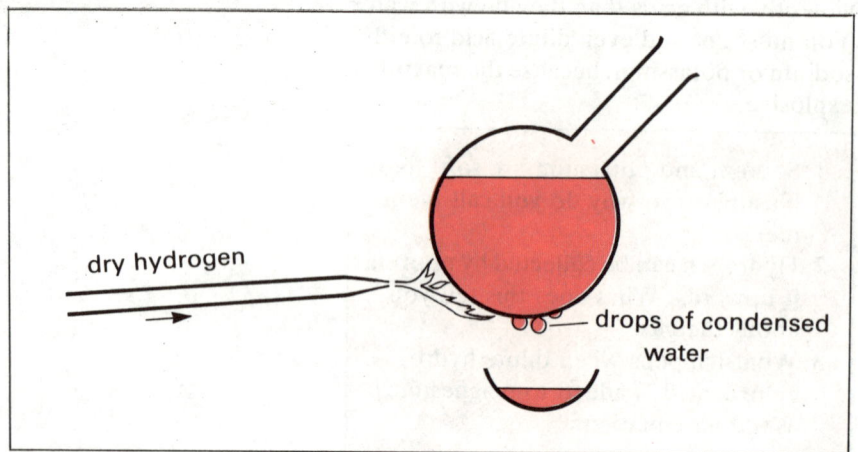

Reactions of hydrogen

It combines with many elements, forming hydrides. With chlorine in light it combines explosively:

$$H_2 + Cl_2 \rightarrow 2HCl$$
hydrogen   chlorine   hydrogen chloride

With nitrogen, under great pressure, it forms ammonia:

$$N_2 + 3H_2 \rightarrow 2NH_3$$
nitrogen   hydrogen   ammonia

This is an important industrial process (the Haber process) for making ammonia and ammonium compounds.

As seen with copper oxide above, hydrogen reduces oxides, that is, it takes away oxygen. It combines with oxygen to form water. The oxide is reduced and at the same time the hydrogen is oxidised. Hydrogen is a reducing agent:

e.g. $PbO + H_2 \rightarrow Pb + H_2O$
lead   lead
oxide

Uses

Hydrogen has many uses, depending on its properties.

1 Being the lightest gas known, it is used to fill balloons, although non-inflammable helium is now preferred.

2 It gives a hot flame when it burns, so it is useful as a fuel.

3 Coal gas and water gas both contain much hydrogen.

4 The hot flame is used in the oxy-hydrogen torch for welding.

5 It is used in making compounds, e.g. ammonia.

6 Hydrogen combines with carbon monoxide under high pressure and in the presence of a catalyst to form methanol.

$$CO + 2H_2 \rightarrow CH_3OH$$

7 Its action as a reducing agent is also used in making margarine from liquid oils. In this case, the atoms of hydrogen adding to the liquid oil molecules turn them into molecules of solid fat.

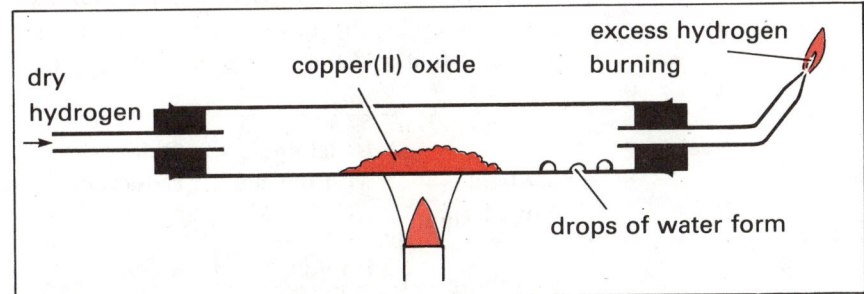

Reaction with copper oxide
Hydrogen can remove the oxygen from copper oxide

Airship
These used to be filled with hydrogen

1 Suppose you were given two black powders. One is carbon, the other is copper(II) oxide, but you do not know which is which. What would you do to find out?

2 Write down the words that complete the following statements:
  (i) magnesium +   *   → magnesium chloride + hydrogen
  (ii) hydrogen + copper(II) oxide → copper +   *
  (iii)   *   + water →   *   + hydrogen.

Reactivity of Metals

40

From what we have already learnt, we can arrange the metals in order of their reactivity. In this table, the most reactive metals are placed first and the most stable ones last.

Metal	Reaction with oxygen	Oxide reduction	Reaction with water	Reaction with acids
Potassium			Violent	Explosive
Sodium	Rapidly oxidised in air	Very difficult to reduce	Violent	Explosive
Calcium			Vigorous	
Magnesium			Burns in steam	
Aluminium	Oxidised slowly	Difficult to reduce	Protected by oxide	Readily gives hydrogen
Zinc			Attacked by steam	
Iron	Rusts in moist air		Action reversible	More slowly
Tin				
Lead	Oxidised when strongly heated	Can be reduced more easily		Needs concentrated acid
Copper			Not attacked by water or steam	
Mercury		Oxide reduced heat alone		Only attacked by nitric acid and hot concentrated sulphuric acid
Silver	Stable			
Gold	Stable			Attacked only by aqua regia (hydrochloric acid plus nitric acid)

Lead tree

Displacement

When a piece of iron is put into a solution of copper sulphate, a deposit of copper is formed on the iron. (So do not stir copper sulphate solution with your penknife!)

If iron powder is stirred into the copper sulphate solution, the iron disappears. A residue of red copper powder is left and the blue solution gradually fades to a pale green liquid.

$$Fe + CuSO_4 \rightarrow Cu + FeSO_4$$
$$\text{blue} \qquad \text{red} \quad \text{pale green}$$

The iron has displaced the copper from the solution of its salt. The solution gets hot while this is happening.

If a strip of magnesium or zinc is put in a solution of lead nitrate, crystals of lead will grow on it. This is called a 'lead tree'.

Copper is below iron in the activity table. Lead is below magnesium and zinc. Any metal higher in the list will displace any metal lower in the list from its salts.

Extraction of metals from their ores

The methods used for extracting metals from ores depend on the nature of the ore and how strongly the metal is combined with other elements in the ore.

Many metals are found as sulphides. The process for obtaining the metal involves roasting the ores in air. For instance, mercury is easily obtained from the ore cinnabar (HgS) by roasting it. The mercury is collected by condensing the vapour:

$$HgS + O_2 \rightarrow Hg + SO_2$$
cinnabar   mercury   sulphur
                                   dioxide

Copper is obtained in a similar way:

$$Cu_2S + O_2 \rightarrow 2Cu + SO_2$$
copper
glance

The copper is made purer by electrolysis (see Topic 53).

Some metallic sulphides are roasted to change them to oxides then the oxides are reduced to the metals, e.g.

$$2ZnS + 3O_2 \rightarrow 2ZnO + 2SO_2$$

The usual reducing agent to change oxides to metals is carbon (coke) or carbon monoxide (CO).

Zinc oxide, for instance, is heated with coke and the zinc metal boiling at 900°C, distils over as vapour and is condensed.

$$ZnO + C \rightarrow Zn + CO$$

To obtain tin, the oxide is heated with coal

$$SnO_2 + C \rightarrow Sn + CO_2$$
tin   tin
oxide

Extracting iron from its ores

Iron ores are chiefly oxides. These are reduced to iron in the blast furnace by a continuous process.

The iron ore together with coke and limestone, is fed into the furnace. A blast of hot air is blown into the furnace. This causes a very high temperature, about 1500°C or more in the middle.

The limestone ($CaCO_3$) breaks up to lime (CaO) and carbon dioxide. The lime combines with the silica impurities in the ore and forms slag which is tapped off. Molten iron is drawn off below the slag. The chief reactions are shown below:

$$CaCO_3 \rightarrow CaO + CO_2$$
$$CO_2 + C \rightarrow 2CO$$
$$2C + O_2 \rightarrow 2CO$$

It is mainly the carbon monoxide that reduces the iron oxide:

$$Fe_2O_3 + 3CO \rightarrow 2Fe + 3CO_2$$

The slag consists of silicates.

The very reactive metals magnesium, aluminium, sodium and potassium cannot be obtained by these reducing methods. Electrical methods are used instead (see Topic 53).

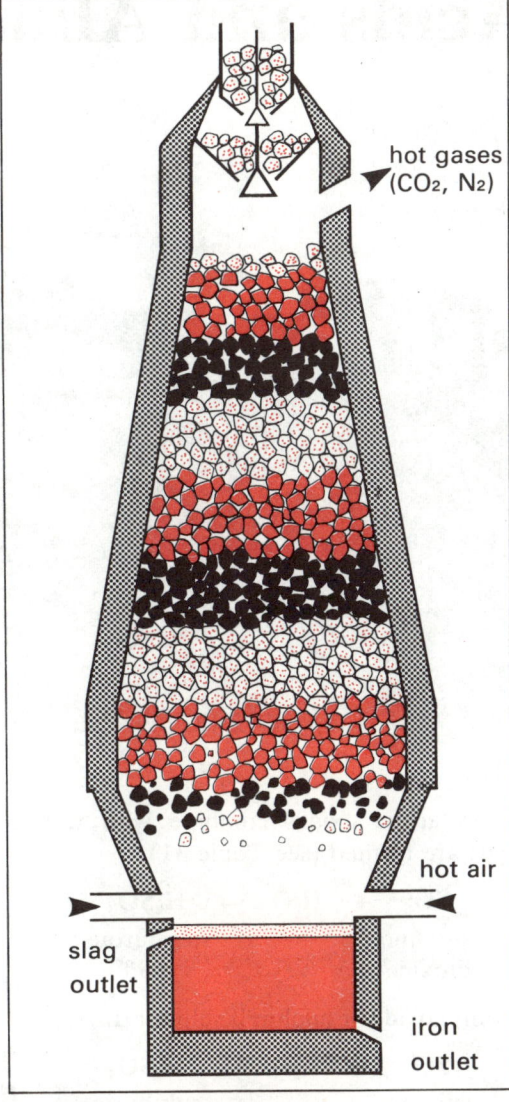

Blast furnace

Iron ores

1 From the list of metals below, select (a) those that react violently with cold water, (b) those that do not react with cold water but give hydrogen with steam, (c) those that do not react with water but react with dilute sulphuric acid. Copper; sodium; magnesium; iron; potassium; zinc; mercury.
2 What would you expect to happen if a strip of polished aluminium were put into some silver nitrate solution? Explain.
3 When zinc is added to dilute hydrochloric acid: (i) water is formed; (ii) magnesium chloride is formed; (iii) no reaction takes place; (iv) hydrogen is set free; (v) oxygen is set free; (vi) zinc chloride is formed. Which, if any, of these statements is/are correct?

Acids and Alkalis

41

Sulphuric acid
Industry uses large quantities of this chemical. Note the 'Hazchem' sign on the back of the lorry

Acids

When acidic oxides combine with water, acids are formed (see Topic 31).

$$SO_2 \;+\; H_2O \;\rightarrow\; H_2SO_3$$
sulphur dioxide   sulphurous acid

Another oxide of sulphur is sulphur trioxide:

$$SO_3 \;+\; H_2O \;\rightarrow\; H_2SO_4$$
sulphur trioxide   sulphuric acid

Warning
Great care must be taken with concentrated acids. Wherever possible dilute solutions are used

Mineral acids

There are three acids used in the laboratory more often than others. They are sulphuric acid (H_2SO_4), hydrochloric acid (HCl) and nitric acid (HNO_3).

They are called mineral acids because they can be formed by reactions with minerals. Sulphuric acid can be made by strongly heating iron(II) sulphate ($FeSO_4.7H_2O$).

Some familiar acids, such as acetic acid (the acid of vinegar), citric acid and tartaric acid are organic acids.

Concentrated solutions of mineral acids are corrosive and they destroy paper, wood and clothing. Special care must be taken when handling them. Any drop falling on flesh can cause a nasty 'burn'. Always use dilute solutions.

Some properties of acids

Acid solutions turn litmus red. Acids usually react with metals (except gold) and sometimes hydrogen is given off. All acids contain hydrogen.

Acids usually react vigorously with carbonates. Hydrochloric acid reacts with sodium carbonate and carbon dioxide is given off.

$$Na_2CO_3 \;+\; 2HCl \;\rightarrow\; 2NaCl \;+\; H_2O \;+\; CO_2$$
sodium carbonate   sodium chloride

An interesting example of this effervescent action happens when 'health salts' are added to water. Health salts contain either citric acid or tartaric acid (or both), and sodium hydrogencarbonate. As long as this mixture of powders is kept dry there is no action, but when water is added, they react with effervescence and give off carbon dioxide.

Acids react with basic oxides. For example, copper(II) oxide put in dilute sulphuric acid seems to dissolve forming a blue solution:

$$CuO \;+\; H_2SO_4 \;\rightarrow\; CuSO_4 \;+\; H_2O$$
copper oxide   sulphuric acid   copper sulphate (blue)

Alkalis

When basic oxides combine with water, alkalis are formed:

$$Na_2O + H_2O \rightarrow 2NaOH$$

sodium   sodium
oxide   hydroxide
(basic oxide)   (alkali)

The common alkalis are sodium hydroxide (caustic soda; $NaOH$), potassium hydroxide (caustic potash; KOH), calcium hydroxide [$Ca(OH)_2$] and ammonium hydroxide (NH_4OH). A weak solution of calcium hydroxide is known as limewater.

Ammonium hydroxide is contained in a solution of ammonia gas (NH_3) in water. The group of atoms NH_4^+ is called ammonium.

Properties of alkalis

They turn litmus blue.

They form precipitates, often coloured, with solutions of metallic salts. Two examples are:

$$CuSO_4 + 2NaOH \rightarrow Cu(OH)_2 + Na_2SO_4$$

copper   copper
sulphate   hydroxide
            (pale blue)

$$FeCl_3 + 3KOH \rightarrow Fe(OH)_3 + 3KCl$$

iron(III)   iron(III)
chloride   hydroxide
            (brown)

When ammonia solution is added to copper(II) sulphate, a pale blue precipitate is formed. But when more ammonia solution is added the precipitate dissolves and a dark blue solution is formed.

Alkalis do not react with metals as a rule, but zinc, aluminium and tin do react. They set free hydrogen when warmed with sodium hydroxide or potassium hydroxide. Magnesium does not give hydrogen with alkalis, so if an unknown liquid reacts with magnesium to give hydrogen, it must be an acid.

Uses of sodium hydroxide

These are many and include the making of other sodium compounds, soap making, paper making and the manufacture of rayon.

ALKALI CISTERNS.

CUTTING INTO BARS

FILLING MOULDS

REFINED SOAP MAKING

STAMPING & PACKING.

Sodium hydroxide
This chemical is used to make soap. It has been used for this purpose for thousands of years

1 Which of the following will form an alkali when added to water? Carbon dioxide; sodium oxide; sulphur dioxide; calcium oxide.
2 Which of the following substances in solution would turn litmus red? Sulphur dioxide; sodium hydroxide; ammonia; hydrogen chloride; citric acid; sodium chloride.
3 Write down the names and formulae of (a) any three acids, (b) any three alkalis.

Salts

42

Neutralisation

Acids turn litmus **red** and alkalis turn it **blue**. What will happen if an acid is added to an alkali?

When hydrochloric acid (dilute) is added slowly to a solution of sodium hydroxide containing some litmus, the litmus can show a balance point at which it is purple. The purple colour shows that the solution is neither acid nor alkaline: it is neutral. If the liquid is now evaporated to dryness, it leaves a solid which is common salt, sodium chloride. The acid has neutralised the alkali:

$$HCl + NaOH \rightarrow NaCl + H.OH$$
$$\text{acid} \quad \text{alkali} \quad \text{salt} \quad \text{water}$$

If dilute sulphuric acid is neutralised with potassium hydroxide the salt formed is potassium sulphate, K_2SO_4:

$$H_2SO_4 + 2KOH \rightarrow K_2SO_4 + 2H_2O$$
$$\text{acid} \quad \text{alkali} \quad \text{salt} \quad \text{water}$$

A salt is a substance made by taking out the hydrogen from an acid and putting a metal or ammonium in its place.

An acid is a compound of hydrogen and an acid-part, e.g. HCl. An alkali is a compound of a metal and a hydroxide-part, e.g. NaOH. The salt is the compound of the metal and the acid part, so that water is also formed from the H and the OH:

$$HX + MOH \rightarrow MX + H.OH$$
$$\text{acid} \quad \text{alkali} \quad \text{salt} \quad \text{water}$$

Methods of preparing salts

The hydrogen of an acid can be replaced by a metal in various ways.

1 Make the acid react with the metal directly:

$$Mg + H_2SO_4 \rightarrow MgSO_4 + H_2$$

2 Neutralising acid with alkali:

$$NaOH + HNO_3 \rightarrow NaNO_3 + H_2O$$

3 Reaction of an acid with a basic oxide:

$$CuO + H_2SO_4 \rightarrow CuSO_4 + H_2O$$

This method is illustrated in detail on the next page.

4 Reaction of an acid with a carbonate:

$$K_2CO_3 + 2HCl \rightarrow 2KCl + H_2O + CO_2$$

5 If the salt required is insoluble, e.g. barium sulphate, it can be precipitated by mixing solutions of two suitably chosen soluble substances:

$$BaCl_2 + CuSO_4 \rightarrow BaSO_4 + CuCl_2$$
| barium | copper | barium | copper |
| chloride | sulphate | sulphate | chloride |

To make good crystalline specimens of salts by the methods **1**, **3** and **4**, it is important to avoid using excess acid, which would prevent crystals forming.

So the oxide or carbonate or metal is added slowly to the acid until it is slightly in excess. It is then filtered off and the liquid evaporated to the point of crystallisation.

Bases

In both reactions **2** and **3** above, the only product besides the salt is water. Both metallic oxides and metallic hydroxides are **bases**.

A base is a substance which reacts with an acid to form a salt and water only:

$$\text{acid} + \text{base} \rightarrow \text{salt} + \text{water}$$

Alkalis are soluble bases.

Acid radicals

A salt is made up of two parts MX where M is a metal (or ammonium) and X a part of an acid, called an **acid radical**.

On the next page are some acid radicals, shown alongside the valency cards according to their valency. The formulae for salts can be built up with the valency cards.

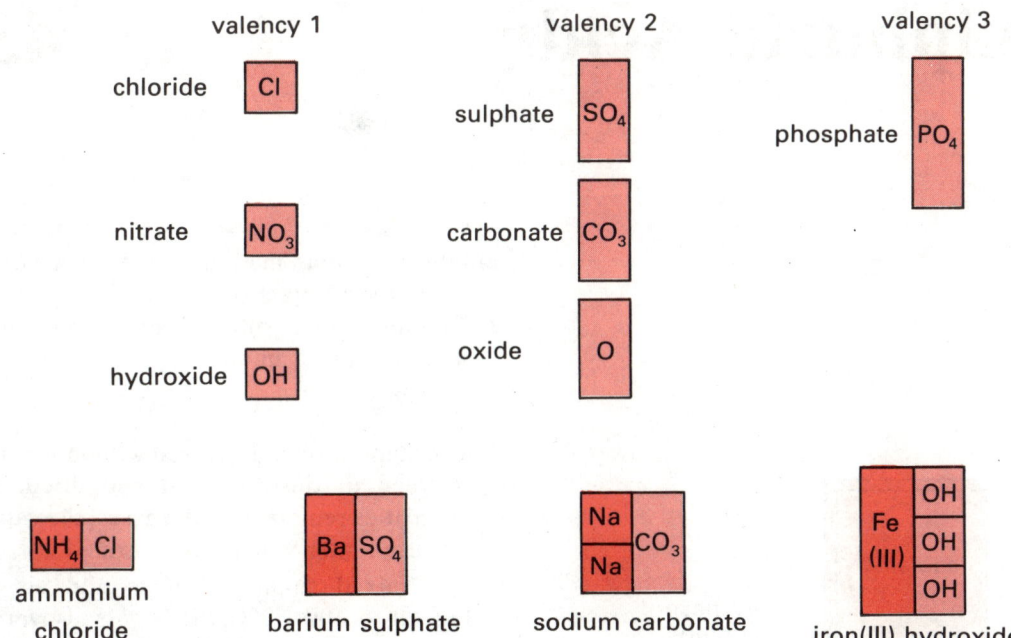

Valency cards

valency 1

chloride Cl

nitrate NO₃

hydroxide OH

valency 2

sulphate SO₄

carbonate CO₃

oxide O

valency 3

phosphate PO₄

NH₄ Cl
ammonium chloride

Ba SO₄
barium sulphate

Na Na CO₃
sodium carbonate

Fe (III) OH OH OH
iron(III) hydroxide

Preparation of copper(II) sulphate

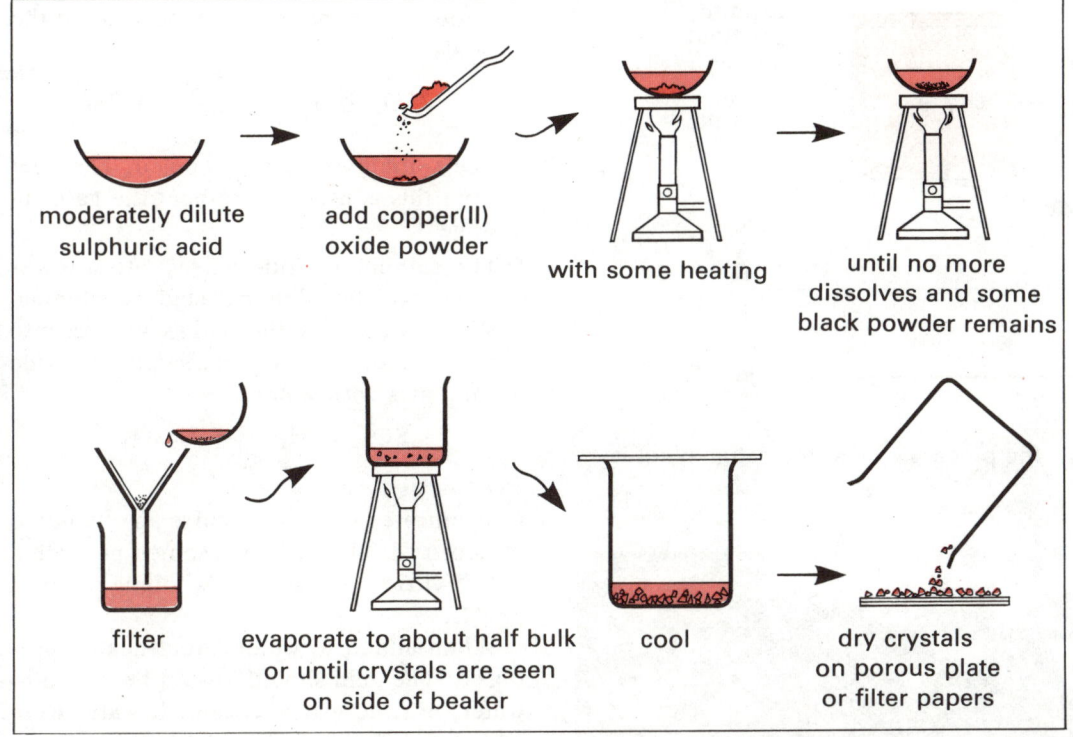

moderately dilute sulphuric acid

add copper(II) oxide powder

with some heating

until no more dissolves and some black powder remains

filter

evaporate to about half bulk or until crystals are seen on side of beaker

cool

dry crystals on porous plate or filter papers

1 Name the salts that would be made by actions between: (a) sodium hydroxide and nitric acid; (b) ammonium hydroxide and sulphuric acid; (c) copper hydroxide and hydrochloric acid.

2 From the list below, choose (i) three alkalis, (ii) three basic oxides and (iii) three salts. Sulphur trioxide; nitric acid; calcium hydroxide; ammonium hydroxide; copper sulphate; magnesium nitrate; potassium hydroxide; copper oxide; zinc oxide; carbon dioxide; sodium carbonate; magnesium oxide.

3 State the methods you think are best for preparing the following salts: sodium sulphate; copper sulphate; zinc chloride; barium sulphate (insoluble); magnesium sulphate.

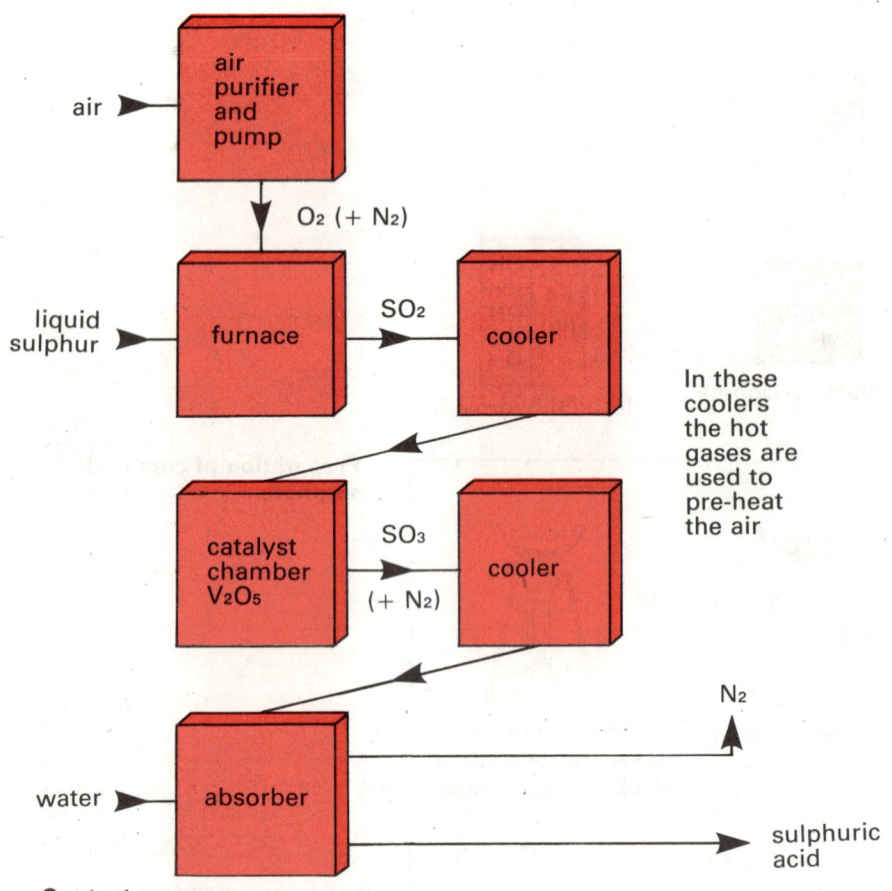

O₂ (+ N₂) — rendered below

$O_2 (+ N_2)$

In these coolers the hot gases are used to pre-heat the air

SO_2 SO_3 $(+ N_2)$

N_2

sulphuric acid

Contact process
The above diagram explains how the process works; the actual plant is shown in the photograph below.

Sulphuric acid is manufactured on a large scale by the Contact process.

1 Sulphur, or a sulphur-containing substance, is burnt in air:

$$S + O_2 \rightarrow SO_2$$

2 The sulphur dioxide, mixed with more air, is freed of dust, washed and dried. If impurities remained in the gases, the catalyst used later would be 'poisoned'.

3 The mixed sulphur dioxide and air is heated to 400–500°C. It is passed over a catalyst, usually vanadium pentoxide. Sulphur dioxide is oxidised to sulphur trioxide. This is not easily done without the catalyst.

$$2SO_2 + O_2 \xrightarrow[\text{V}_2\text{O}_5]{\text{catalyst}} 2SO_3$$

4 The sulphur trioxide is cooled. The heat from this is used to pre-heat the gases in stage **3**.

5 The sulphur trioxide passes into a tower down which sulphuric acid is running. Water is added to the acid as it passes into storage tanks. In effect the sulphur trioxide combines with water.

$$SO_3 + H_2O \rightarrow H_2SO_4$$

Properties

Concentrated sulphuric acid is an oily liquid of density 1.84 g/cm³. It is sometimes called 'oil of vitriol'. It has a high boiling point of 333°C.

When added to water, great heat is produced. The acid should always be added to water, not the water to acid. If water were added to the acid, the liquid that might splash out owing to the violence of the reaction would be strong, corrosive acid.

Dehydrating action. Sulphuric acid has a great affinity for water. You have already seen how it can remove water vapour from a gas. It will remove the elements of water from substances.

It takes the water of crystallisation out of blue hydrated copper(II) sulphate, leaving the white anhydrous salt.

$$CuSO_4.5H_2O \quad - 5H_2O \rightarrow CuSO_4$$
blue crystals   white powder

When added to cane sugar, frothing takes place and carbon is left.

$$C_{12}H_{22}O_{11} - 11H_2O \rightarrow 12C$$
cane sugar
(sucrose)

When sulphuric acid is heated with excess ethanol (ethyl alcohol) at 140°C, diethyl ether is produced and this can be distilled off.

$$2C_2H_5OH - H_2O \rightarrow (C_2H_5)_2O$$
ethanol   ether

At a higher temperature, 170°C and with sulphuric acid in excess, ethene (ethylene) is obtained instead. Ethene, C_2H_4, is an important chemical. It is prepared from petroleum in large quantities to make other substances. Polythene (or poly-ethene) a familiar plastic, has long molecules made up of ethene molecules – $(CH_2 - CH_2)_n$ –.

Sulphuric acid is an oxidising agent. This is why metals do not set free hydrogen from the concentrated acid. Instead the sulphuric acid is reduced and sulphur dioxide is formed, e.g.

$$Cu + 2H_2SO_4 \rightarrow CuSO_4 + SO_2 + 2H_2O$$

Dilute solutions of sulphuric acid show the usual properties of an acid solution (see Topic 41).

Making other acids

Hydrochloric acid is formed when sulphuric acid is heated with a chloride.

Nitric acid is formed when sulphuric acid is heated with a nitrate.

Acid salts

Some acids have more than one replaceable hydrogen atom in the molecule. Sulphuric acid, H_2SO_4, has two. So the acid is said to be dibasic. The basicity of an acid is the number of replaceable hydrogen atoms contained in one molecule of the acid.

When both hydrogen atoms are replaced by a metal, a normal salt is formed, e.g.

$$2NaOH + H_2SO_4 \rightarrow Na_2SO_4 + 2H_2O$$
normal
sodium
sulphate

The alkali has exactly neutralised the acid.

But if half as much alkali is added to the same amount of acid only one of the hydrogen atoms is replaced. A solution of this acid salt turns litmus red.

$$NaOH + H_2SO_4 \rightarrow NaHSO_4 + H_2O$$
sodium
hydrogen-
sulphate
(acid salt)

Uses of sulphuric acid

In addition to the uses already mentioned, its action as an acid is useful in making the salt, ammonium sulphate, which is used as a fertilizer.

$$2NH_4OH + H_2SO_4 \rightarrow (NH_4)_2SO_4 + 2H_2O$$
ammonium
sulphate

It reacts with oxides. It is used for 'pickling' steel before galvanizing or plating. The acid removes the oxide covering.

It is the acid used in accumulators.

It is used in making rayon, plastics, paints, detergents, weed-killers, explosives, drugs and dyes.

Uses of sulphuric acid
In the above photograph explosives are being made by remote control. Another use of sulphuric acid is in accumulators (rechargeable electric batteries)

1 Which of the following metals does not react with dilute sulphuric acid: magnesium, iron, copper, zinc, sodium?
2 How can sulphur be changed into sulphuric acid?
3 State two different ways of changing blue copper(II) sulphate crystals into white copper sulphate (anhydrous) powder.

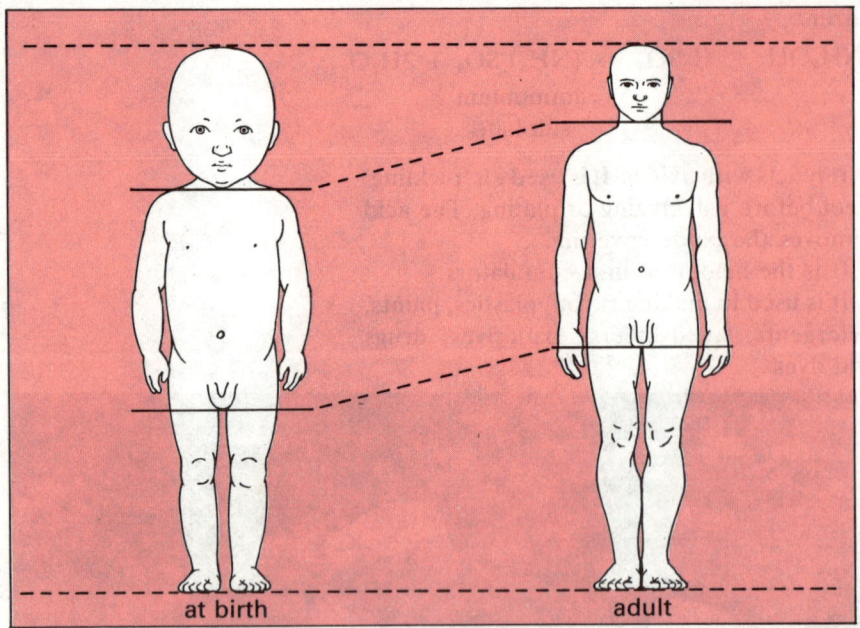

Growth
Not all parts grow at the same rate

Even under these conditions, seeds do not always germinate successfully. Some seeds collected in autumn will not germinate until the spring. They remain dormant for some months and must be kept at low temperatures before they will start growing. This can be done by leaving the seeds in a refrigerator for a time.

Structure of seeds

It is easy to dissect a seed, e.g. broad bean or pea. Inside the tough outer coat (**testa**) there are two fleshy 'seed leaves'. These are called **cotyledons**. Between these can be seen a miniature root (**radicle**) and miniature shoot (**plumule**).

A drop of iodine solution put on the cotyledon gives a dark blue colour. This means that starch is present. It is a reserve food present in the cotyledons.

Growth

Living organisms are made up of cells. They can grow larger in two ways: by cells becoming larger, or cells dividing into more cells.

The process of growth happens in some regions of the body more than in others. In plants, rapid division of cells and increase in size take place in parts called **meristems**. These are found at the tip of the root and the stem. Some parts of the plant do not grow, e.g. woody tissues on the outside of the stem.

For animals, growth does not go on for ever. Some parts, such as bones, stop growing. Not all parts of the body grow at the same rate.

Germination

Suppose seeds are grown under different conditions, some in the cold, others in the warm; some on dry fibre others on wet and so on. It is found that for seeds to grow into seedlings there must be warmth, water and air present.

Growth of seedlings

The stages in germination are easily studied if some broad beans are put on damp fibre or blotting paper and held between two sheets of glass or perspex.

Drawings can be made of the stages. Some seeds such as sycamore and onion, grow rather differently from the pea and broad bean. The cotyledons are pushed up above ground and form the first two leaves of the seedling.

Cotyledons

Most plants have seeds with two cotyledons inside, like the bean. But some seeds, such as maize, have only one cotyledon.

It is striking that the plants with two cotyledons in their seed are quite different in structure from those with only one.

Compare some dicotyledonous plants such as buttercup, wallflower, rose, daisy with some monocotyledonous plants like onion, tulip, lily, daffodil and iris.

Structure of broad bean seed

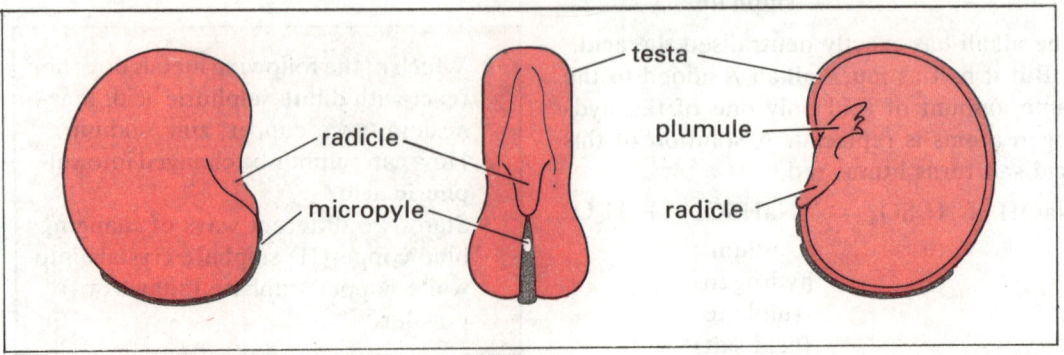

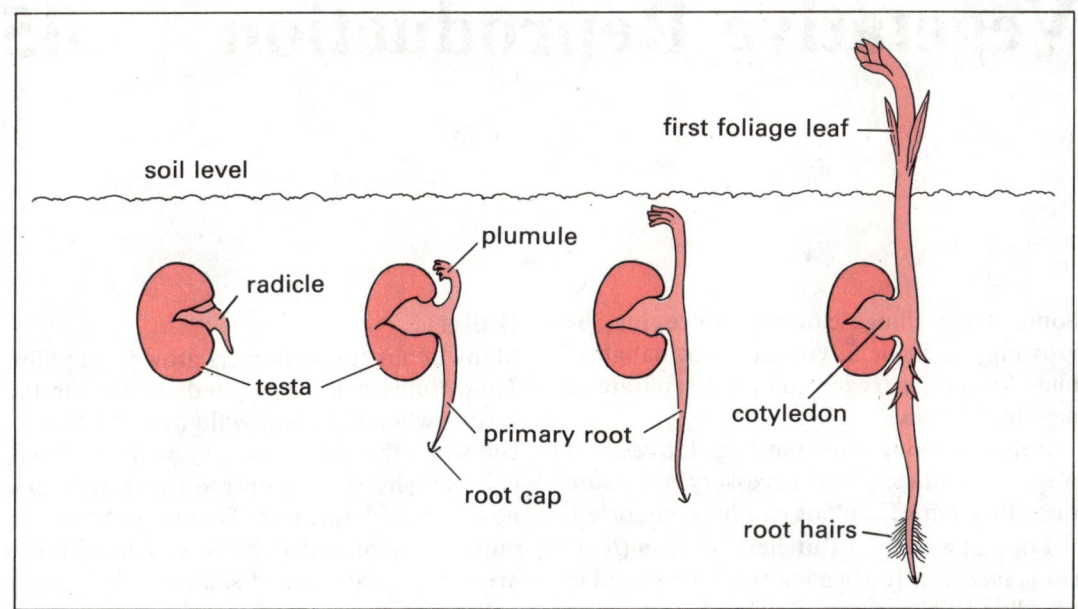

Germination of broad bean seed

soil level

first foliage leaf

plumule

radicle

testa

primary root

root cap

cotyledon

root hairs

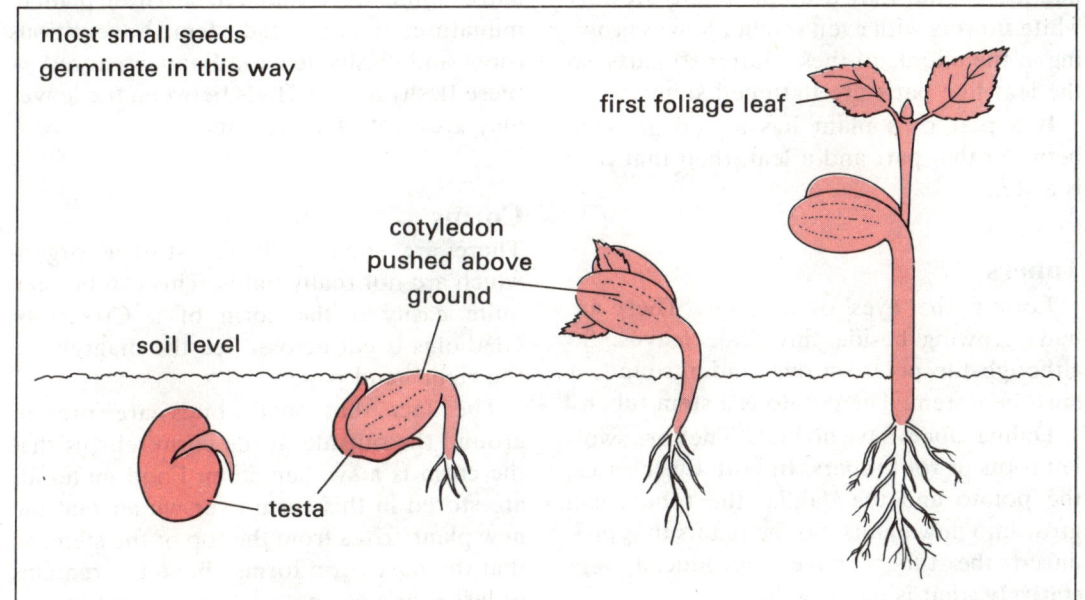

Germination of sycamore seed

most small seeds germinate in this way

first foliage leaf

cotyledon pushed above ground

soil level

testa

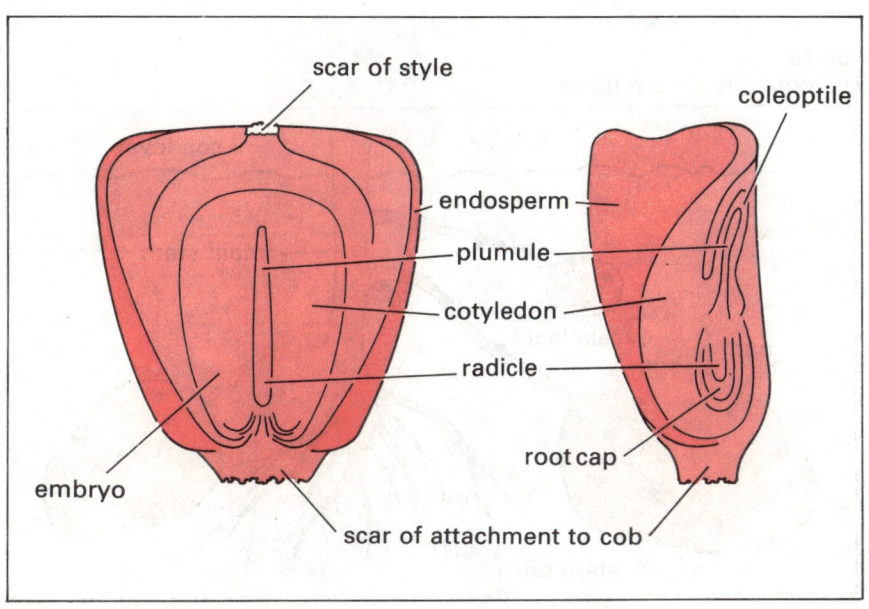

Structure of seed from a monocotyledon (maize)

scar of style

coleoptile

endosperm

plumule

cotyledon

radicle

root cap

embryo

scar of attachment to cob

1 Can you think of ways in which the leaves and flowers of monocotyledons, like lily, tulip and daffodil, are quite different from those of dicotyledons like wallflower, rose, buttercup?

2 Write down a list of seeds that we use as food or use in cooking.

3 When a seed germinates, chemical changes are taking place inside. So what would you expect to notice as a sign of these chemical changes?

4 State three conditions needed for a seed to germinate.

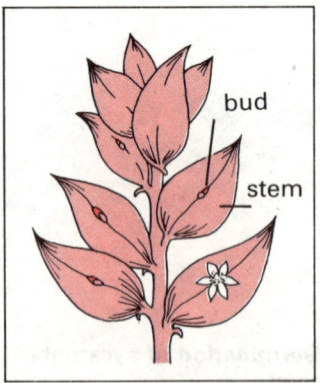

Butcher's Broom

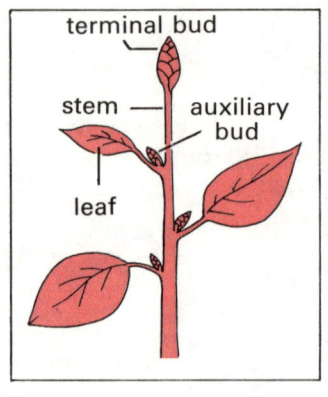

Stem
A stem is between a leaf and a bud

Some living things can reproduce just by growing. In Topic 5, you saw that daughter buds form on a *Hydra* and later separate as new individuals.

Before looking into some similar cases of plants reproducing, it is necessary to be sure that the stem of a plant can be recognized.

Look at a sprig of Butcher's Broom (*Ruscus aculeatus*). It can sometimes be found in woodland. The green flattened parts look like leaves, but they are not. There are tiny white flowers with even smaller leaves growing in the middle of these flattened parts, so the leaf-like parts are flattened stems.

If a part of a plant has a bud growing between that part and a leaf, then that part is a stem.

Tubers

Look at the 'eyes' on a potato. These are buds growing beside tiny scale leaves. So although the potato grows underground, it must be a stem. The potato is a stem tuber.

Dahlia tubers have no buds. They are swollen roots or root tubers. In both these cases, the potato and the dahlia, the tubers can grow into new plants. So the plants that produced the tubers have reproduced **vegetatively** (that is by growth).

Bulbs

Many of our spring flowers grow from bulbs. Tulip bulbs can be planted, ready for the spring, when the plant will grow and flower. But when the plant seems to be dying down, it is actually storing up food materials in a new bulb underground. There may be two or more bulbs formed in this way, so more tulips are ready to grow next season.

When a tulip or onion bulb is cut across and examined, it is found to be a little plant in miniature, it has a flat stem, adventitious roots and fleshy leaves. Food is stored in these fleshy leaves. Buds between the leaves may grow into fresh plants.

Corms

There are some bulb-like storage organs which are not really bulbs. This can be seen quite easily if the corm of a Crocus or Gladiolus is cut across. See the diagram on the right-hand page.

The fact that small buds are present around the outside of the corm tells us that the corm is a swollen stem. Food materials are stored in this corm over winter and the new plant arises from the top of the stem, so that the next corm forms above the remains of last year's corm.

Potato
The potato is a stem tuber

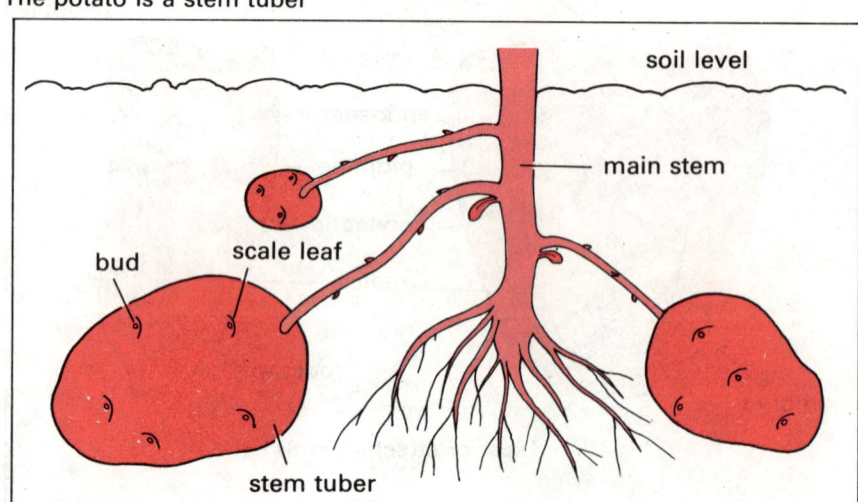

Dahlia
The dahlia has a root tuber

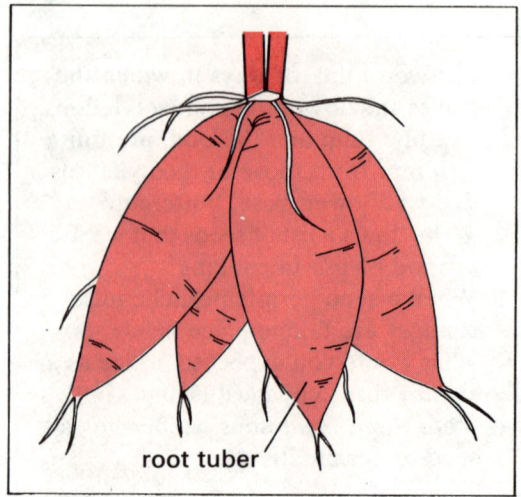

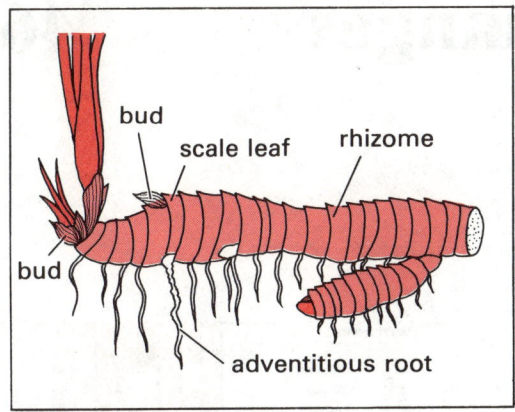

bud
scale leaf
rhizome
bud
adventitious root

Couch grass (right)
Iris (left)
Both iris and couch grass have underground stems called rhizomes

scale leaf
rhizome
adventitious root

Couch grass (right)
Iris (left)
Both iris and couch grass have underground stems called rhizomes

Rhizomes

Sometimes stems are to be found underground. The diagrams above show the rhizomes (underground stems) of Iris and Couch grass (*Agropypron repens*). Scale leaves and buds are found on these rhizomes. The roots are called adventitious roots to distinguish them from true roots that grow directly from a seed.

Couch grass is sometimes called 'Twitch'. It is an unwelcome weed in gardens. If a small piece is broken off and left in the ground a new plant will grow. The grass is reproducing by vegetative means.

Runners or stolons

Some aerial stems can give rise to new plants. A good example is the runner of the strawberry. Fresh plants are produced by roots forming along these runners. In Blackberry *Jasminum nudiflorum*, and other plants, the trailing stem will form adventitious roots when it touches the ground. A new plant grows up and can be separated from the parent plant.

Strawberry
This plant spreads by aerial stems called runners

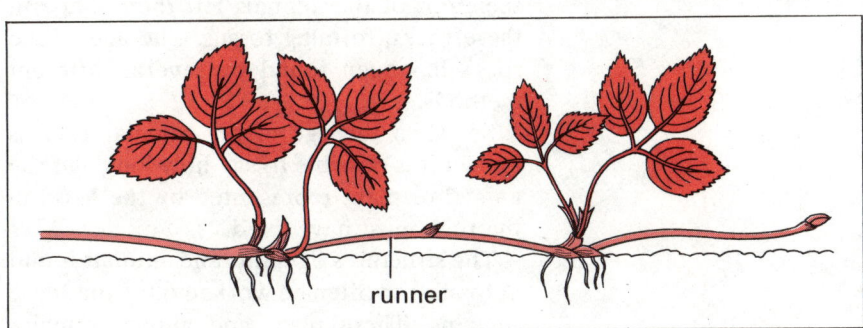

runner

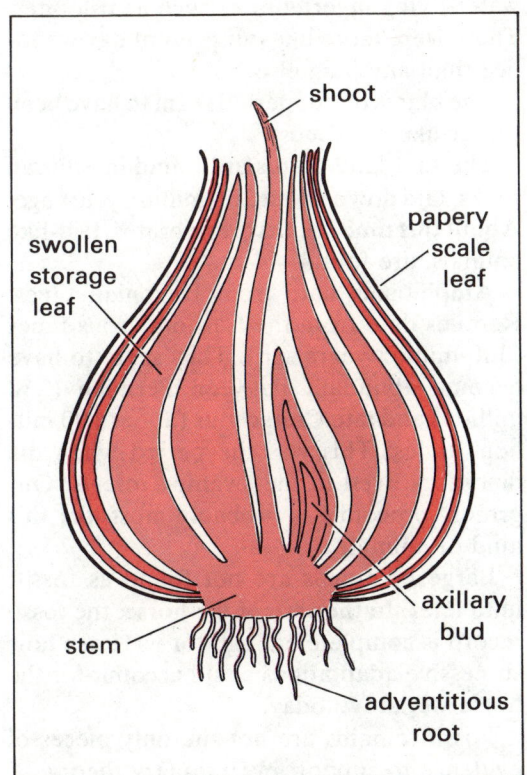

shoot
swollen storage leaf
papery scale leaf
stem
axillary bud
adventitious root

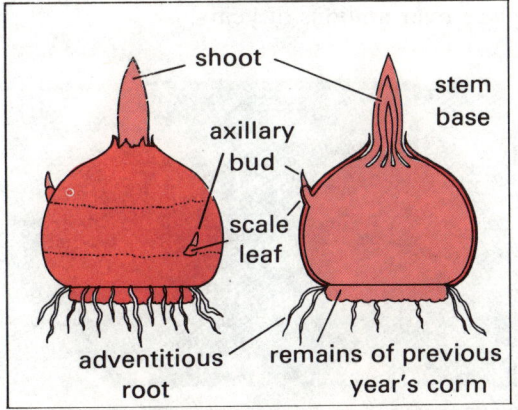

shoot
axillary bud
scale leaf
adventitious root
stem base
remains of previous year's corm

Tulip bulb (left)

Gladiolus corm (right)

1 Make a diagram of a section through an onion bulb and label the parts seen.
2 Plants such as tulip and daffodil that grow from bulbs flower in spring earlier than other plants grown from seed. Can you explain why this should be?
3 Name plants that store food during the winter in (a) a swollen stem, (b) a swollen root, (c) a bulb, (d) a corm.

101

The basis for the modern theory of evolution was put forward by Charles Darwin in 1859 in his **Origin of Species**.

In this book Darwin suggested that the animals and plants living today were not created exactly as they are now, but they have descended from ancestral types as a result of gradual changes over several million years.

There is now much evidence to support this theory.

Fossils

Some of the organisms that were living many years ago were buried in layers of mud and silt after they died. These layers of mud and silt were slowly changed into rocks. The skeletons of the animals left their shape in these rocks, forming fossils. The age of the rocks has been found by several different methods.

So when the age of a particular rock is known it is possible to say how long ago the animal or plant represented by the fossil in the rock must have lived.

The structures of plants and animals found as fossils can often be worked out from these remains. These plant and animal remains form a sequence in time. This shows the slow but astonishing changes that have taken place over millions of years.

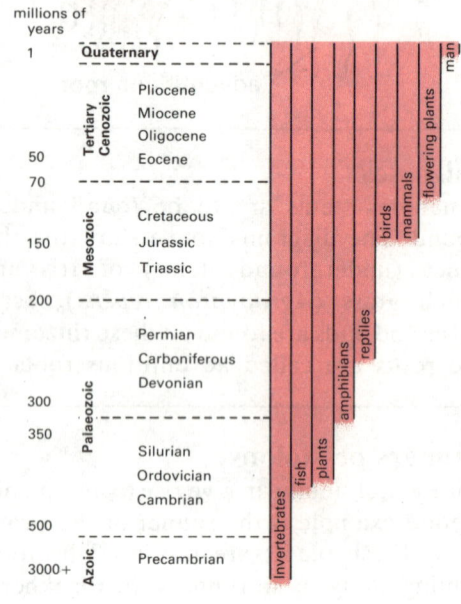

Geological time scale

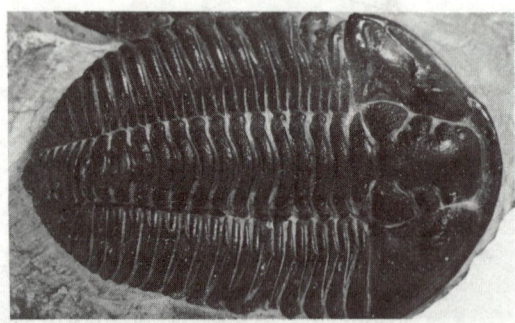

Fossils
A trilobite (above) and an Ichthyosaurus (below)

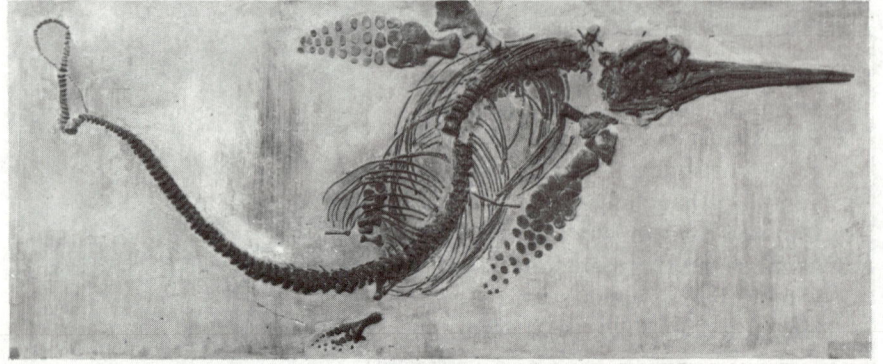

For example, the animals found in Cambrian rocks (about 500 million years old) are water-living invertebrates such as trilobites. These were more like our present day woodlice than anything else.

The plants of this period seem to have been rather like seaweeds.

The first land plants are found in Silurian rocks, laid down about 350 million years ago. About this time the first vertebrates, fish-like animals, are found.

Amphibians date from Devonian times. Reptiles date from the Carboniferous times (300 million years ago). They seem to have become abundant between Permian (250 million) and late Cretaceous (about 100 million) times. This was the period when the dinosaurs lived in the swampy forests. Our present crocodile is probably most like this kind of animal.

Large mammals are not found as fossils until later. In the case of the horse, the fossil record is complete enough for us to see how successive adaptations could account for the horse we know today.

Fossil remains are not the only pieces of evidence to support evolutionary theory.

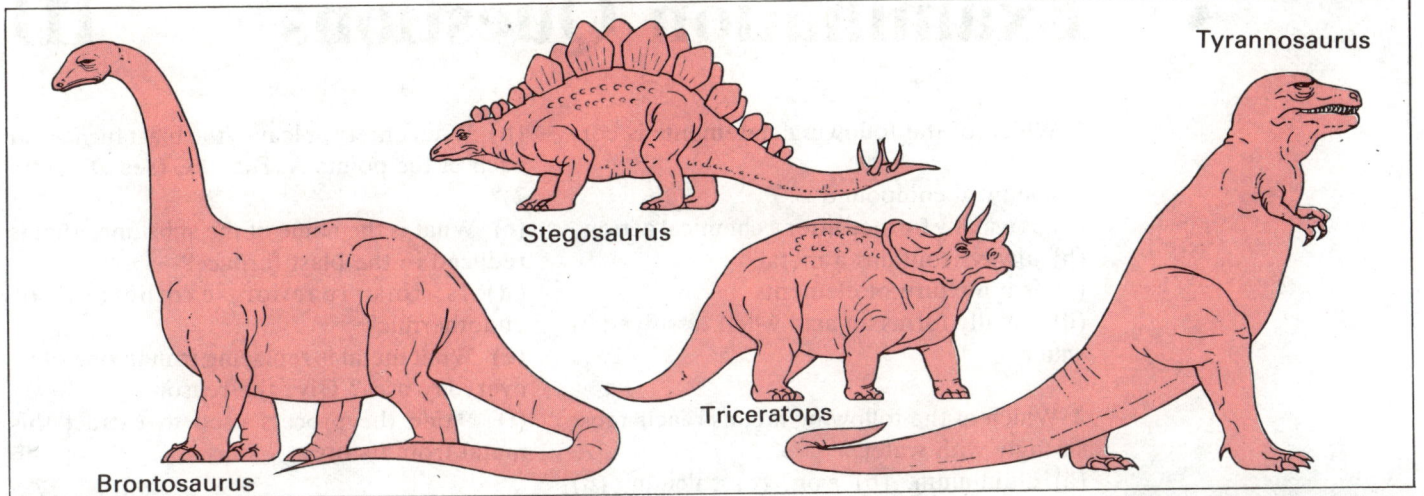

Stegosaurus
Tyrannosaurus
Triceratops
Brontosaurus

Embryos

The idea that different animals have descended from a common ancestor is supported by a study of embryos. In the early stages of development from eggs, the embryos of a fish, a reptile, a bird and a mammal are all very much alike. All have gill slits, like a fish, but the bird and the mammal embryos develop further than the fish or reptile.

Vestigial structures

Some animals have structures that seem useless today, but might have been useful to an ancestor. Some snakes, for instance, have tiny bones where limbs might be. We ourselves are covered with hair which stands on end when we are cold. This might have been very useful to our ancestors before clothing was used.

Perhaps our appendix was more useful to our distant ancestor than it is to us today.

Adaptive radiation

If evolution does take place, can changes in the environment be the cause? If different individuals of one particular species of a 'prehistoric' animal lived in different places, subject to different temperatures and climates, would the animals descended from them be similar?

Darwin, in his voyage in the 'Beagle', actually found several examples of divergencies of this kind. These are examples of what is called adaptive radiation. So the birds on the Galapagos Islands were found to be different in several small features from those on the American mainland. But they were alike enough to suppose that they had all descended from some common ancestor.

Natural selection

No two animals of a species are exactly alike in every detail. Those having variations that enabled them to adapt better to changing conditions might give rise to offspring that could survive whereas others would die in the 'struggle for survival'.

Evolutionary change can sometimes be seen to happen in measureable time. A good example of this is the Peppered moth. This moth exists as a dark form and also a light form. In industrial areas with smoky atmospheres, more of the dark form are found. It has replaced the light form. This is presumably because the dark form is not so easily spotted against the drab background by birds that feed on the moth.

Peppered moths

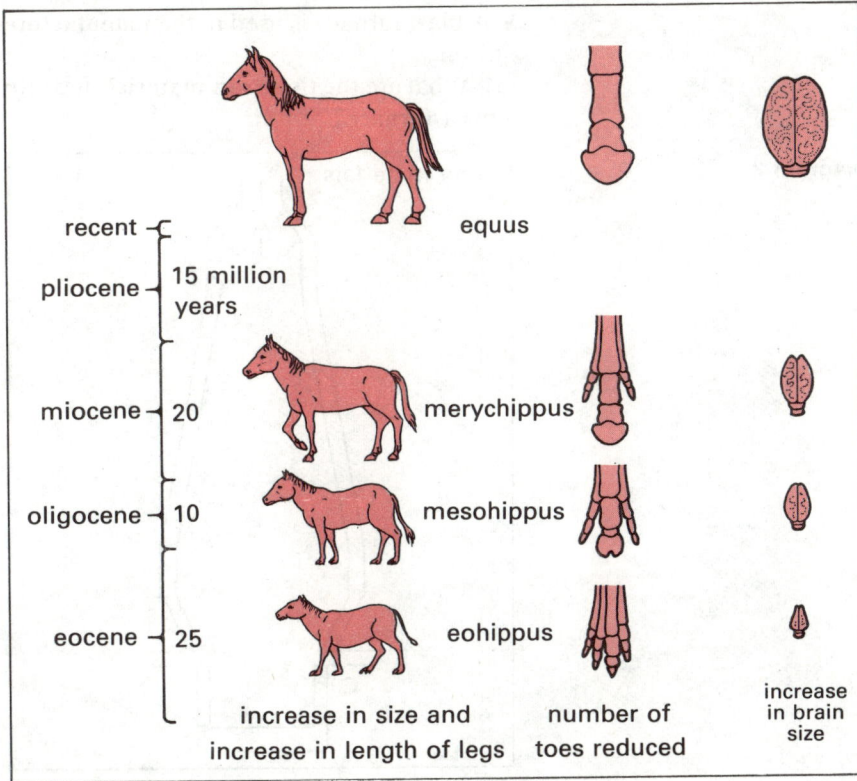

recent — equus

pliocene — 15 million years

miocene — 20 — merychippus

oligocene — 10 — mesohippus

eocene — 25 — eohippus

increase in size and increase in length of legs | number of toes reduced | increase in brain size

Examination Questions III

1 Which of the following statements is correct?
A chemical compound . . .
(a) is usually formed after a chemical action.
(b) always contains a metal.
(c) is a mixture of elements.
(d) usually forms an acid when dissolved in water.   WM

2 Which of the following metals reacts most strongly with water?
(a) aluminium, (b) iron, (c) calcium, (d) sodium, (e) copper.   WYL

3 Two nails were set up in water as shown in Diagram 1. Which will rust. Why?   WYL

Diagram 1

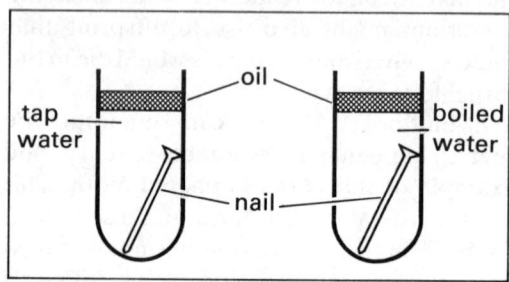

4 Litmus paper or litmus solution will turn red when treated with (a) alkalis, (b) water, (c) ammonia, (d) limewater, (e) lemon juice. Which answer is correct?   M

5 A blast furnace is used in the manufacture of iron.
(a) What are the three raw materials fed into a blast furnace?

Diagram 2

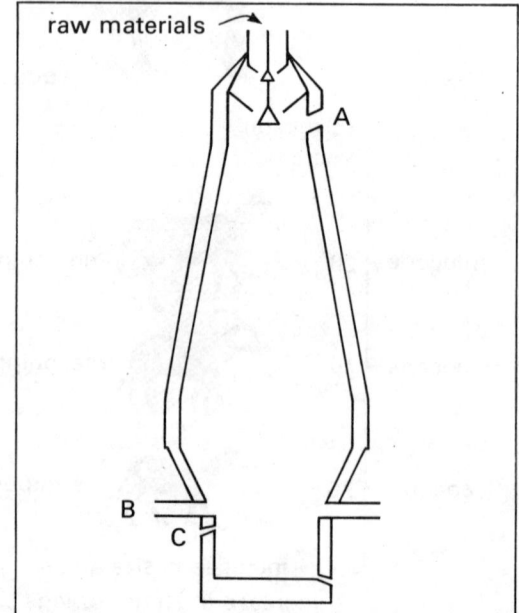

(b) What enters or leaves the blast furnace at each of the points A, B and C (see Diagram 2)?
(c) What is the name of the substance that is reduced in the blast furnace?
(d) Is this reaction exothermic or endothermic?
(e) What metal is replacing iron in one of its everyday uses? Give the reason.
(f) Name the process used to extract this metal from its ore.   SE

6 A typical example of an alkali is (a) lead, (b) hydrogen chloride, (c) sulphur dioxide, (d) sodium hydroxide, (e) calcium carbonate. What is the correct answer?   EM

7 Natural gas or town gas is passed over heated black copper oxide as shown in Diagram 3.
(a) What will you see happening to the copper oxide?
(b) What substance will remain in the tube?
(c) What happens to the mass of the substance in the tube during the experiment.
(d) Explain your answer to part (c).   EM

8 Two of the following will react to produce copper sulphate and water only: copper; copper oxide; copper nitrate; sulphuric acid; nitric acid; hydrochloric acid; ammonia; water; copper carbonate.
What are the two correct substances?   M

9 Which of the following is essential for germination to take place? Soil; light; water; carbon dioxide; darkness.   ALS

10 In a bulb such as an onion, most of the food reserve is stored in the (a) root, (b) leaves, (c) flower, (d) stem, (e) seed.
Which answer is correct?   EM

11 A mixture of iron filings and sulphur was placed on a watch glass.
(a) How could you remove the iron from the mixture leaving the sulphur?
(b) Using a different method, how could you remove the sulphur from the mixture leaving the iron?
(c) When the mixture was heated in a test tube it began to glow. After the bunsen was removed the glow spread throughout the whole mixture. What does this observation indicate?
(d) What is the name of the grey compound left in the test tube after heating?   EAN

Diagram 3

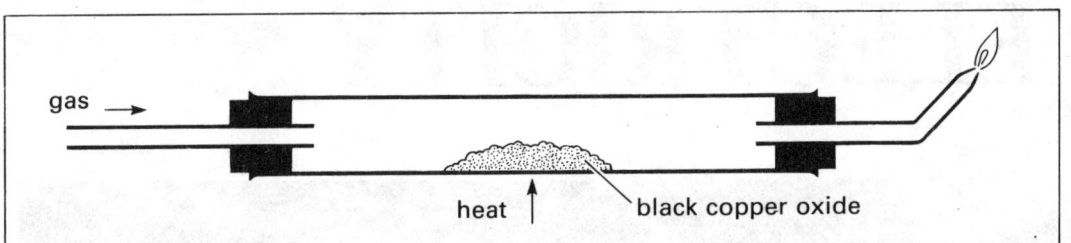

gas →

heat ↑   black copper oxide

12 (a) Draw a labelled diagram showing the main parts of a seed.
(b) What are the essential conditions for seed germination?
(c) Describe **two** ways in which man makes use of food stores of seeds.   W

13 (a) Why are metals rarely found in a pure state in the earth's crust?
(b) Arrange the following metals in order of chemical activity: copper; magnesium; iron; zinc.
(c) (i) In the blast furnace, which element has to be removed from iron ore to leave iron? (ii) Describe the chemical action that takes place in the blast furnace to remove the element named above.
(d) Why are aluminium and tin suitable for food packaging?
(e) Why is copper a suitable material for the base of saucepans?   SW

14 Write down four of the following items that are properties of hydrogen: (a) Less dense than air; (b) denser than air; (c) colourless; (d) green; (e) pungent smell; (f) odourless; (g) non-inflammable; (h) flammable.   NW

15 The apparatus shown (Diagram **4**) was assembled and left for a few days. (i) What would you expect to happen to the roots of each bean seedling? (ii) What feature of plant growth is this experiment designed to show?   SE (part)

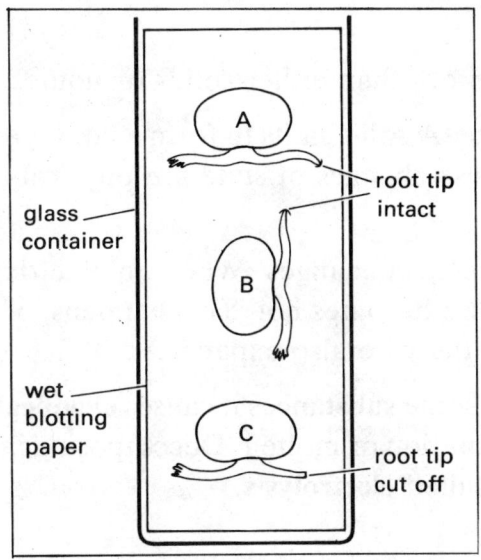

A

root tip intact

glass container

B

wet blotting paper

C

root tip cut off

Diagram 4

16 (a) Complete the following word equations:

heat
(i) Calcium carbonate → * + *
(ii) Sodium hydroxide + hydrochloric acid → * + *
(iii) Zinc + sulphuric acid → * + *
(iv) Sulphur + oxygen → *
(b) Write down the **formulae** for the following compounds: hydrochloric acid; calcium carbonate; sodium hydroxide. ALS

17 (a) Describe carefully an experiment to make a sample of pure sodium chloride from dilute hydrochloric acid and sodium hydroxide solution.
(b) Draw a diagram to show the shape of the sodium chloride crystals you would expect to obtain.
(c) State three precautions you should take when diluting concentrated sulphuric acid.
(d) Give one use in industry for each of the following.
(i) sulphuric acid, (ii) nitric acid, (iii) lime, (iv) caustic soda.
(e) Name one acid that is used regularly in the home and state its use.
(f) An old wives' tale states that you should put vinegar on the skin to remove the pain from a wasp's sting. This is found to work. State why this might be so.   EAS

18 (i) Name **one** element of which you have studied the extraction. (ii) Is the element you have chosen a metal or non-metal? (iii) Give **two** properties of the element. (iv) Give **one** use of the element. (v) What is the name of the raw material from which the element is extracted? (vi) Name **one** other raw material used in the extraction stage and state the function of its addition.   SE (part)

HEAT & ELECTRICITY

Expansion can cause spectacular damage

Heating can cause chemical changes. Heating can also make substances change in size and shape. Such changes are **physical** changes.

Solids become slightly longer when they are heated. They expand. Since they become longer in every direction, their volume also increases. Liquids expand more than solids.

All gases expand at the same rate, much more than either solids or liquids.

Heating a substance can also change its state. A solid melts to form a liquid; a liquid boils to form a gas (or vapour). These changes of state are physical changes; they can easily be reversed.

Electricity can cause both physical and chemical changes. When an electric current passes through a metal wire, the wire becomes hot. This happens, of course, in an electric fire. Because of this the wire also expands.

When an electric current is passed through some substances it causes **chemical** changes. The substances may be either in solution or molten. Decomposing in this way by passing an electric current is called **electrolysis**.

Expansion

Expansion of solids

This is so very small that the increase in length of a solid when it is heated must be magnified before it can be measured. The diagram shows this can be done.

An iron rod, 1 metre long, heated by 100°C would become longer by only 0.12 centimetre (just over 1 mm).

The following decimal fractions show how much a unit length of a solid expands for every 1°C. These fractions are called **coefficients of linear expansion**.

Aluminium	0.000 023
Copper	0.000 017
Iron	0.000 012
Platinum	0.000 0089
Glass	0.000 009
Invar	0.000 000 1

Expansion of liquids

Flasks are filled completely with liquids and fitted with glass tubes. When the flasks are heated, the expansion in volume shows by the level of liquid in the tubes rising.

Results show that different liquids expand by different amounts. 1 dm³ (1 litre = 1000 cm³) of water, heated from 15°C to 65°C would increase in volume by about 10 cm³.

The following decimal fractions show how much unit volume of a liquid expand for every 1°C rise in temperature. These fractions are called **coefficients of volume expansion**.

Water	0.000 21
Paraffin	0.000 9
Mercury	0.000 18

Expansion of gases

A flask filled with air and fitted with a glass tube, is inverted with the end of the tube under water in a beaker. There is now a definite volume of air enclosed in the flask.

When the flask is warmed, any air expansion shows itself as bubbles escaping. When the flask cools again, the water will rise up in the tube to take the place of the air that has escaped.

It has been found that all gases expand by 1/273 part of their volume at 0°C for each degree Celsius.

Galileo in 1638 used an inverted flask and tube like this to measure temperature. This was one of the earliest thermometers.

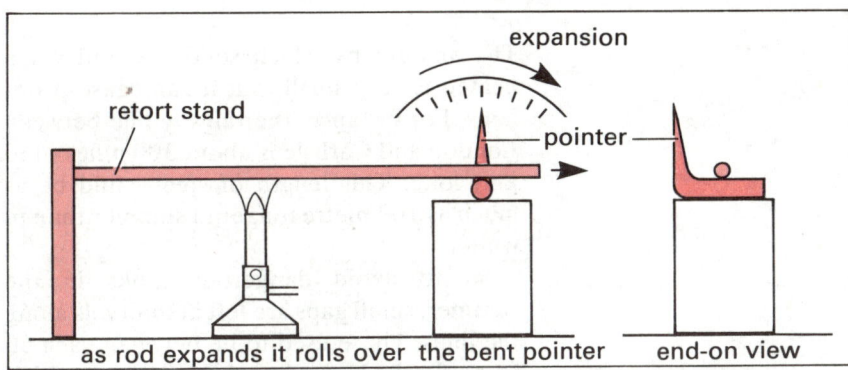

Expansion of solids

Molecular ideas on expansion

Molecules are continually moving about.

Molecules in a solid do not move far, they just vibrate. They stay more or less in the same place. This is because they attract and repel each other.

When the solid is heated, the molecules 'get excited' because they receive more energy. They vibrate more and take up more space. So the solid as a whole expands.

In a liquid the molecules are even more energetic. They flow past each other and are not kept in one place. Their energy overcomes the attractive forces.

Molecules in a gas are not held together by the small forces between them. They move about quite freely. When heated, gases expand a great deal more than liquids.

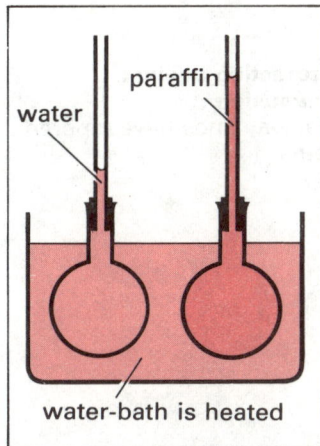

Expansion of liquids

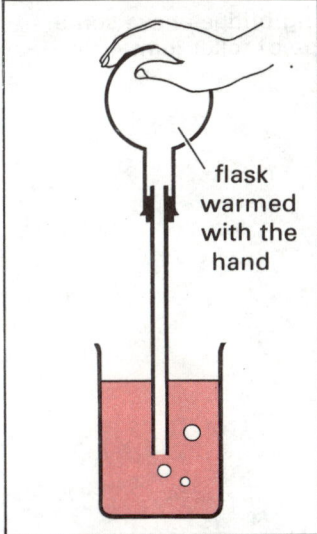

Expansion of gases

1 From the following metals, select the metal that (i) expands most, (ii) expands least. Iron; copper; invar; aluminium.
2 Which of the following gases expands most when heated from 0°C to 100°C? Air, oxygen, carbon dioxide, nitrogen.
3 When a flask of water is heated suddenly, the level of liquid in the tube falls slightly at first, before it goes up. Can you suggest an explanation? Remember that the heat reaches the glass flask first before it reaches the water.

The amount by which solids expand when heated is very small, but it can cause problems. For instance, the railway line between London and Carlisle is about 300 miles (480 km) long. This length of steel could be as much as 100 metre longer in summer than in winter.

So to avoid dangerous kinks in the summer, small gaps are left at intervals along the lines. These used to be between each 60 feet of rail. Nowadays lengths are welded together to make sections of about 100 metre. Tapered ends are left between these sections to allow for expansion.

Protection against expansion
Railway lines have tapered joins

In the case of a 'fly-over' road or a long bridge, allowance is made for expansion by letting one end of the road slide on rollers or by having teeth-like ends that slide in and out.

For smaller lengths of solid, expansion usually does not matter.

But what about the pendulum of a grandfather clock? There may be only a frac-

Long bridges have some type of roller joint

tion of a millimetre difference in length over the year, but a longer pendulum swings more slowly. In hot weather this small difference means that the clock will be slow, unless the expansion is corrected for.

If very hot liquid is poured into a thick glass vessel, the sudden expansion may make the vessel crack. Glass does not conduct heat well, so the heated inside expands more quickly than the outside. The strain may crack the glass.

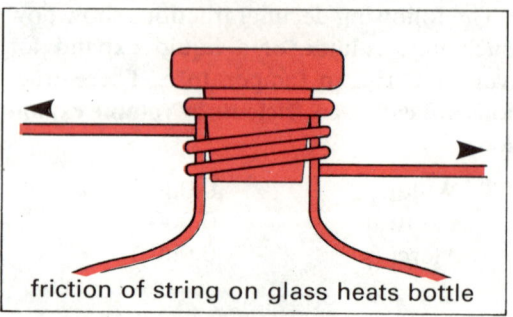

friction of string on glass heats bottle

But expansion is sometimes useful. When a glass stopper is stuck in the neck of a glass bottle, it can be released by warming the neck. One way of doing this is by pulling a length of string wrapped round the neck to and fro several times.

When hot metal cools it contracts and the force of this contraction is quite strong. When the outside rim (tyre) of hard steel is to be fitted on the wheel of a railway engine, it is made slightly too small. It is then put over the wheel when hot. As it cools, it contracts and binds to the wheel very tightly.

Bimetallic strip

This consists of equal lengths of two different metals, e.g. copper and iron, riveted together. When this double bar is heated, the strip bends because the copper expands more than iron.

Such a strip can form part of an electric circuit. If the strip gets heated, it will bend and so break contact and break the circuit.

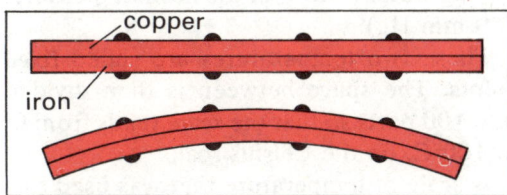

copper
iron

Thermostats

These are devices to control temperature. They often use the different expansion of two metals. A good example is the thermostat used in regulating the temperature of a cooking oven heated by gas.

The brass tube shown below is in the oven. The control knob is outside. When the oven gets hot the brass tube expands, but the invar steel rod does not expand very much. So it pulls the valve along and cuts down the flow of gas. The temperature at which this happens can be chosen by turning the control knob to the required setting.

An interesting thermostat is found in the cooling system of a motor car. The engine of a car gets hot when running and needs to be cooled. The engine block is surrounded by water. When the engine has been running for some time, the water itself gets hot. It is allowed to pass into the radiator where it is cooled by air, sometimes helped by a fan.

But it is not a good idea to have too much cooling while the engine is starting. So a thermostat is put between the engine block and the radiator. The bellows in this thermostat contains liquid which when cold

keeps the valve in its setting. As the liquid gets hot, it expands and pushes the valve up. Water is then able to pass into the radiator.

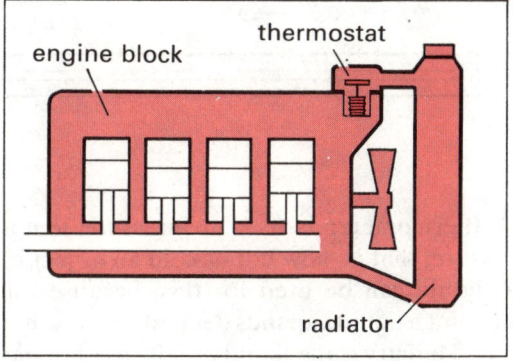

engine block   thermostat

radiator

Thermostat
A thermostat is used to control the flow of water through a car engine

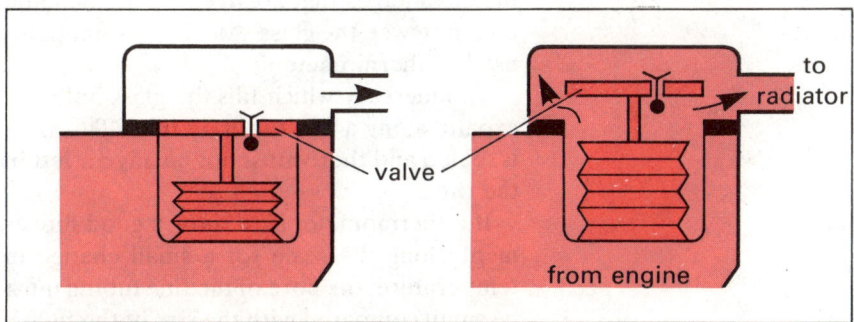

to radiator

valve

from engine

1 Why are pipes that may carry steam sometimes looped?
2 A bimetallic strip is included in the electric circuit of a car flashing indicator. It also forms part of one of the lamps in a set of flashing lights on the Christmas tree. In each case, the effect is that the lights flash on and off again. Can you suggest how this might happen?
3 Draw a diagram to show how you could use a bimetallic strip in an electric circuit with a bell so as to act as a fire alarm.
4 How would you expect a bimetallic strip to be used as part of the circuit of an electric iron? In this case the electricity supply is cut off when the iron reaches the required temperature.

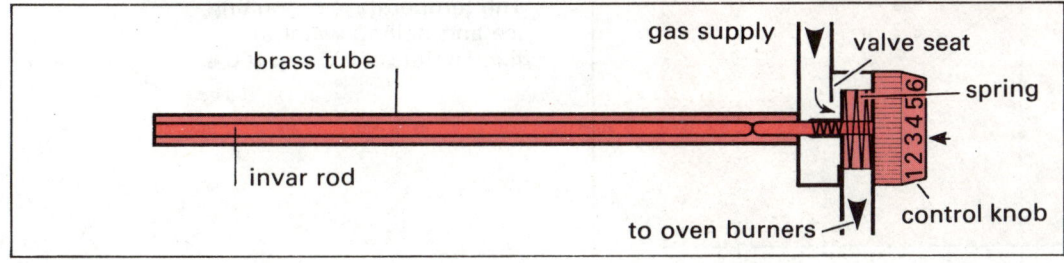

brass tube

invar rod

gas supply   valve seat

spring

1 2 3 4 5 6

to oven burners   control knob

Thermostat
A thermostat is used to control the flow of gas to an oven

Thermometers

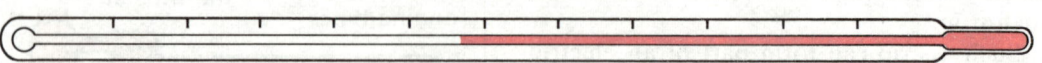

A thermometer is used to measure temperature, that is, how hot or cold an object is. A liquid can be used for this, because the amount a liquid expands depends on how hot it is. Mercury is the liquid usually used. It has the advantages that it can easily be seen and does not 'wet' the glass. Alcohol is sometimes used in thermometers.

The mercury which fills the glass bulb can expand along a fine capillary tube. This tube is sealed and there must not be any air left in the tube.

If a thermometer is to show a good movement along the scale for a small change in temperature, the bore of the fine tubing must be small compared with the size of the bulb.

The scale marked along the stem shows the temperature. This is done by noting the mercury levels when the thermometer is put in melting ice (0°C) and held in steam from boiling water (100°C at the normal pressure, 760 mm Hg).

These two temperatures are called **fixed points**. The space between is then divided into 100 parts so that the scale reads from 0° to 100°C, on the Celsius scale.

A scale of temperature that was used formerly is the Fahrenheit scale. It is still used sometimes in cookery recipes, weather reports, gardening and so on. On this scale, ice melts at 32°F and water boils at 212°F.

It is easy to plot a graph to show the relation between the two scales. Plot degrees Celsius against degrees Fahrenheit. If the point representing 0°C = 32°F is joined to the point 100°C = 212°F, the line will give equivalents. So, body temperature is about 98°F on the Fahrenheit scale, but about 37°C on the Celsius scale.

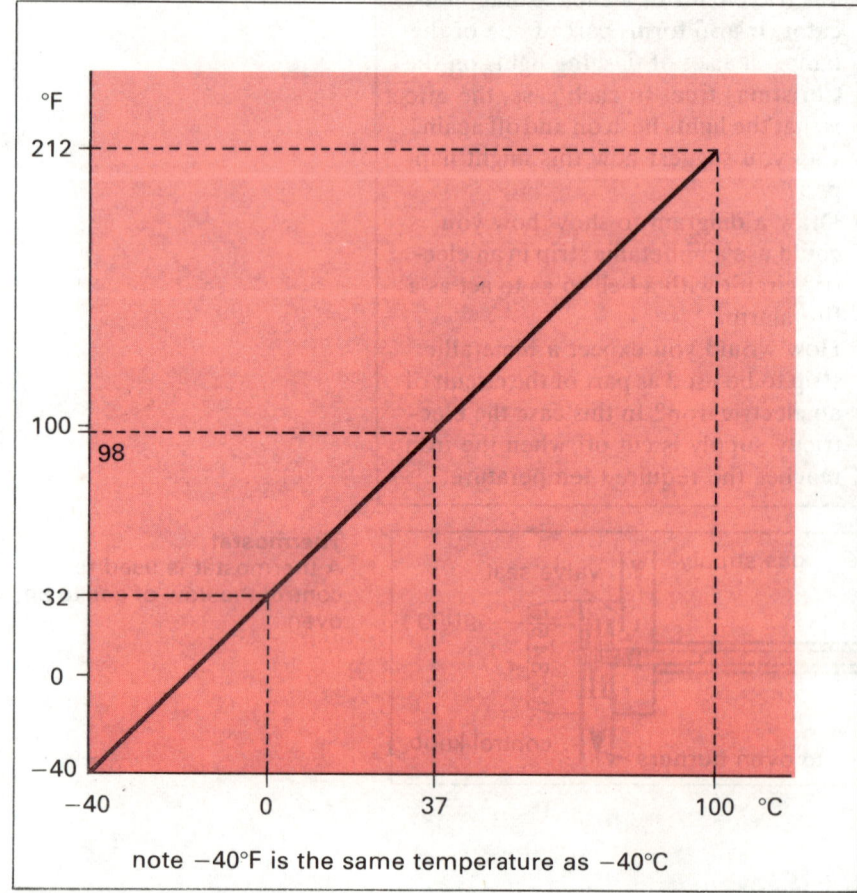

note −40°F is the same temperature as −40°C

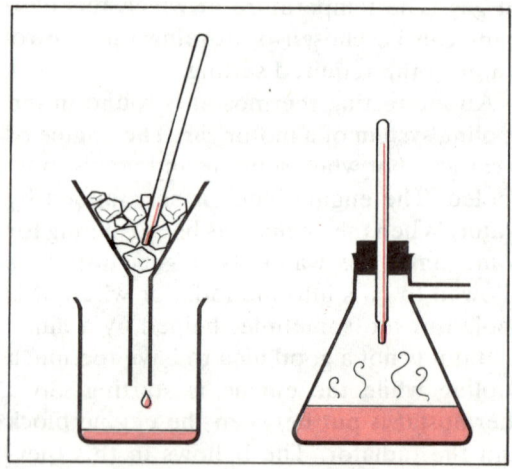

Fixed Points (above)
The temperature of melting ice and boiling water are fixed points of 0°C and 100°C

Conversion graph (left)
This graph converts the Fahrenheit scale and the Celsius scale

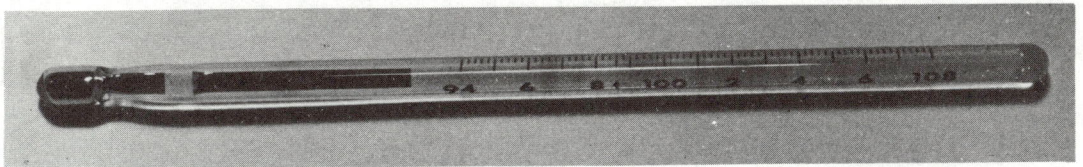

Clinical thermometer

One special kind is the clinical thermometer. This is a small thermometer only about 10 cm long. It is used for measuring human body temperatures. So it must read from about 35 to 42°C (or 95 to 110°F). It has some special features.

1 For the small range of temperature measurement to be possible, there is a very fine bore capillary tube.

2 The bulb must be small, because it must go easily under the patient's tongue.

3 The glass of the bulb is thin, though strong, so that heat can pass to the mercury quickly. This means that the thermometer will show the reading quickly.

4 There is a constriction between the bulb and the fine part of the tubing. This causes the mercury thread to break when the thermometer is taken out of the patient's mouth. But the mercury above the constriction stays in place. The nurse can take the reading before shaking the thermometer to make the mercury go back to the bulb.

5 The front edge of the stem is usually curved so that it acts as a magnifying glass. The fine thread of mercury can be more easily seen against the background of white enamel in the glass.

Maximum and minimum thermometer

It is sometimes necessary to know the lowest and highest temperatures reached during a certain time, but it is not possible to stay looking at the thermometer all the time.

Alcohol is sometimes used in a thermometer instead of mercury. It can be coloured with a dye so that it is more easily seen. Alcohol is used in this thermometer. Alcohol in bulb A expands when the temperature increases. This pushes the mercury along so that the light steel index, X, is pushed along.

A light spring on this index, holds it in position when the mercury falls back. But as the temperature falls, the mercury in the left hand part of the tube pushes the index N further up so that it shows the minimum reading. Index X shows the maximum. Bulb V contains alcohol vapour which 'cushions' the expansion. A magnet can be used to re-set the indices after readings have been taken.

There are many ways of measuring temperature besides the thermometer. Some depend on measuring an electrical property such as the resistance of a coil of wire, which changes with temperature. One of the most reliable ways depends on measuring the pressure of a fixed volume of gas.

You may think you could depend on your hands to judge temperature. If you do, try this experiment.

You need three bowls of water, A cold, B tepid and C quite hot. Put your left hand in A, your right hand in C. Hold them there for about 30 seconds. Take both out and put

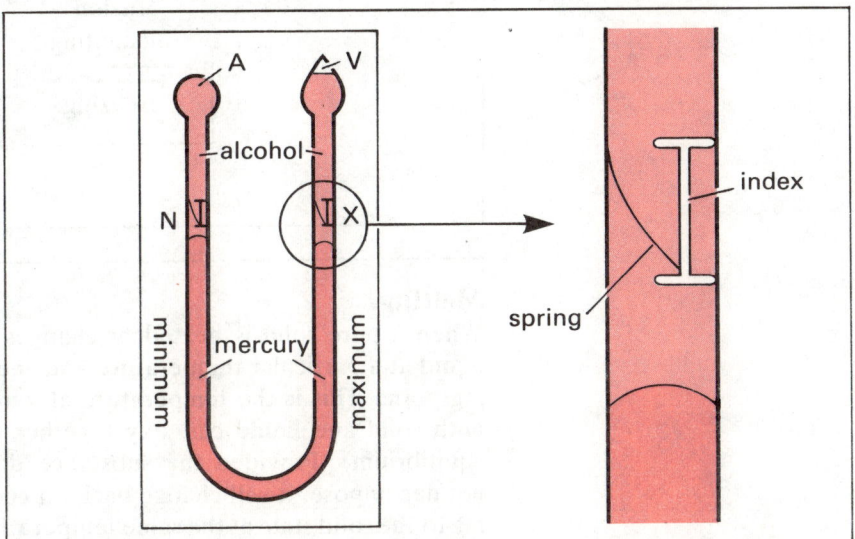

them straight away in B. What do you feel? Is your feeling a reliable guide to how hot or cold a thing is?

1 Why must you not sterilize a clinical thermometer by putting it in boiling water?

2 Select the temperatures from the list below to correspond with the following: surface temperature of the sun; boiling point of mercury; temperature of the human body; freezing point of water; boiling point of water; freezing point of mercury.
   −39°C; 0°C; 37°C; 100°C; 357°C; 6000°C.

3 Describe how the device shown in the diagram on the right indicates temperature.

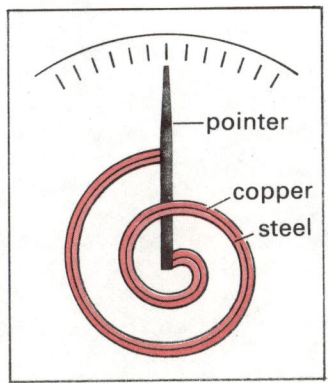

A substance can usually exist in three different states. These are solid, liquid and gas (or vapour). They are the **states of matter**. Changes from one state to another are physical changes. They can be reversed.

A vapour is different from a gas because a vapour can be condensed to liquid by increasing the pressure, without lowering the temperature.

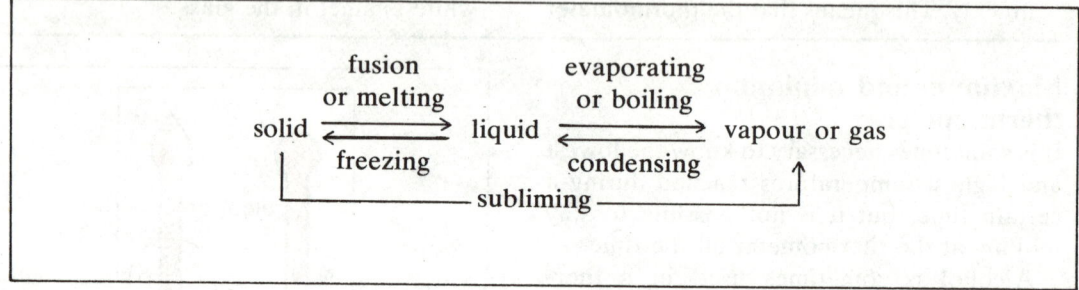

Melting

When a pure solid is heated, it changes to liquid at a particular temperature – its melting point. This is the temperature at which both solid and liquid can stay together (in equilibrium). Provided the substance does not decompose, it will change back on cooling to the solid state at the same temperature (freezing point).

Sublimation

When iodine is heated it does not melt, but forms a violet vapour. When a solid changes directly to vapour without forming liquid, it is said to **sublime**.

Another substance that sublimes is ammonium chloride. This case is rather different though. The ammonium chloride when heated, breaks up chemically into two gases, ammonia and hydrogen chloride. But these gases on cooling re-combine to form ammonium chloride. So the solid appears to sublime:

$$\text{ammonium chloride} \underset{\text{cooling}}{\overset{\text{heating}}{\rightleftharpoons}} \text{ammonia} + \text{hydrogen chloride}$$

If you want to separate a mixture of sodium chloride and ammonium chloride, you can heat the mixture in a crucible, with a lid. The ammonium chloride will settle on the lid, leaving the sodium chloride in the crucible.

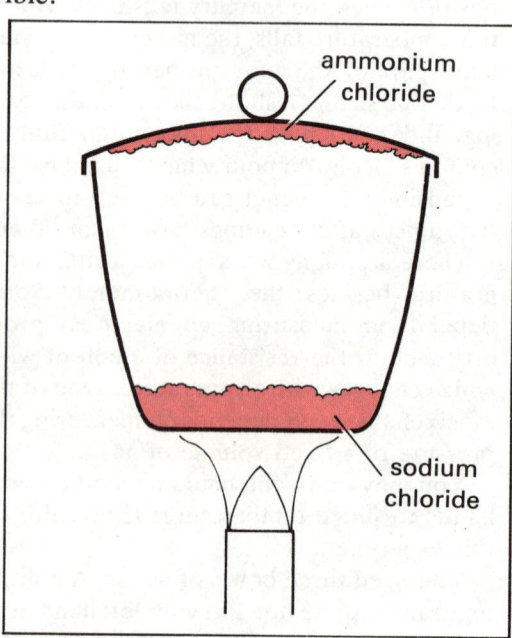

Separation of sodium chloride and ammonium chloride

Effect of pressure on melting point

The melting point of a substance depends on the pressure. If the atmospheric pressure above ice was doubled, the melting point would be lowered by 0.0075°C. When snow is moulded in the hands to make a snowball, the pressure applied is enough to melt the outside layer of ice and so help to compact the snow.

An ice-skater exerts great pressure on the narrow blade edges of the skates so that the melting point of ice is lowered slightly (about 0.1°C). The skater moves on a thin film of water, with little friction.

Evaporation

Some liquid ether left out in a watch glass quite quickly disappears. This can be explained by using the idea of molecular motion. Some ether molecules are moving quicker than others. Some move so fast that they escape into the air, away from the rest of the molecules in the liquid. The ether is **evaporating**.

The process can be speeded up by blowing air across the surface of the ether. The evaporation rate is increased if the liquid is hot or if it has a large surface area.

Boiling

When evaporation becomes so rapid that bubbles of vapour arise from all parts of the liquid, the liquid is boiling. Liquid and vapour exist together at the boiling point.

At the boiling point, the vapour pressure of the liquid is saturated and is equal to the atmospheric pressure.

Each pure substance has a definite melting point and/or boiling point. These are used to identify a substance.

In the case of a mixture of liquids, the different liquids can often be separated by distilling the mixture and condensing each liquid at its particular boiling point. This method of separating substances is known as **fractional distillation**. This is how crude oil is separated into different fractions at the oil refinery.

Effect of pressure on boiling point

The vapour pressure of a boiling liquid is equal to the external pressure at the boiling point. So if the external pressure is increased, the liquid will need to reach a higher temperature for its vapour pressure to equal the new external pressure.

The boiling point of a liquid is raised by increasing the external pressure and lowered when the pressure is reduced.

Pressure cooker

The steam formed by the water boiling inside the sealed cooker increases the pressure inside. This can be almost double the normal atmospheric pressure. So the water reaches a temperature well above 100°C and the food is cooked much more quickly. A safety valve prevents too great a pressure developing. An autoclave for sterilising surgical instruments works in the same way.

1 Iodine crystals are grey and metallic in appearance. How would you separate iodine from charcoal if they were mixed together?
2 What is the vapour pressure of ethanol at its boiling point (78°C) on a day when atmospheric pressure is normal (760 mmHg)?
3 Can you explain why it is more difficult to cook an egg at the top of a high mountain, than it is at sea level? (Think of the air pressure at the top of a high mountain.)

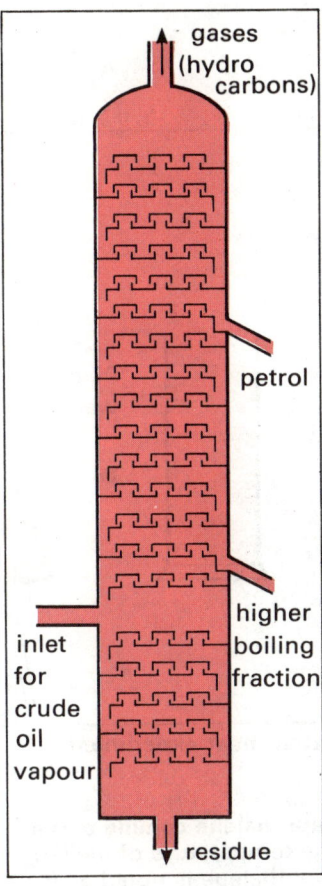

Fractional distillation column

Latent Heat

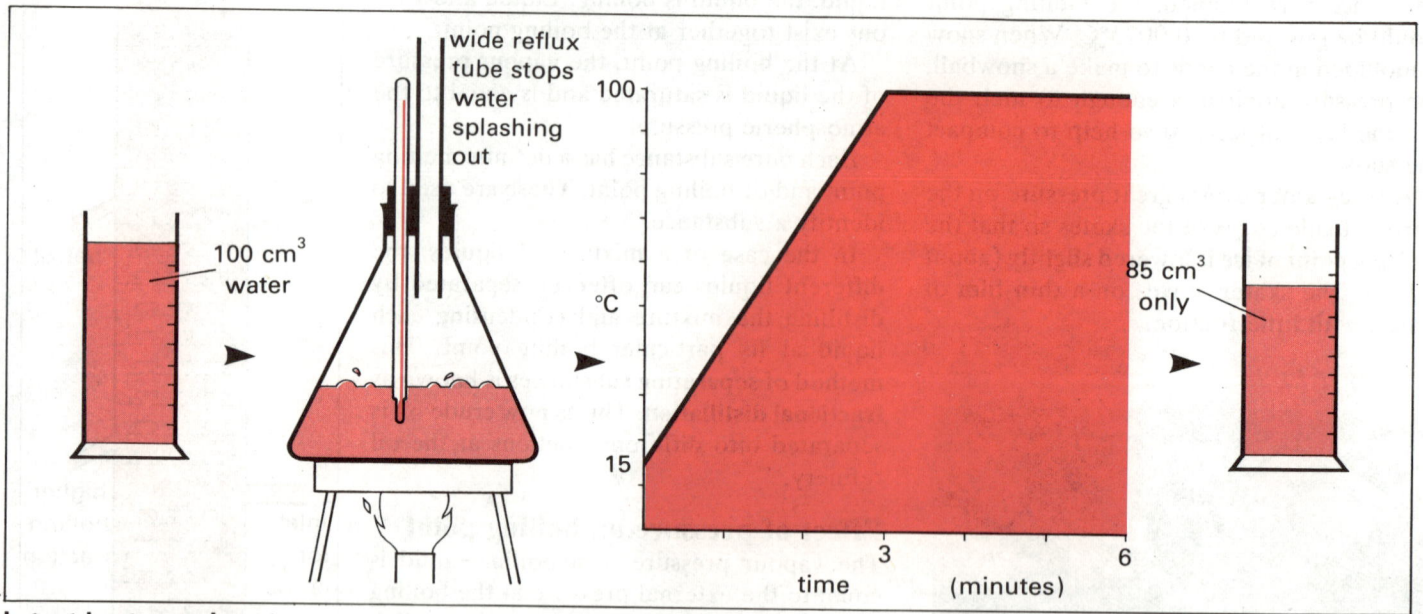

Latent heat experiment

Naphthalene cooling curve
The temperature of melted naphthalene is noted as it cools and graph plotted

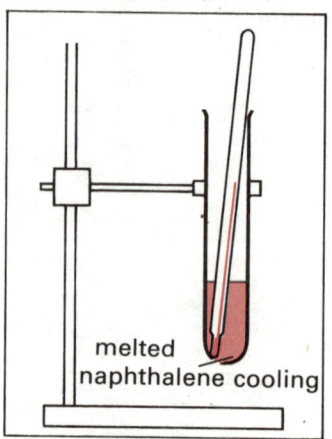

melted naphthalene cooling

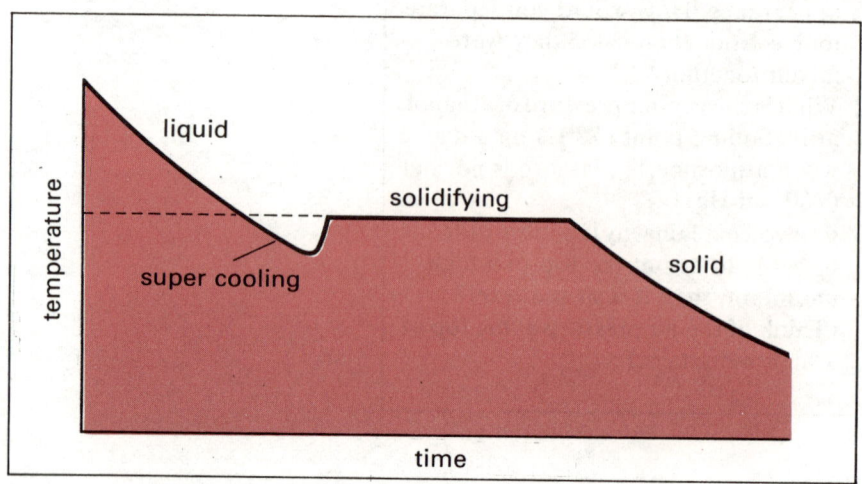

When some water is heated at a steady rate its temperature rises to boiling point around 100°C. But further heating does not make the temperature rise above 100°C. All that seems to happen is that some of the water changes to steam.

In the experiment shown above, it took 3 minutes heating for the water, originally at 15°C, to reach boiling point. A further 3 minutes heating changed 15 cm³ of water into steam. The temperature did not rise above 100°C.

It is not difficult to think out what happens. Even when the temperature has reached 100°C, the molecules of water need energy to escape from the liquid into the vapour space. This energy is known as **latent heat**. As long as both liquid and vapour are present together the temperature must be boiling point.

The latent heat concerned in changing water to steam is called the **latent heat of evaporation** of water.

As another example, imagine that the temperature of some melted naphthalene is noted (wax can be used instead of naphthalene). The temperature is recorded every minute as the liquid cools.

When the temperatures are plotted against time, a cooling curve is obtained. The temperature falls slowly until the naphthalene begins to solidify. Then it stays at this point until all the liquid has solidified, that is, at the melting point.

Latent heat is being given out by the cooling liquid as it changes to solid. This is called the **latent heat of fusion**. To change 1 kg of ice at 0°C to water at 0°C takes 336 kilojoules.

It sometimes happens, if the liquid is not stirred as it cools, that it may become supercooled – below the melting point. But any disturbance will make the temperature rise quickly to the melting point. This happens with the release of heat.

The latent heat of evaporation of water is quite large. To heat 1 kg of water from 0 to 100°C takes 420 kilojoules of heat. To change 1 kg of water at 100°C into steam at 100°C takes 2260 kJ of heat, that is, over five times as much.

This means that if it takes 5 minutes to boil some water in a kettle, it will be at least another 25 minutes before the kettle 'boils dry'.

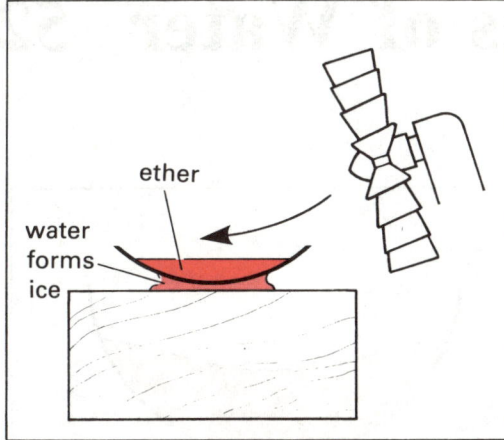

Evaporation causes cooling
When ether evaporates it gets cold enough to freeze water

Cooling by evaporation

Some ether is forced to evaporate in a fume cupboard, by having air blown across the surface. The ether needs quite a lot of latent heat to evaporate, so it takes it from its surroundings. If a drop of water is first put under the dish containing the ether, the water may lose so much heat that it changes to ice.

Perspiration. Our skin is kept cold when we perspire by the evaporation of sweat. This comes from the sweat glands (see Topic 110). There is more sweat when the body gets hot, so it acts as an automatic cooling system.

Refrigerator. The losing of heat by the evaporation of liquid is made use of in the familiar 'fridge'.

A pump makes the refrigerant liquid boil under reduced pressure. This takes heat away from the inside of the refrigerator. The vapour condenses and is circulated round again.

The refrigerant liquid can be ammonia or sulphur dioxide. Modern domestic refrigerators use a chemical called Freon, which boils at about −29°C.

In a gas refrigerator, cooling is brought about by heating! A gas flame takes the place of the pump and causes the refrigerant to circulate.

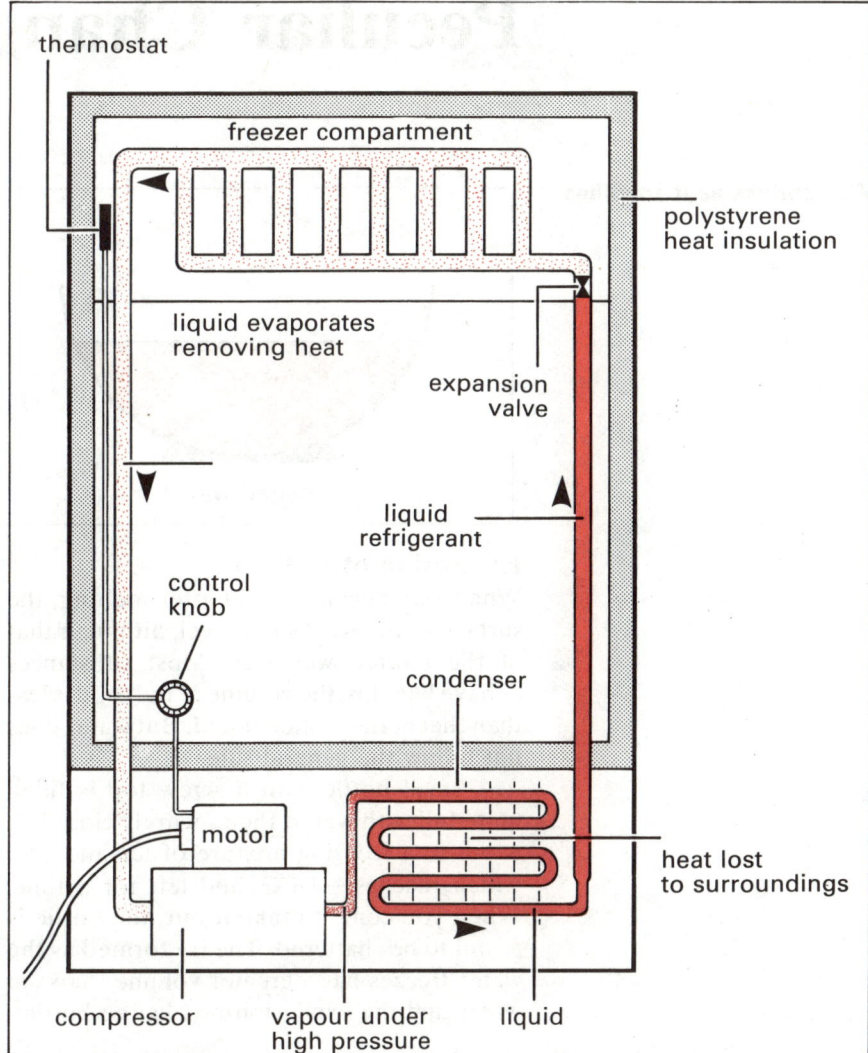

1 Why is a burn with steam usually more severe than a burn with boiling water?
2 Can you explain why, after a fall of snow, it takes a long time to melt the snow, even though it is warm and sunny?
3 Why is it more uncomfortable on a hot day when the air is humid (saturated with water vapour) than on a day when the air is dry, even though the temperature is the same?

Wax shrinks as it solidifies

melted wax

solidified wax

Expansion of water

When wax becomes solid after melting, the surface of the wax is not level, although that of the molten wax was. Most substances behave like this, the volume of the solid is less than that of the molten liquid. But water does not follow the general rule.

A small bottle with a screw top is filled quite full with water, then securely closed. It is put in a freezing mixture of ice and salt, which reaches −18°C, and left for a time. When you come to take it out, the bottle is found to be shattered. The ice formed as the water freezes has a greater volume than the water and the expansion breaks the bottle.

The reverse change can be studied. A few small pieces of ice are carefully dried, then dropped into a measuring cylinder containing some paraffin. Ice sinks in paraffin. The combined volume of paraffin and ice is noted and the ice allowed to melt. When the new volume of paraffin and melted water is noted it is found to be less than the first reading.

The expansion of water when it freezes makes itself unpopular in winter! During severe frosts, the water may freeze in the pipes. If the pipes are not free to expand, they will crack because of the greater volume of the ice. But this is not discovered until the thaw melts the ice and the water escapes through the crack.

Weathering of rocks

Another result of water expanding on freezing is that rocks are broken up in winter.

Water may seep into fine cracks in rocks. It then freezes and makes the cracks larger so that even a large rock can be broken. This is only one way in which rocks break. Uneven expansion in hot weather can cause the initial cracks.

Water shrinks when it melts

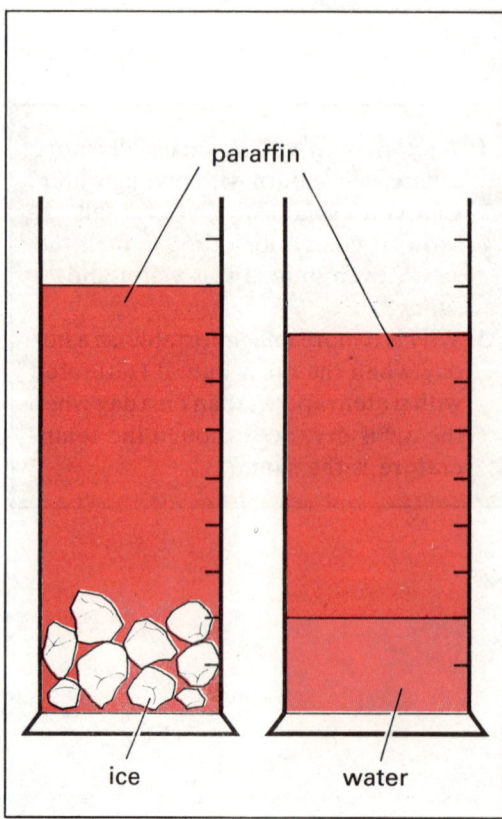

paraffin

ice

water

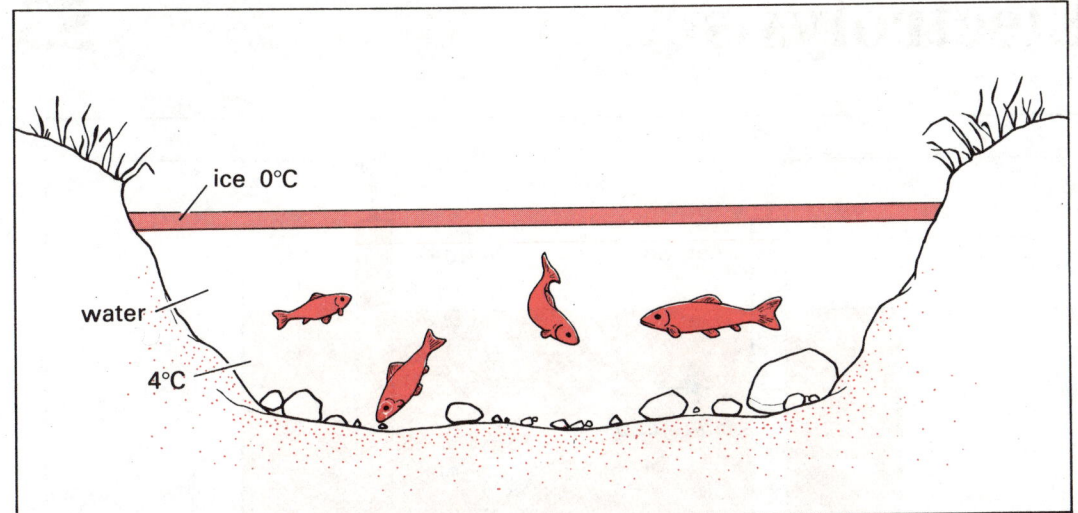

Frozen pond

Maximum density of water

Water is unusual in another way. Water at 0°C might be expected to expand when it is heated. But up to 4°C, the water actually decreases in volume. Above 4°C it expands. So water is more dense at 4°C than at any other temperature.

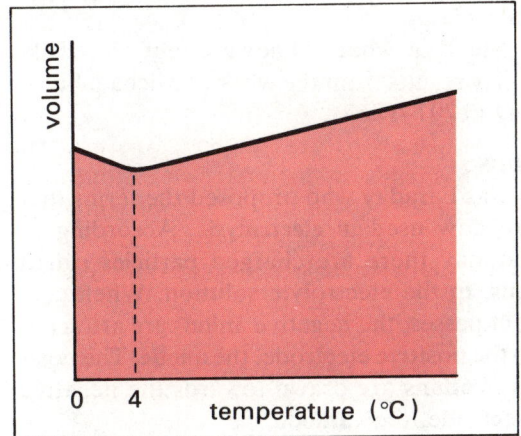

Volume changes
This graph shows the change in volume of water with temperature

When ponds begin to freeze in winter the first ice is formed on the surface by contact with the cold air. But this ice, at 0°C is not likely to sink because water at 4°C is dense and will occupy the bottom of the pond.

This means that fish can survive the cold weather below the ice instead of becoming frozen in solid ice.

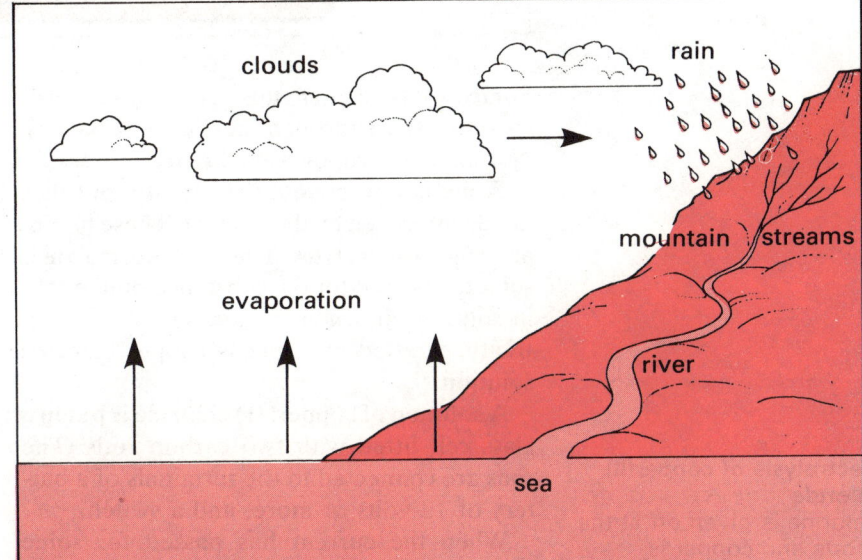

Water cycle

Water cycle

From what you have learnt, it is not difficult to see that water circulates. The sun over the seas makes the water evaporate into the air. Here it condenses into small droplets as it cools, forming clouds.

Under some circumstances, such as meeting warmer air or meeting mountains, the clouds release the water as rain. The rain falling down forms streams that join to form rivers, which run back to the sea again.

1 Gardeners like to rough-dig the garden in autumn before the frosts. Can you suggest what will happen to the soil when rain is followed by frost?
2 Could water, possibly coloured with a dye, be used as the liquid in a thermometer? What problems would there be near 0°C?
3 After severe frosts, there may be leaks from water pipes when a thaw takes place. Does this mean that when ice melts the expansion bursts the pipes?

Electrolysis
Anions move to the anode; cations move to the cathode

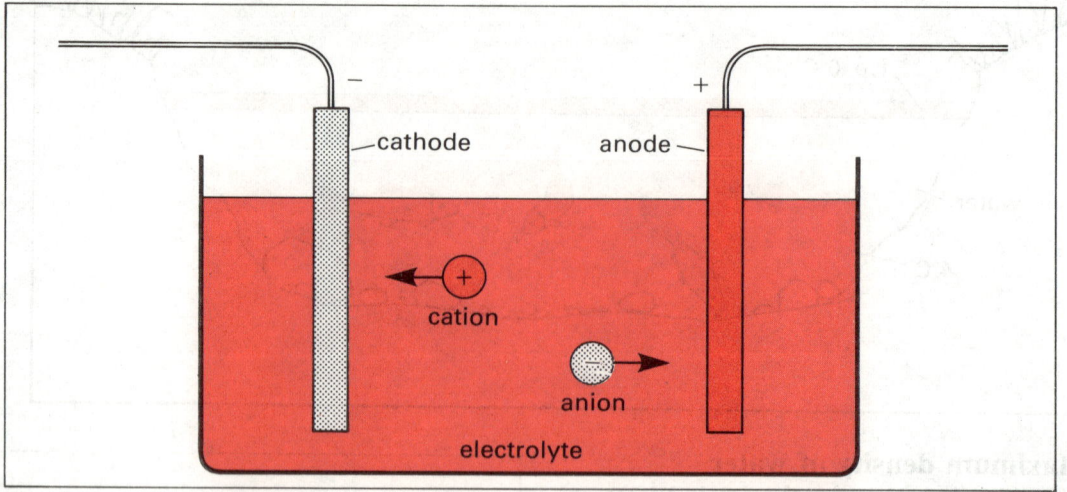

Some solids, e.g. metals, allow an electric current to pass through them (see Topic 67). Not all solids conduct electricity.

Some liquids conduct electricity, but they are decomposed by the current. These liquids are called **electrolytes**. The term electrolyte is sometimes used for the substance that, either in solution or when melted, conducts electricity. A good example is copper chloride solution.

A solution of copper(II) chloride is put in a glass cell fitted with two carbon rods. The rods are connected to the terminals of a battery of 12 volts or more, and a switch.

When the current has passed for some minutes, a reddish-brown powder is seen on the carbon rod connected with the negative pole of the battery.

A yellow gas comes off around the other carbon rod. So the current has separated the copper and chlorine of the copper(II) chloride.

Much of what is known about electricity today results from the work of Michael Faraday (1791–1867).

Electrolysis of copper(II) chloride
Chlorine is given off at the anode and copper is deposited at the cathode

copper(II) chloride solution

electron flow

Ions
It was Faraday who proposed the terms that are now used in **electrolysis**. According to Faraday there are charged particles called **ions**, in the electrolyte solution. When current passes, the negative **anions** are attracted to the positive electrode, the **anode**. The positive **cations** are drawn towards the negative electrode, the **cathode**.

It is now known that in a solution of copper(II) chloride, even before current is passed, there are copper ions, Cu^{2+}, and chloride ions, Cl^-. A chloride ion is a chlorine atom that has gained an electron

$$Cl + e^- \rightarrow Cl^-$$
atom   electron   ion

A copper ion is a copper atom that has lost two electrons

$$Cu - 2e^- \rightarrow Cu^{2+}$$
atom   electron   ion

When the carbon rods are connected to the battery, the ions are attracted towards the electrodes.

Ions are quite different from atoms. When the copper ions reach the cathode, they form copper atoms. These are seen as a reddish-brown powder. The chlorine atoms formed at the anode are seen as a yellowish gas, but the chloride ions cannot be seen in the solution.

During this process of electrolysis, electrons are taken off the cathode to change the copper ions into atoms. Electrons are given to the anode by the chlorine ions. So electrons move round in the circuit through the battery from anode to cathode.

An electric current is a flow of electrons.

Copper sulphate

When copper sulphate is electrolysed, copper is formed at the cathode. If a sheet of copper is used as cathode, it becomes coated with more copper.

Copper can also be used as the anode. This copper becomes thinner and is gradually used up. In this case, copper ions from the solution become atoms at the cathode.

$$Cu^{2+} \ + \ 2\,e^- \ \rightarrow \ Cu$$
$$\text{ion} \qquad \text{electrons} \qquad \text{atom}$$

At the anode, copper atoms lose electrons and form ions. These go into the solution.

$$Cu \ \rightarrow \ Cu^{2+} \ + \ 2e^-$$
$$\text{atom} \qquad \text{ion} \qquad \text{electrons}$$

Valency cards

You may have wondered why an ion of copper is Cu^{2+} and not Cu^+. The valency of copper in both copper(II) chloride and copper(II) sulphate is two. The electric charge on ions depends on the valency of the element or group.

You can use the valency cards to show ions.

1 There are no ions present in chloroform. What do you think would happen if you tried to pass a current through chloroform?
2 Write down the symbols for: (a) a chlorine atom; (b) a chlorine ion; (c) a copper atom; (d) a copper ion; (e) the ions present in sodium chloride; (f) the ions present in copper sulphate solution.
3 Use your valency cards to show the ions that together make (a) sodium nitrate; (b) calcium chloride.

Electrolysis of water

Pure water does not conduct electricity. It is a **non-electrolyte**. Your teacher may show you this striking experiment.

A suitable electricity supply is connected in series through electrodes in a beaker of distilled water and a switch, to an electric lamp. When the switch is closed, nothing happens. The lamp does not light.

But if a little dilute sulphuric acid is added to the water, the lamp glows then lights up brightly. It can be seen that bubbles of gas are produced at the electrodes in the dilute acid in the beaker.

These bubbles can be collected in the apparatus shown. Very dilute sulphuric acid is electrolysed, with pieces of platinum foil as electrodes. Hydrogen comes off at the cathode and oxygen at the anode. There is twice as much hydrogen as oxygen by volume.

These facts can be remembered by thinking of the middle letter of the electrode names:

$$\text{an}\mathbf{O}\text{de} \qquad \text{cat}\mathbf{H}\text{ode}$$

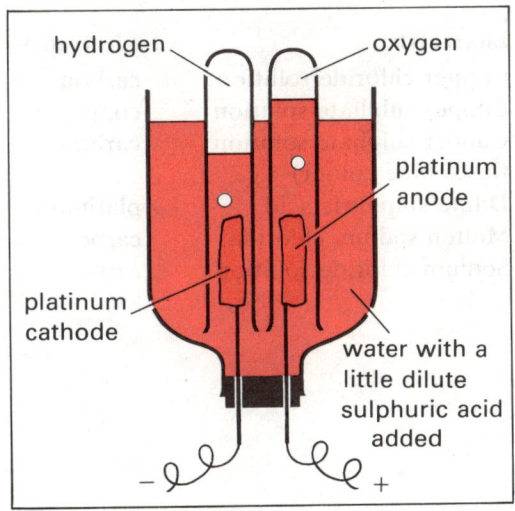

Electrolysis of acidified water
Hydrogen is given off at the cathode and oxygen at the anode

Electrolysis can give information about the composition of a solution. It can also be used to make some chemicals that cannot easily be made in other ways.

Electrolysis of molten substance

Although solid sodium chloride does not conduct a current, it can be heated strongly to melt it and make a molten liquid.

Electrolysis of molten sodium chloride

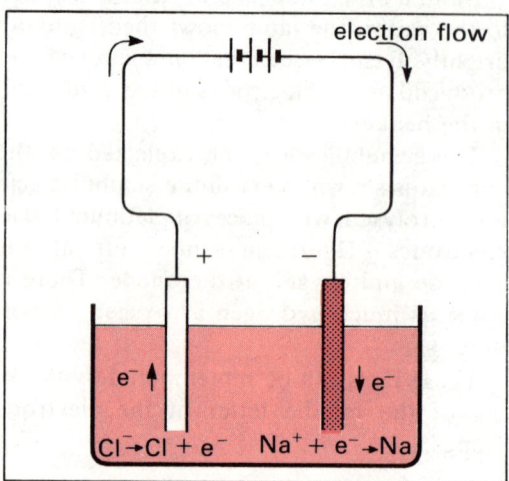

When molten sodium chloride is electrolysed, sodium forms at the cathode and chlorine is given off at the anode. This was the way in which the metal sodium was first obtained, by Sir Humphry Davy in 1807.

If the sodium chloride is in solution, when it is electrolysed, chlorine is still given off at the anode, but no sodium is formed at the cathode. This is not surprising when you remember how violently sodium reacts with water. The liquid near the cathode, tested with litmus, proves to be alkaline. Sodium hydroxide is formed.

This method is used on a large scale for making sodium hydroxide (caustic soda) from common salt.

The table below is a summary of the examples of electrolysis we have dealt with. Note that the electrodes are chosen so that they do not react with the products formed. Neither copper nor platinum may be used when chlorine is given off, because chlorine attacks both these metals.

Electrolyte	Electrodes	Product at cathode	Product at anode
Copper chloride solution	carbon	copper	chlorine
Copper sulphate solution	copper	copper deposited	copper removed
Copper sulphate solution (loses blue colour)	carbon	copper deposited	oxygen gas
Dilute sulphuric acid	platinum	hydrogen	oxygen
Molten sodium chloride	carbon	sodium	chlorine
Sodium chloride solution	carbon	sodium hydroxide in solution and hydrogen gas	chlorine

Electrolytic extraction of metals

Some metals are not easily separated from other elements in their ores. Electrical methods can often be used.

Magnesium. This is obtained by electrolysing fused magnesium chloride. The metal forms on the cathode.

Aluminium. The ore bauxite is first purified to give alumina. The alumina is dissolved in molten cryolite. This liquid is then electrolysed using a carbon cathode. Aluminium is formed at the cathode. Oxygen is given off at the anode.

Copper. Another example of the use of electrolysis is the purifying of copper. The block of impure copper is made the anode. A thin strip of pure copper is the cathode. The electrolyte is acidified copper sulphate. The copper moves from the impure anode to form a thicker pure copper cathode. The impurities are left behind.

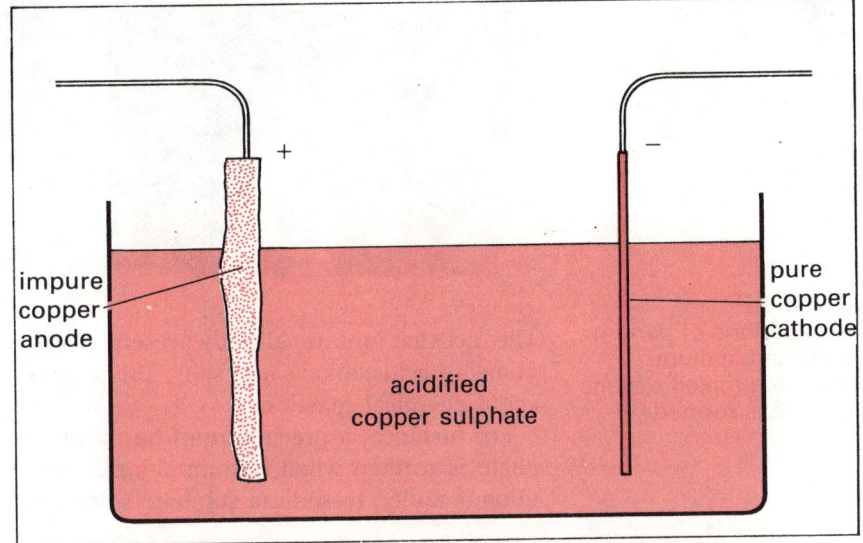

impure copper anode

pure copper cathode

acidified copper sulphate

Electrolysis of Copper(II) sulphate
This method is used to purify copper metal

Aluminium production
Alumina is electrolysed in this massive array of cells

1 It is often argued that the production of hydrogen and oxygen when water with a little acid is electrolysed shows that water is a compound of hydrogen and oxygen. Do you think this is a good argument remembering that pure water is a non-electrolyte?
2 State which of the following will conduct an electric current and which of them will be decomposed by a current: copper; distilled water; dilute sulphuric acid; silver; molten sodium chloride; solid sodium chloride.
3 Aluminium is a conductor, but it is not an electrolyte. Explain.

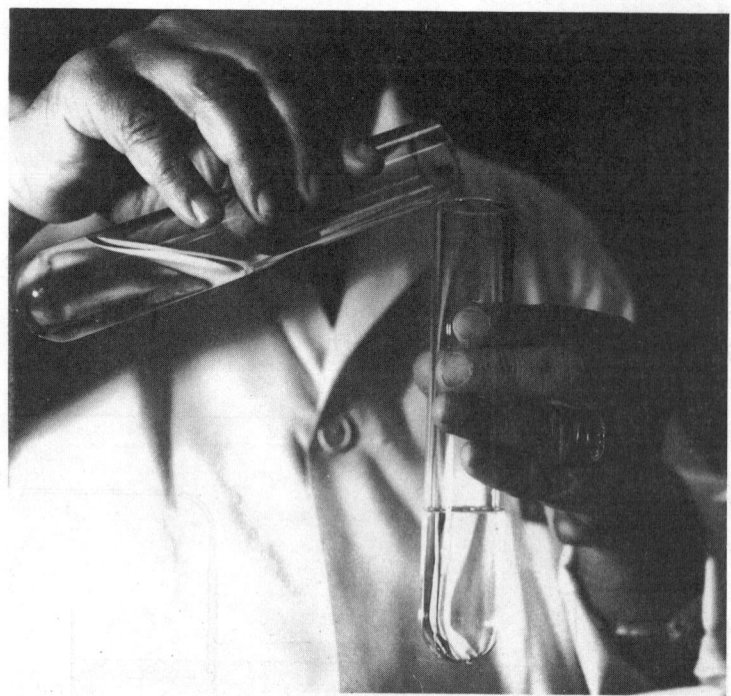

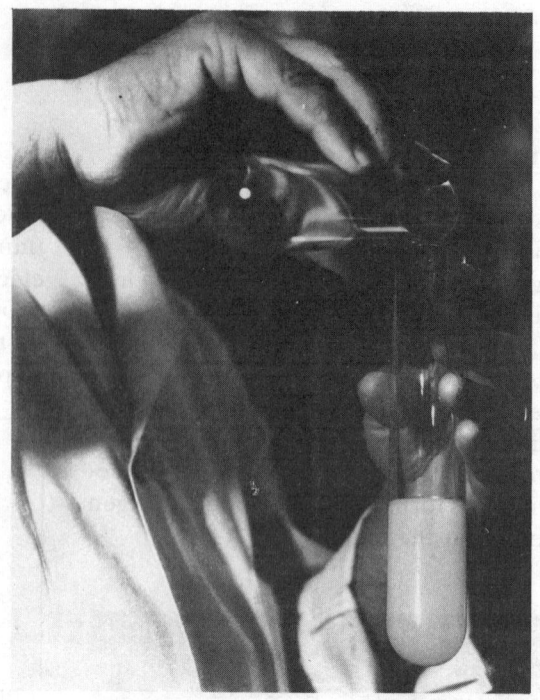

Precipitation
When solutions of barium chloride and sodium sulphate are mixed a white precipitate is formed

The fact that ions are already present in solutions of acids, alkalis and salts, alters ideas about chemical reactions.

For instance, a precipitate of barium sulphate is formed when barium chloride solution is added to sodium sulphate solution:

$$BaCl_2 + Na_2SO_4 \rightarrow BaSO_4 + 2NaCl$$

barium	sodium	barium	sodium
chloride	sulphate	sulphate	chloride

This precipitate could also have been formed by adding barium nitrate to sodium sulphate or potassium sulphate or ammonium sulphate.

$$Ba(NO_3)_2 + (NH_4)_2SO_4 \rightarrow BaSO_4 + 2NH_4NO_3$$

In fact the essential reaction is between barium ions and sulphate ions:

$$Ba^{2+} + SO_4^{2-} \rightarrow BaSO_4$$

barium	sulphate	barium
ions	ions	sulphate

Any solution containing barium ions mixed with any solution containing sulphate ions gives a precipitate of barium sulphate.

This is useful. Suppose you have a substance that you think could be a sulphate. Make it into a solution, then add a solution of a barium salt (usually barium chloride). If there is a heavy white precipitate it is most likely that it is a sulphate.

Acids
Sulphuric acid reacts with zinc to give hydrogen:

$$Zn + H_2SO_4 \rightarrow ZnSO_4 + H_2$$

Hydrochloric acid and zinc also give hydrogen:

$$Zn + 2HCl \rightarrow ZnCl_2 + H_2$$

The essential reaction is between zinc and hydrogen ions:

$$Zn + 2H^+ \rightarrow Zn^{2+} + H_2$$

Hydrogen ions
So acids have hydrogen ions available in solution. This gives us a very good definition of an acid.

An acid is a substance that gives hydrogen ions in solution. The more ions present in a given volume, the stronger the acid.

Water has a very few hydrogen ions, but for every H^+ ion there is a hydroxyl ion OH^-.

$$H_2O \rightleftharpoons H^+ + OH^-$$

H^+ and OH^- ions cannot exist together in great numbers. In an alkali there are many OH^- ions so there is only a small number of H^+ ions. So the concentration of hydrogen ions in a solution can be used as a measure of degree of acid or alkaline character.

pH values													
1	2	3	4	5	6	7	8	9	10	11	12	13	14

acidic

litmus turns red

neutral
litmus turns purple

alkaline

litmus turns blue

The pH scale.

This is used to state the concentration of hydrogen ions. pH 7 is neutral (it stands for the concentration of hydrogen ions in pure water: water has a pH of 7). Solutions with a pH greater than 7 are alkaline and those with pH less than 7 are acidic.

Indicators

A substance like litmus that shows different colours at different pH values, is called an indicator. Many substances act as indicators, though they do not all change colour at pH 7.

The BDH Universal Indicator shows several different colour changes at different pH values:

3 4   5   6   7   8   9   10   11   12   13
red orange yellow green   blue   purple

Importance of pH

There are many ways in which pH is important. For instance, in the human body, the blood must keep to a pH of about 7.4. If it drops below 7.0 or above 7.8, death is likely. Fortunately, the body has its own way of controlling the pH of blood.

Our food needs to have a particular acidity for the digestive juices to act properly. The stomach juices are very acidic, pH about 1.7, but the pancreatic juice is alkaline, pH about 8.

Some food preservatives are successful because they are acidic, e.g. sulphur dioxide. Moulds may develop on food at neutral pH, so if the pH is kept at 1–2 or 8–9, moulds are not likely to grow.

Gardeners need to know the acidity of the soil. The pH can be found by mixing a little soil with a few drops of Universal Indicator and the colour will show the pH.

Some plants such as heathers and azaleas do not grow well in an alkaline soil. If lime-loving plants like Clematis are to be grown successfully, it may be necessary to add lime to the soil to make the pH alkaline.

Clematis
This plant will only grow in an alkaline soil

1 State what colour you would see if litmus were added to each of the following substances in solution: hydrogen chloride; sodium hydroxide; ammonia; sulphuric acid; carbon dioxide.
2 Silver chloride, AgCl, is insoluble. Write down an ionic equation to show the essential reaction to produce a precipitate of silver chloride.
3 State what you understand by the equation:
$$2H^+ + CO_3^{2-} \rightarrow H_2O + CO_2$$

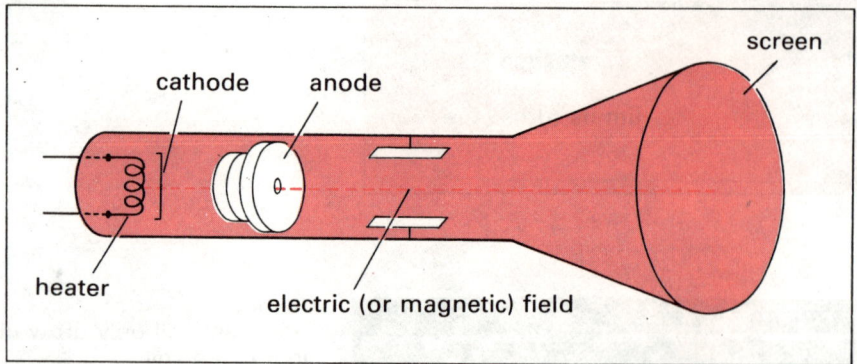

Cathode ray tube

No matter what metal was used for the cathode, electrons were produced. This shows that there must be electrons present in all these metal atoms. Atoms of all elements contain electrons.

But if this is so, there must be a positive part in the atom to balance the negative charge of the electrons, because all atoms are neutral. So every atom consists of a **nucleus** which has a positive charge. This is surrounded by a number of negatively charged electrons. The mass of electrons is very small. So the nucleus can be thought of as having all the mass of the atom.

The nucleus is made up of two kinds of particles; **protons** are positively charged; **neutrons** have no charge.

The electrons outside the nucleus form into groups or shells. The first shell holds only two electrons, the second eight and so on.

The diagrams show the kind of structures that can be used when thinking of some of the atoms. Since atoms are too small to be seen, these diagrams are really just pictures of ideas!

The atoms of different elements have different atomic numbers, and usually different mass numbers.

Although electrons are extremely small particles, they have been isolated and studied. If a metal cathode in an evacuated tube is heated by an electric current, electrons are pushed off the cathode and can be guided by an electric charge on the anode.

This stream of electrons can be deflected by an electric field and by a magnetic field. If a suitable substance is painted on the end of the tube it will act as a screen and a flash of light will be seen when an electron hits the screen.

This kind of cathode ray tube enabled the scientists, Sir J. J. Thomson, R. A. Millikan and others to find out what electrons were like. The television tube is a well-known modern version of a cathode ray tube.

Structure of some atoms

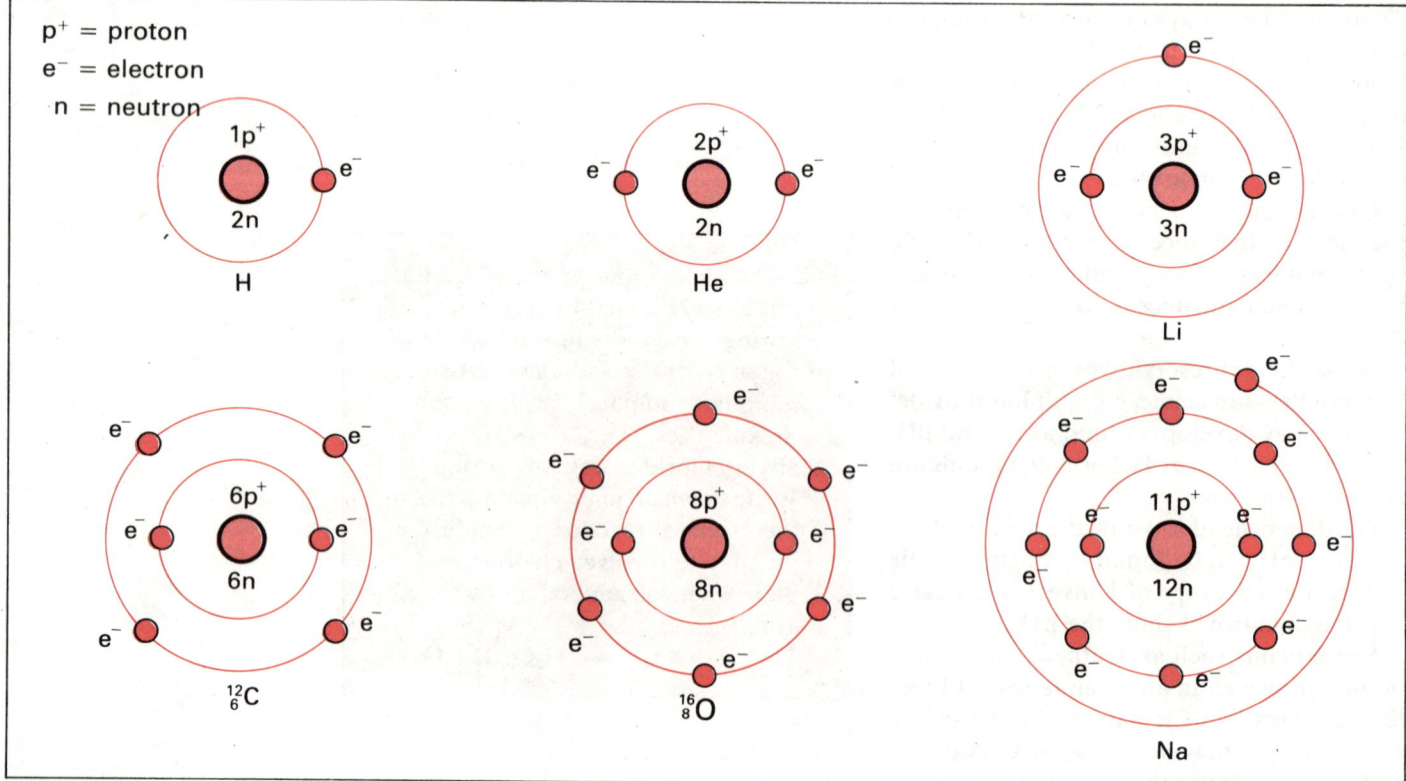

p^+ = proton
e^- = electron
n = neutron

H   He   Li $^{12}_{6}C$ $^{16}_{8}O$ Na

The **atomic number**, Z, of an element is the total number of charges on the nucleus, that is the number of protons in the nucleus. So it is also the number of electrons outside the nucleus in the neutral atom (not ions).

The **mass number** is the number of nucleons (protons plus neutrons), in the nucleus.

To get used to these ideas you should now draw diagrams of the atomic structures on the reverse side of your valency cards for some of the elements.

Isotopes

The chemical properties of an element depend only on the electrons in the outermost shell. These are the least strongly attracted by the positive charge on the nucleus and are the easiest to remove in order to form ions. These loosely-held electrons are called valency electrons because they determine the valency of the element.

The top diagram shows the three forms of hydrogen. They all have one valency electron. But the masses are different: (a) has no neutrons in the nucleus; (b) is **deuterium**, with one neutron and (c) is **tritium**, with two neutrons. All these have the same atomic number, but different mass numbers. They are **isotopes** of hydrogen.

The diagram below shows two carbon isotopes $^{12}_{6}C$ and $^{14}_{6}C$. [Note that the lower number (6 in this case) is the atomic number. The top number gives the total number of nucleons (protons + neutrons).]

Ordinary chlorine gas is a mixture of two isotopes, about three quarters of $^{35}_{17}Cl$ with about one quarter of $^{37}_{17}Cl$.

The **relative atomic mass** of an element is the average mass of the atom of the element on a scale on which the carbon atom ^{12}C has a

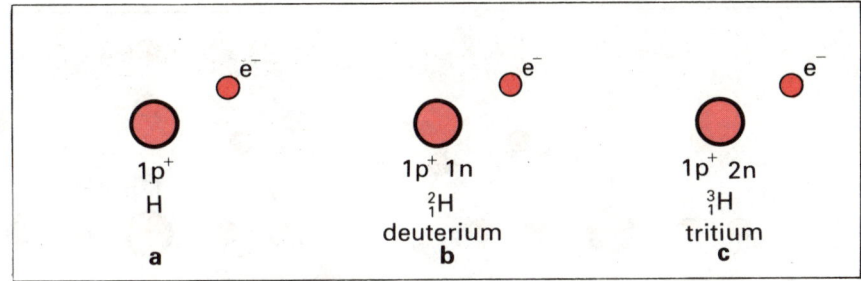
Isotopes of hydrogen

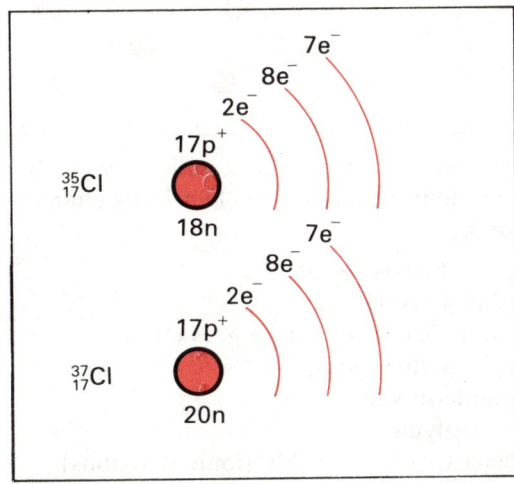
Isotopes of chlorine

mass of 12 units. For chlorine this works out to be 35.5.

Chemical bonds

Electrons are involved when atoms combine together. There are two distinct ways in which this happens.

In an **electrovalent** or **ionic** bond, one or more valency electrons pass from one atom to another so as to make a complete group of electrons. The result of this giving and taking of electrons is that two ions are formed.

In a **covalent** bond, one or more pairs of electrons are shared, an equal number com-

(Continued on following page)

Isotopes of carbon

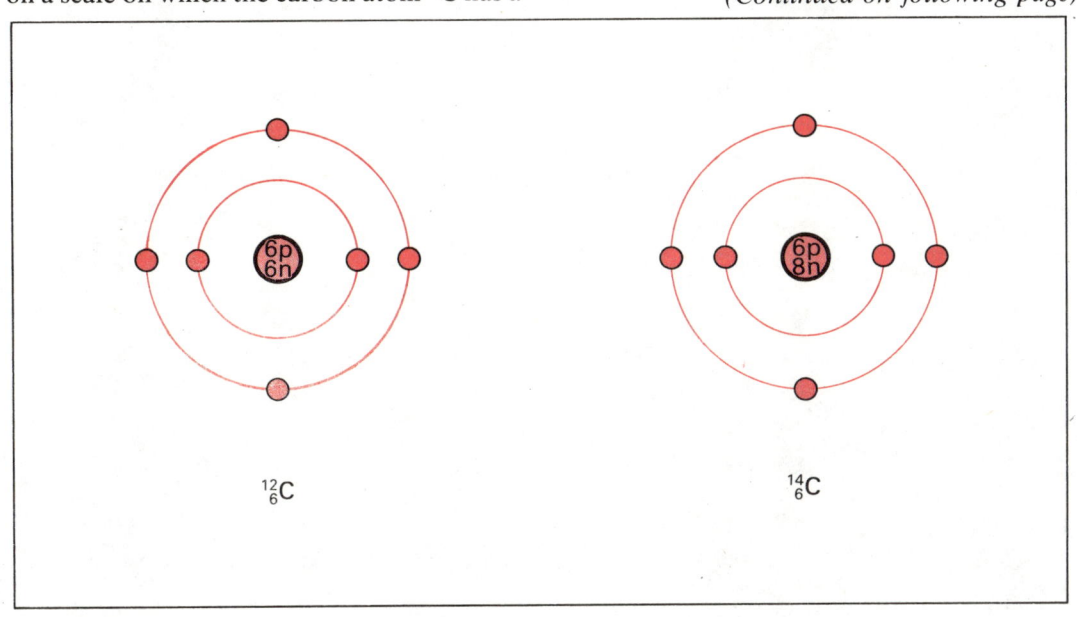

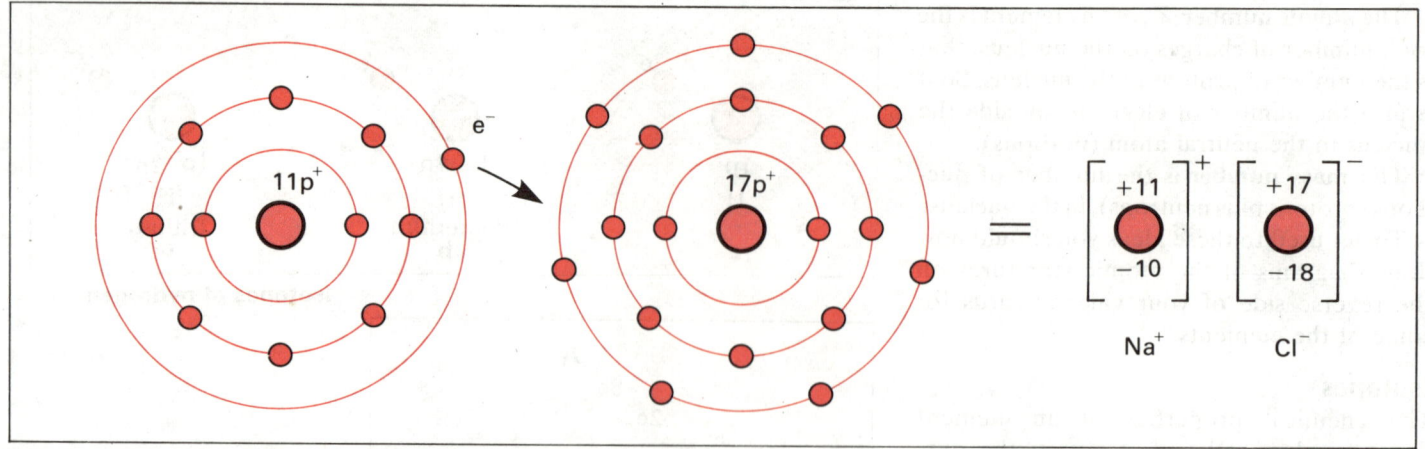

Electrovalent compound
Sodium 'gives' one electron to chlorine to form sodium chloride

ing from each atom. The resulting compound is not ionised and has no electrically charged bond.

Electrovalent
Usually crystalline
Inorganic compounds, e.g. Na^+Cl^-
High melting point
Soluble in water
Electrolytes
React together quickly (ionic reactions)

Electrovalent compounds have different properties from covalent compounds.

Covalent
More often liquids or gases
Organic compounds, e.g. ethanol, C_2H_5OH
Low melting and boiling points
Soluble in organic solvents
Non-electrolytes
React slowly

Covalent compound
Hydrogen and carbon share electrons to form methane

×   electrons from H atoms

○   electrons from outer shell of C atom

1 Which atomic particle has the smallest mass? What particle is positively charged? What particle has no charge?
2 How many (a) electrons; (b) neutrons; (c) protons are there in each atom of (i) $^{14}_{6}C$; (ii) $^{131}_{53}I$.
3 Draw diagrams to show how the electrons involved in the bonds between atoms are arranged in (a) lithium chloride, (LiCl); (b) chlorine, Cl_2.

Examination Questions IV

1 Water is at its most dense at: (a) 18°C; (b) 0°C; (c) 10°C; (d) 4°C; (e) −4°C. Which answer is correct. WYL

2 (a) What is the arrangement of metals shown in Diagram 1 called?
(b) Draw a diagram to show what you would see this piece of equipment do after you had heated it for a few minutes in a bunsen flame.
(c) Why does this happen?
(d) Name one practical use of this piece of equipment. M

3 Copy the following and fill in the missing words.
(i) An atom is the smallest part of an * that can take part in a chemical reaction.
(ii) An * is the smallest part of an element or compound that can have a separate existence.
(iii) An atom is made up of three parts, which may be relatively light or heavy, and positively or negatively charged. M

particle	mass	charge	other property
*	light	*	orbits around the nucleus
*	heavy	*	part of the nucleus
*	*	neutral	part of the nucleus

4 Explain the following:
(i) A bottle of milk may be kept cool on a warm day if it is covered with a damp cloth, placed in a bucket of water and stood out of doors.
(ii) Boiling water poured into a milk bottle will crack the bottle.
(iii) Small gaps are often left between railway lines.
(iv) Mercury cannot be used in thermometers that will be used on an Antarctic expedition. M

5 Atoms of elements are often represented by simple diagrams similar to the one shown opposite:
(a) (i) Name the particles represented by the symbol   ; (ii) What charge do these particles carry?
(b) How many protons will there be in the nucleus of the atom shown?

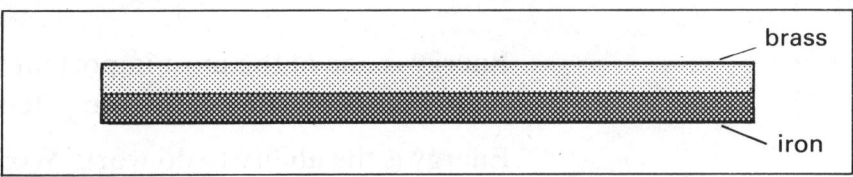

Diagram 1

(c) Name one other type of particle in the nucleus.
(d) What is the charge on the nucleus? Y

6 (a) If a beaker of water is cooled sufficiently, where will the ice start to form? How has this effect helped to preserve life in the sea?
(b) What is the latent heat of vaporization of water?
(c) Draw a simple diagram of the water cycle. SE

7 During the electrolysis of acidified water, hydrogen is released at (a) anode; (b) anode and cathode; (c) electron; (d) cathode; (e) nucleus. What answer, if any, is correct? EM

8 A clinical thermometer differs from an ordinary laboratory thermometer in that (a) it contains alcohol; (b) it has a constriction in the tube; (c) it has a longer stem; (d) it must be sterilized in boiling water before use; (e) its scale covers only a few °C; (f) the bulb is made of thicker glass. Which two statements are correct? EM

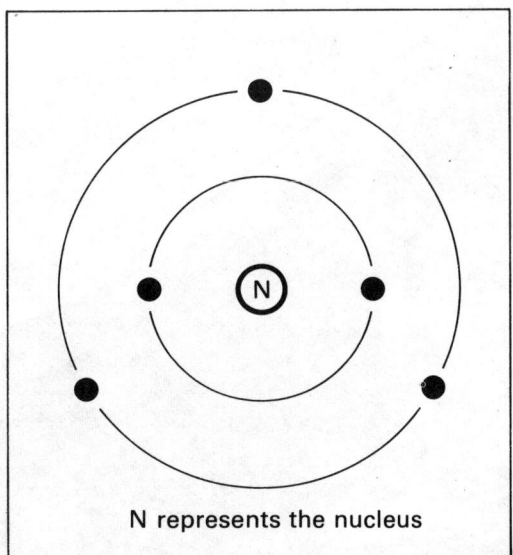

Diagram 2

N represents the nucleus

ENERGY & FOOD

Energy is one of the most important topics in science. Coal, gas and oil which supply most of our energy are often in the news. So what is energy?

Energy is the ability to do work. Work is done when a force moves something through a certain distance (see Topic 14). Work is measured in joules: 1 joule is the work done when a force of 1 newton moves through 1 metre.

There are different forms of energy; **kinetic energy** is the energy of motion; **potential energy** is energy due to position. Heat is a form of energy. These forms can be changed into one another. Work can be done by changing one form of energy to another.

Energy is never 'finished'! It cannot be created nor can it be destroyed. In changing from one form to another, part of the energy changes to heat. This heat cannot generally be put to good use. In effect it is a 'loss' of energy. So it is important to have energy sources, such as stocks of coal, gas and oil.

The energy source for our own bodies is food. You use energy even when you are asleep, because the heart, lungs and other organs are still working and need energy. The energy within the food you eat is not available until the food is digested. That is, the complex chemical substances in the food must be broken down. The simpler, soluble substances formed can be taken up by the blood.

Plants do not eat food. They are able to make their food from simple chemical substances. The process is called **photosynthesis**. It needs the energy from the sun. Animals depend on plants for their food.

Concorde
Concorde changes the chemical energy in its fuel to kinetic energy of motion

Forms of Energy

Kinetic energy

Moving things have energy. Think of a hammer hitting a nail: the nail is pushed in a small distance. The moving hammer head has kinetic energy due to its motion. Just resting the hammer head on the nail would not drive it home.

Motor cars, railway engines, aeroplanes and rockets all have kinetic energy when they are moving. Moving water has kinetic energy, for example, when it flows over a mill wheel. Moving air has energy; it can drive windmills.

The formula:

$$\text{kinetic energy} = \frac{1}{2} \text{ mass} \times (\text{velocity})^2$$

shows that if the speed of a moving thing is doubled, the kinetic energy is increased fourfold. When tackling a rugby player it is a good idea to do so before he has gained speed.

Rotating things such as electric drills, spin driers and turbines also have kinetic energy.

A washing machine changes electrical energy into rotational kinetic energy

Potential energy

Imagine a man standing on a platform a few metres above ground. He has potential energy due to his position. He can do work that he could not do if he were on the ground.

Suppose he held on to a rope passing over a pulley above his head. If there is a load on the other end of the rope, the man can use his weight to lift the load by just jumping down holding on to the rope.

A spring in a clock or watch gains potential energy when coiled up. As it is released, it does work. In a clock, this release of energy is controlled by the escapement.

Chemical energy is a kind of potential energy. Some chemical substances have a great deal of energy inside their molecules; part of it is released when these substances change into others. Magnetism is a kind of potential energy. Suppose you have a piece of iron attached to a magnet. You pull it off, the energy from your finger muscles passes into potential energy of the magnet. The magnet can now draw the iron piece back to it.

Electrostatic energy

Rub each of two balloons against some woollen material. The balloons become charged with an electric charge. If they are hung side by side, from long threads, they repel each other. You will have to do work to push them together.

Other forms of energy discussed in later topics are sound, electricity and radiant energy such as light, heat, X-rays, also atomic energy.

Potential energy
The man can use his potential energy to lift the sack

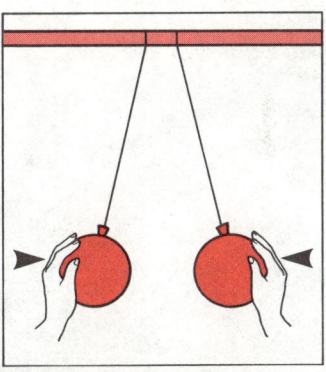

Electrostatic energy
Electric charge is a form of energy

1 State the kinds of energy a football has when (a) it is kicked; (b) the ball is at the top of its flight, during a goal-kick.
2 (a) Find how much work is done when a weight of 500 N falls through 10 m. (b) How much potential energy (in joules) has a weight of 500 N when it is 10 m above the ground?
3 Work out the kinetic energy of a man of mass 80 kg moving with a speed of 10 m/s.

Potential and kinetic energy

One kind of energy can be changed into another. When an object falls from a height its potential energy changes into kinetic energy. Some of the energy is 'lost' as heat.

Think of a ball bouncing up and down. As it falls, potential energy is changing to kinetic. On the upward movement, kinetic is changing into potential energy. The gradual loss of some of the energy as heat makes the ball slightly warm. The loss of energy is seen because each bounce is not quite so high as the one before.

This exchange between the two kinds of energy is well shown in a swinging pendulum. The swinging bob has maximum potential energy when it is pulled aside. It slowly changes this for kinetic energy. It has maximum kinetic energy at the lowest point and then loses kinetic for potential as it swings upwards again.

Hydroelectric schemes of power supply depend on water being kept in a reservoir by damming it in a high position in a mountain. This water has potential energy. It is allowed to flow down the mountain side in pipes. When it reaches the power house at the foot of the mountain its potential energy has changed to kinetic energy. The moving water is used to drive turbines of the turbo-generator and so generate electricity – another form of energy.

Potential energy of
water in high reservoir
↓
Kinetic energy
in turbine
↓
Electrical energy

Electricity

Everyone knows that electricity can be changed into heat, e.g. in the electric heater or electric iron, or into light in the electric lamp. It can be changed into kinetic energy, in the washing machine.

Chemical energy

Chemical energy stored in substances like gunpowder or dynamite can be changed into kinetic energy. Exploding dynamite does useful work in blasting rocks.

Petrol is a familiar example of a substance with potential (chemical) energy. Sparking a mixture of petrol and air provides the kinetic energy of cars and other internal combustion engines.

Hydroelectricity

The photograph shows the building of the turbine hall in a man-made cave at the Dinorwic power station. Energy is stored by pumping water into a lake high on a mountain side

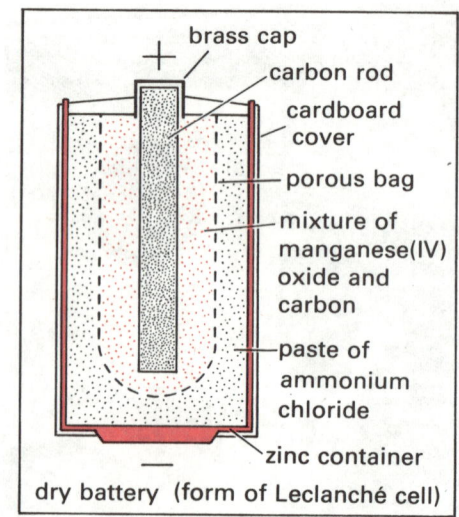

dry battery (form of Leclanché cell)

Dry battery

Batteries convert chemical energy into electrical energy

Labels on dry battery diagram:
- brass cap
- carbon rod
- cardboard cover
- porous bag
- mixture of manganese(IV) oxide and carbon
- paste of ammonium chloride
- zinc container

shows itself as energy. Mass has changed to energy. A small mass gives rise to a very large amount of energy.

When released suddenly in the atomic bomb this energy had a devastating effect. But when controlled in a nuclear reactor, it can be usefully changed into electricity.

The change of mass into energy explains the tremendously high temperature of the sun.

Conservation of mass–energy

The possibility of mass changing to energy means that we must revise our statement that energy cannot be created or destroyed: the total of **mass** and **energy** does not change.

Nuclear power station
Mass is turned into energy in the reactor core

Energy stored in muscles

When an archer pulls back the strings of his bow, energy from his muscles changes to potential energy in the stretched string. This changes to kinetic energy in the arrow as the bowstring is released. A working person needs food to put back the energy-giving chemicals in the muscles.

Measuring heat

James Joule (1818–1889) proved that a definite amount of kinetic energy could be changed into a definite amount of heat. He did this by churning water with paddles, knowing the amount of work done and measuring the amount of heat produced. This gives us a method of measuring heat in terms of the energy unit, the joule.

It takes 4.2 joules to heat up 1 g of water by 1°C. It takes 4 200 joules to heat 1 dm³ (litre) of water by 1°C. The unit of energy formerly used (still sometimes seen) was the **calorie**.

A calorie is the heat needed to raise the temperature of 1 g of water by 1°C. So 1 calorie = 4.2 J.

The kilocalorie (kcal or Cal), which is 1 000 cal or 4 200 J (Joules) is still used in calculating energy values of foods (see Topic 62).

Atomic energy

Energy can be released from some atoms. Briefly, when some large atoms break into two other atoms, the total mass of the two atoms formed is a little less than the mass of the original atom. This slight difference

1 The diagram shows some possible energy changes. One arrow has been labelled. Give examples of the changes shown by a, b, c and d.

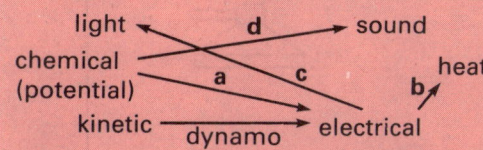

2 What energy changes are there when a boy throws a ball high into the air and the ball then falls to the ground?

3 How much energy is needed to heat 4 dm³ of water from 20°C to boiling point?

Heat

Heat is a form of energy. It is not the same as temperature. To say that the heat of the oven is so many degrees Celsius or Fahrenheit is quite wrong because it confuses heat with temperature.

The distinction can be shown very neatly by an experiment. A small copper rivet of mass about 1 g is made red-hot by heating in a bunsen flame. At the same time a lump of copper of about 200 g mass is heated to 100°C. This is done by keeping it in boiling water for a few minutes. Then both rivet and large mass are dropped separately into 100 cm³ of cold water in beakers. The rise of temperature in each case is noted. Here is a typical set of results.

Each mass of water	=	100 g
Initial temperature	=	15°C
Final temperature of water with rivet	=	16°C
Final temperature of water and large mass	=	30°C

The red-hot rivet must have been at the temperature of the bunsen flame, at least 700°C. The larger mass of copper was only at 100°C. But the larger mass gave up more heat to the water, because its mass was much greater.

The heat given up can be worked out in each case from the rule:

heat   = 4.2 × mass × temperature
(joules)   (g)   difference
                                       (°C)

The rivet gave up:

4.2 × 100 × (16 − 15) = 420J

The larger mass gave up:

4.2 × 100 × (30 − 15) = 6 300J

that is fifteen times as much.

Temperature means **degree of hotness** (e.g. in degrees Celsius). **Heat** is **energy** (measured in joules) that passes from a body at a high temperature to one at a lower temperature.

Hot metal gives up heat to cold water

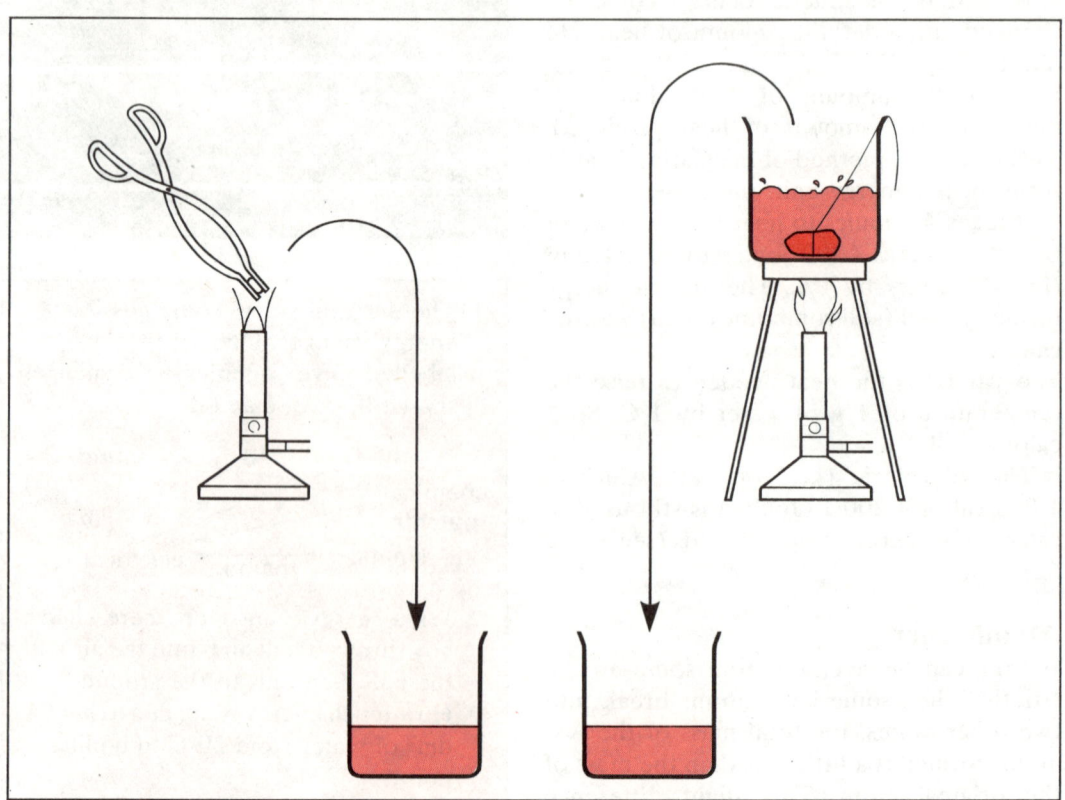

Latent heat
Extra energy is needed to
change water into steam

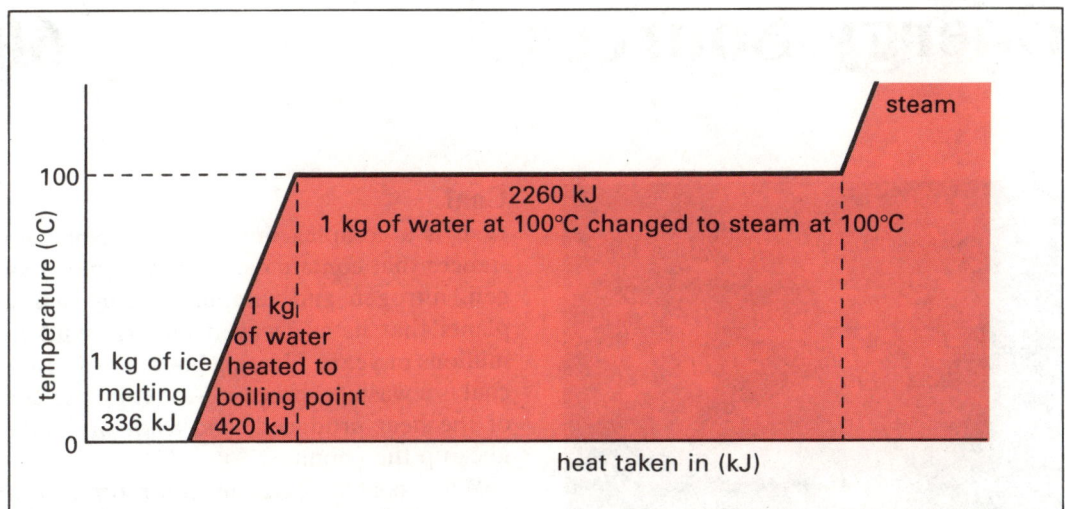

Specific heat capacity

Suppose you heat a pan containing 1 kg of water and a mass of 1 kg of copper both on the same hot plate for 1 minute. You find you can still put your fingers in the water without pain, but you would get your fingers burnt if you tried to handle the mass of copper. The same amount of heat has just warmed the water, but has made the copper very hot.

It takes 4 200 J to heat 1 kg of water through 1°C, but it takes only 400 J to heat 1 kg of copper through 1°C. The heat that would increase the temperature of water by only 30°C would heat the same mass of copper through $\frac{30 \times 4\,200}{400} = 315°C$

The heat needed to increase the temperature of 1 kg of a substance through 1°C is called the **specific heat capacity** of the substance.

Specific heat capacity of water = 4 200 J/kg
Specific heat capacity of copper = 400 J/kg

When an object is gaining or losing heat:

heat = specific × mass × temperature
(J)   heat   (kg)   difference
         capacity   (°C)

The fact that water has a large specific heat capacity can be useful. Suppose you use a hot water bottle, holding 500 g of water at a temperature of 90°C. When this water cools to 20°C, it gives up:

$4\,200 \times 0.5 \times (90 - 20) = 147\,000\,J$ or $147\,kJ$

Imagine someone decides to use a lump of hot copper wrapped in cloth, instead. To have the same amount of heat from 500 g of copper, he would need to make the copper red hot!

What is heat?

When a substance is heated, energy passes into it. What happens to this energy? The molecules of the substance get 'hotter', which means that they move more quickly and their average kinetic energy increases. You can think of heat as the energy possessed by a substance in the form of kinetic energy of molecules.

Latent heat

By looking back to Topic 51, the latent heat of water is seen to be quite large. The specific latent heat of ice is 336 J/g (or 336 kJ/kg); the specific latent heat of steam is 2260 J/g (or kJ/kg). The graph shows how these amounts compare with the heat needed to boil the same amount of water.

Dissipation of energy

When one form of energy is changed into another, heat is always produced, although it is not always wanted. Friction is the usual cause, e.g. the heat produced in the running of a car engine.

Although we can convert heat into useful energy, e.g. kinetic or electrical, as in the steam engine there is no way in which we can use all the heat. There is always wastage.

1 'Metals have a lower specific heat capacity than water'. Explain what this means and describe what difference you would find if you heated 1 kg of water and 1 kg of metal on the same heater for the same time.
2 After sawing a piece of wood, the saw feels very hot. How can you explain this?
3 How much heat is needed to change 1 kg of ice at 0°C into steam at 100°C?

Water wheel

How long will our present energy sources last? What energy sources will become available in the future? These are questions of world-wide importance.

Centuries ago all work had to be done by man's muscle. Heavy jobs, like building pyramids in about 2 500 BC, were possible only because there were large numbers of slaves. Water mills were used about 20 AD for grinding corn and windmills probably came into use after about 600 AD.

Fuels

Most of the energy sources we depend on today are fuels. Fuels are substances that release their potential (chemical) energy when they are burnt (oxidised) to produce energy. The earliest was wood. Nowadays we use coal, petrol, oil and gases, such as North Sea gas, butane and propane.

Coal

Coal is a complex mixture of organic substances that contain carbon, hydrogen, oxygen, nitrogen and sulphur. It comes from plants that have decayed underground for millions of years. Heating houses by burning coal is a wasteful process. About 80 per cent of the heat produced, and a lot of smoke, goes up the chimney.

When coal is heated in iron retorts, without air being present, a process called dry distillation takes place. Coal tar, ammonia and coal gas are driven off. Coke is left in the retorts. Many valuable chemicals can be made from coal tar.

Gas

Coal gas or town gas contains about 50 per cent hydrogen and 30 per cent methane. Both these gases produce much heat when they are burnt. Natural gas now replaces town gas as a domestic fuel. It occurs naturally with petroleum. Since 1967 it has been obtained from under the North Sea. This gas, after treatment at points on the east coast, is distributed by a system of pipes throughout the country. It is mainly methane and gives greater heat than town gas.

Butane (C_4H_{10}), known as Calor gas, is another gaseous fuel. It is found with methane in natural gas or can be obtained from petroleum.

Petroleum

Petroleum is really a fossil fuel. It is a thick oil formed by the decay of plants and animals

Fractional distillation of crude oil
Fractions of different boiling points are collected

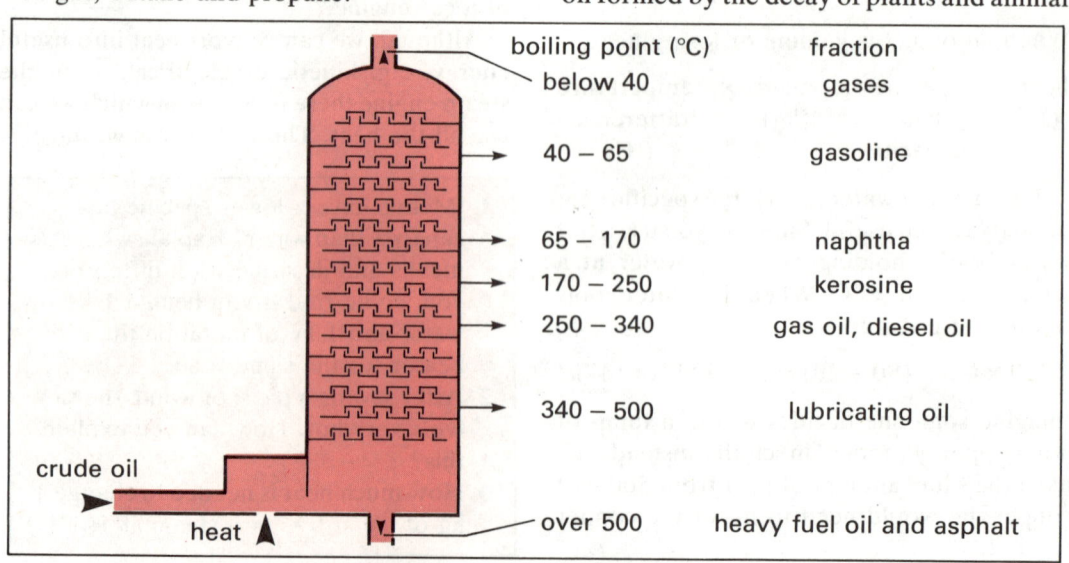

	boiling point (°C)	fraction
	below 40	gases
	40 – 65	gasoline
	65 – 170	naphtha
	170 – 250	kerosine
	250 – 340	gas oil, diesel oil
	340 – 500	lubricating oil
crude oil		
heat	over 500	heavy fuel oil and asphalt

buried under rocks many million years ago. It was first drilled in the USA in 1859. Oil was first obtained from the North Sea oil fields in 1970. It is now pumped to refineries at several places in the country. Petroleum is a mixture of many hydrocarbons, (compounds that contain carbon and hydrogen only). These have different boiling points, so they can be separated by fractional distillation.

The liquid that distils over between 40 and 65°C is gasoline from which the various grades of petrol are obtained.

All the above fuels are energy sources because they have potential (chemical) energy.

Food

Food is a source of energy for man and other animals (see Topic 61). Food is a kind of fuel.

Solar energy

When you sun-bathe you know you can get very hot in direct sunlight. This shows that energy must be reaching you from the sun. It is radiant energy, or **radiation**, and travels through empty space. When it falls on an object it changes to heat. This radiant energy is responsible for warming up the soil during spring. Nowadays, some people have solar cells in panels on their roofs. These trap the solar energy and use it for heating the house.

Tidal energy

The rise and fall of tidal water represents a good amount of potential energy. It should be possible to change it to useful forms of energy.

The sun

If we neglect nuclear energy for the moment, all the sources of energy mentioned here come in the first place from the sun. Coal and petroleum are stores of energy obtained from the sun. Animal food depends on green plants being able to make carbohydrates from carbon dioxide and water, with the help of energy from the sun. Wind depends on the sun's energy. Hydroelectric power depends on reserves of water at a height. This water only gets to this height after evaporation caused by the sun's energy.

Further outlook

By the time our present sources of energy, gas, coal and oil, are exhausted, it is to be hoped that we shall be able to make greater use of nuclear, solar, wave and tidal energy.

Oil rig

Solar panels

Wave energy

1 What is meant by 'fuel'? Give three examples of fuels.
2 Suggest one advantage that natural gas has as a fuel.
3 What reasons are there for thinking that the sun's energy is responsible for our stocks of coal?

The more active you are, like these sportsmen and women, the more energy you need. The energy comes from the food you eat

Every living thing needs food to provide energy, just as a motor car needs petrol. Eating food enables you to move about. When more energy is used, as for example, when playing games, you become hungry. Food is also needed to make things grow. When an animal or plant grows, it increases in substance. Body-building needs food. Food is also needed to replace worn and damaged tissues.

Vitamins are needed for us to avoid certain diseases. Vitamins help to make substances that are essential if the organs of the body are to work properly. We eat a great variety of foodstuffs, but they all contain various amounts of three important classes of chemical substances.

They are:

carbohydrates to provide energy

proteins for body-building and replacing tissues

fats for energy storing

But we cannot use the potential energy of these substances as they are. Their complex molecules must be broken down into simpler ones. These simpler substances are soluble and can dissolve and pass into the blood stream. This breaking down process is called **digestion**. It involves the use of some active catalysts called **enzymes**.

Carbohydrates

All carbohydrates contain the elements carbon, hydrogen and oxygen. There are always twice as many hydrogen atoms as oxygen in the molecules. For example glucose has a formula $C_6H_{12}O_6$. It is one of the sugars found in honey. Glucose is present in our blood. Glucose gives a reddish-brown precipitate when it is warmed with Benedict's solution (or Fehling's solution). This can be used as a test.

If a sample of urine shows glucose to be present, it means that glucose is escaping from the patient's blood. He may be suffering from diabetes.

Sucrose (cane sugar, $C_{12}H_{22}O_{11}$), is obtained from sugar cane and from sugar beet. It will not give a precipitate with Benedict's solution. But if cane sugar is boiled for a few minutes with dilute acid, it is changed to two sugars, glucose and fructose.

sucrose + water → glucose + fructose

Starch is a complex carbohydrate. It is present in many of our foods such as cereals. It is also in potatoes. Starch does not dissolve, but it can be easily spotted because it gives a dark blue stain with iodine solution.

Cellulose is another complex substance. It is present in the cell walls of plants, wood, paper, cotton and cotton wool. It cannot be digested in the human body, so when present in food it just acts as 'roughage'. It stimulates the action of the muscles in the digestive canal and passes out of the body as waste matter in the faeces.

Percentages of carbohydrates in foods

cane sugar	100	cabbage	3
boiled rice	30	white bread	50
banana	19	potato	20
porridge	8	peanuts	8

Fats

A **fat** is a kind of salt of glycerol and a fatty acid. Mutton fat, for example is mainly glyceryl stearate. A simple test for a fat is that when rubbed on a filter paper it makes a mark that is translucent. A drop of water will form the same kind of spot, but when the spot dries it disappears. The fat spot will remain. Also the fat dissolves in ether.

Percentages of fat in foods

olive oil	100	egg	12
margarine	85	lard	99
butter	83	peanuts	49
almonds	54	cheese	35

Proteins

Proteins must be included in our food. Our bodies contain nitrogen, sulphur and phosphorus besides carbon, hydrogen and oxygen. So carbohydrates alone cannot form body tissue. Proteins do contain these elements. The protein molecule is very complex. Egg albumen and all enzymes are examples of proteins.

Test for protein. The liquid containing protein is treated with some sodium hydroxide solution, then a few drops of very dilute copper(II) sulphate solution. The liquid becomes purple in colour.

Percentages of protein in foods

cheese	25	egg	13
corned beef	25	peanuts	28
beefsteak	30	brown bread	10
fish	16	lamb	23

1 What elements are present in: (a) a hydrocarbon; (b) a carbohydrate; (c) a protein?
2 What types of substances are, A, B and C in the following:
   (i) A deep blue colour is produced when A is added to iodine solution;
   (ii) B turns purple when treated with sodium hydroxide solution and a few drops of copper(II) sulphate are added.
   (iii) When warmed with Benedict solution C gives a reddish-brown precipitate.
3 Name items of food that contain large amounts of: (i) carbohydrates; (ii) fat; (iii) protein. Give three examples in each case.

Carbohydrate-rich food

Fat-rich food

Protein-rich food

A Balanced Diet

Carbohydrates, fats and proteins account for most of our food. It is not correct to say that meat is a protein or that bread is a carbohydrate. Proteins are present in lean meat. Starch, the carbohydrate, is present in flour used in making bread. Some other chemical substances in food are required only in small amounts.

Inorganic salts

Calcium salts are needed for bone formation. Iodine (as iodide) is needed for the proper working of the thyroid gland. Fluorides, in minute amounts, seem to be needed for the enamel of teeth to form properly. Potassium, sodium, magnesium and traces of cobalt, copper, aluminium, boron, nickel and zinc are also needed. These are all present in the food we usually eat.

Vitamins

Animals do not keep healthy on a diet that contains carbohydrates, fats, proteins and mineral salts and nothing else. Certain extra substances or 'accessory factors' are missing from the diet. These are known as vitamins. The first ones discovered were given letters, A, B, C, D, but nowadays many more are known. Scurvy and beri-beri are diseases caused by the absence of vitamins C and B_1 from food.

Do not link vitamins too closely with the effects connected with them. The action of a vitamin depends on other factors. For example, vitamins A and D cannot be absorbed into the body unless fat is present in the diet. The absence of a vitamin does not automatically mean that a disease will develop.

Roughage

Is it possible that in the future we shall take the food we need in the form of tablets, instead of meals? It is not likely, because some bulk is necessary in food. For example cellulose that is present in cell walls of cabbage, lettuce and other plants, cannot be digested by the body, but the bulk is needed, as roughage, to stimulate the muscles of the digestive system to work (see Topic 63).

Food values

To decide on the value of a particular foodstuff, it is necessary not only to know what substances it contains, but it is also important to know how much energy it can provide. To do this we measure the total amount of heat that is set free when the foodstuff is completely oxidised. This can be shown in a rough way by burning a weighed amount of starch, or sugar, beneath a known mass of water in a beaker. For example, 10 g of bread will give about 100 kJ of energy.

Balanced diet

To get the best results from your food you need carbohydrates, fats and proteins in the right amounts, besides vitamins and the essential mineral salts. Further you need enough food to give the energy needed.

For basic needs you should have about 7 500 kJ per day. This would be enough if you were lying down doing nothing. A man doing heavy work might need about 20 000 kJ.

1 gram of carbohydrate or 1 gram of protein yields about 17 kJ. 1 gram of fat yields about 38 kJ.

Protein is needed for growth and repair of tissues. It might be thought that a large amount of fat would be good. It is useful to include about 25 per cent fat in your diet. But excess fat cannot easily be digested and can be bad for you.

A typical diet might be:

400 g of carbohydrates giving	6 800 kJ
100 g of protein	1 700 kJ
100 g of fat	3 800 kJ
Total	12 300 kJ

The amount of energy available from different foodstuffs can be found by reference to tables of food values. The following table gives an analysis of a single meal. The carbohydrate content is high. An apple (100 g) instead of the rice pudding would keep carbohydrate content down.

Food item	Amount	Content (g)			Energy
	(g)	Carbohydrate	Fat	Protein	(kJ)
Beef hamburgers	100	0	20	24	1150
Chipped potatoes	50	18	7	2	570
Peas	50	7	0.2	3	180
Rice pudding	100	27	3	4	640
	Total	52	30	33	2540

Figures for 100g of food	protein (g)	fat (g)	carbo hydrate (g)	vitamins				energy value kJ
				A (µg)	thiamin B1 (µg)	C (µg)	D (µg)	
brown bread	8.9	2.2	44.7	0	240	0	0	991
ham	24.7	18.9	0	0	440	0	0	1119
chicken	26.5	4.0	0	0	80	0	0	599
milk	3.3	3.8	4.7	40	40	2000	0.2	272
cheddar cheese	26.0	33.5	0	412	40	0	0.26	1682
carrots	0.7	0	5.4	2000	60	6000	0	98

This meal contains about 2000 kJ

1 What class of food substances serves as body-builders? Why are carbo-hydrates not suitable for body-building?

2 Describe tests you could use to find out if a banana contains: (a) starch, (b) fat.

3 Look back to the tables in Topic 61 and find out how much energy could be obtained from 100 g of peanuts. (Remember that 1 g of carbohydrate or of protein yields 17 kJ, but 1 g of fat yields 38 kJ.)

The Path

The food you eat needs treatment before you can get the benefit from it. The energy in the food is used in the body tissues by a process called respiration (see Topic 81). But first the food must travel to the tissues in the bloodstream.

Starch, fat and proteins cannot pass into blood vessels because the molecules are too large. They must be changed into soluble substances that can diffuse into the blood. This process of changing complex, often insoluble, substances into simpler, soluble substances which can reach the tissues of the body is called **digestion**. This takes place in the alimentary canal (gut).

In a simple animal, like the earthworm, the gut is simply a long tube from mouth to anus. In the human body the tube is very long and coiled.

Food is broken down in two ways, **chemically**, it is changed into smaller molecules by hydrolysis. But first it is broken **physically**, into smaller pieces by the teeth. It then has a larger surface for chemical action.

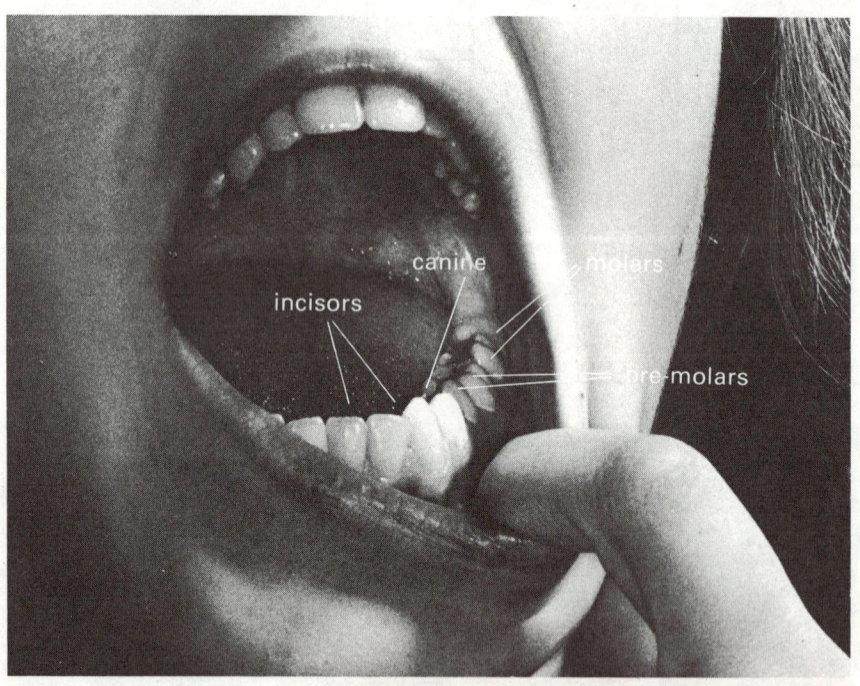

Identify the teeth in your own mouth

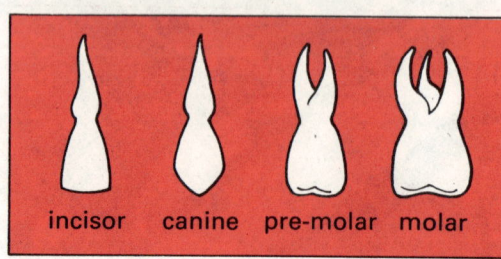

incisor   canine   pre-molar   molar

Teeth

Look at your own teeth with the aid of a mirror. You can see different kinds.

Incisors, chisel-shaped; used for cutting and gnawing

Canines, conical; used for gripping and tearing flesh

Pre-molars, broader and thicker at the end. Used for crushing and grinding

Molars, like pre-molars but larger. They are not present in the child's (milk) teeth but grow in the second (permanent) set.

The kind of teeth an animal has can be linked with its feeding habits. Grass-eaters usually have no canines. Ruminants, like the cow and sheep, which chew the cud, have no upper teeth in front. A horny pad is present and the grass is ground against it with the jaw moving sideways, instead of up and down.

Whatever the shape, the structure of all teeth is the same. A firm substance, dentine, surrounds the pulp cavity in which there are blood vessels and nerves.

The outer surface of the exposed part of the tooth, the crown, is protected by a hard, smooth layer of enamel. This is the hardest substance in the whole body. The rest of the tooth, the root, is fixed in the jaw by a material called cement.

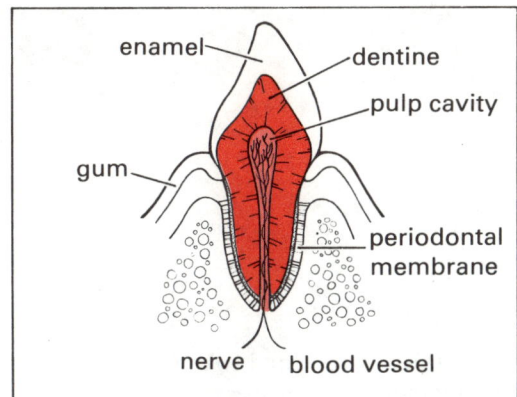

Structure of a tooth

Care of the teeth

Teeth are liable to decay by dental caries. This disease is less likely if there is enough vitamin D in food and traces of fluoride, either in food or in toothpaste to strengthen the enamel.

It seems that the most likely cause of decay is formation of acids, particularly lactic acid, from carbohydrates in food. This is produced by bacteria. The acid attacks the enamel, making a small pit and this allows acid to reach the dentine, then the pulp. If the attack spreads further into the surrounding tissues, a gum-boil or abscess results.

The most vulnerable places for attack are the hollows of the molars and the small gaps between the teeth. Careful brushing after meals is necessary to remove waste food on which bacteria can thrive. This brushing, up and down, to deal with particles between teeth and across the top of the molars, is perhaps most needed just before going to bed.

During the day the presence of fluid in the mouth and movement of the tongue make the teeth less liable to attack. It is certainly not wise to eat sweets after brushing teeth at night-time. Bacteria flourish on sugar!

Tongue

The tongue serves two useful purposes. It pushes food towards the teeth. Also it pushes each morsel of food to the back of the mouth ready to be swallowed.

A reflex action (see Topic 106) then causes muscular action called **peristalsis**, to squeeze the food along the oesophagus into the stomach.

Digestive tract

The path taken by the food during digestion can be seen in the simplified diagram. The oesophagus leads into the middle of the stomach. At the lower end of the stomach is a long tube, not as shown in the diagram, but very coiled. This is the small intestine. This, in turn, leads into a wider tube, the large intestine, which ends at the anus.

Digestive glands

All along the digestive path juices are added to the food. Saliva comes from glands in the mouth; more juice is secreted from the walls of the stomach.

The first loop of the small intestine, where it leads from the stomach, is the duodenum. Small pancreatic ducts lead into this part from a special organ called the pancreas.

Another small tube, the bile duct, also leads into the duodenum. Along this bile duct, a juice called bile passes from the gall-bladder which is close to the liver.

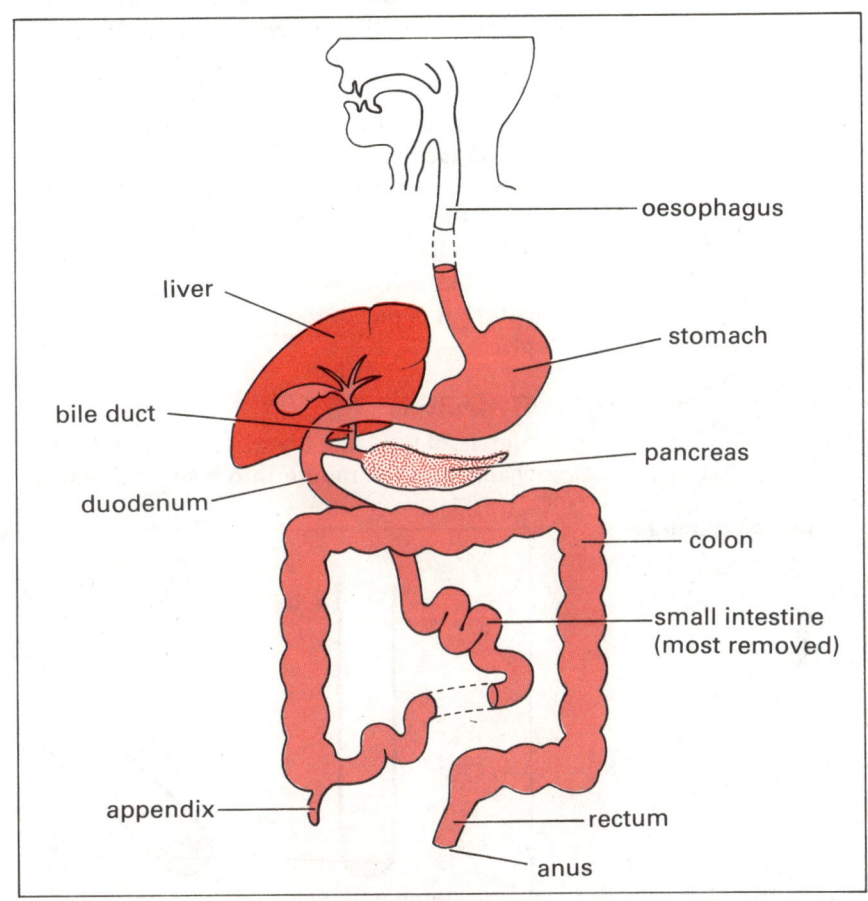

The digestive tract

1 Why is it thought necessary to include (i) a trace of fluoride, (ii) some calcium salts, in the food we eat?
2 Draw a diagram of a cross section through a molar tooth. Label the parts seen in the section.
3 What differences would you expect to find between the teeth of a dog and the teeth of a sheep?

Chemical action

A small amount of starch mixed with water is put in a test tube. A little saliva from the mouth is added. The mixture is kept at about body temperature for a few minutes. A control experiment without saliva is also set up.

After a minute, a drop taken out of each mixture is tested with iodine solution. A blue colour shows that starch is present. The test is repeated each minute. It is found that in a few minutes there is no starch in the test tube that contained saliva. Starch is still present unchanged in the control test tube.

Saliva has changed the starch into something that does not give the deep blue colour with iodine solution.

If the saliva is added to boiling water before it is put in the starch no action takes place.

Hydrolysis

Starch, in food like bread, rice or potatoes, is changed in the mouth into a sugar, maltose. But maltose is further broken down in the small intestine, to glucose.

Starch is a carbohydrate, with a long chain of sugar units in its molecule. Water molecules in saliva attack this chain and break it up into molecules of glucose. This is called **hydrolysis**. Glucose is soluble and diffuses into the blood in the small intestine.

Enzymes

In saliva there is a 'chemical helper' or **enzyme**, called amylase. This changes starch to maltose. Enzymes are proteins that act as catalysts in living organisms. A small amount brings about a change in a lot of material.

Different enzymes act only on certain reactions (i.e. they are **specific**). They act best at a particular temperature and are destroyed by excessive heat. They act best at a specific acidity or alkalinity. For example, amylase no longer acts when it reaches the stomach which is acidic.

Hydrolysis of starch by saliva

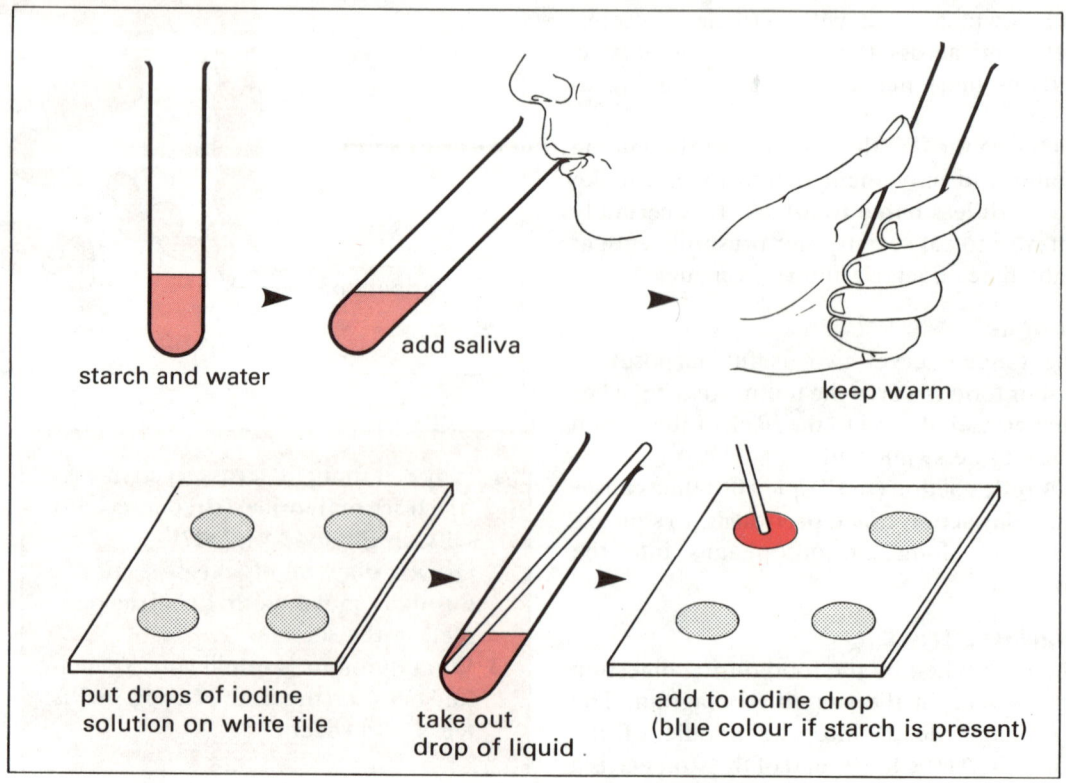

starch and water

add saliva

keep warm

put drops of iodine solution on white tile

take out drop of liquid

add to iodine drop (blue colour if starch is present)

Proteins

Proteins are made of chains of amino acids. The links that hold the amino acids together can be broken by water (hydrolysed) in the presence of enzymes. The individual amino acids are then released.

There is no digestion of proteins in the mouth. Enzyme action resulting in the break up of proteins takes place in the stomach and the small intestine. Amino acids are soluble and are absorbed into the blood in the small intestine.

So the food, in its passage along the digestive tract, is broken down by water molecules, helped by several different enzymes, each having a specific action.

Absorption of food

In the small intestine the solutions of simple sugars, amino acids, fatty acids and glycerol (all soluble) pass into the tissues of the body.

A section of the small intestine shows many projections called **villi** (singular: villus). These greatly increase the surface area of the intestine. Each villus has a network of tiny blood-vessels into which these solutions diffuse. Also inside, there is a loop of the lymph vessel, known as a lacteal. Tiny fat droplets pass into the lacteals and are taken round the body in the separate lymph system.

In the large intestine water is taken into the blood stream.

Summary

The process of feeding can be divided into stages.

1 **Ingestion** The food is taken into the mouth

2 **Digestion** The insoluble substances in the food are changed into simpler, soluble ones

3 **Absorption** These solutions are taken into the blood system

4 **Assimilation** This is the process by which the substances now in the blood are used. Some of them actually become part of the body

5 **Egestion or Elimination** Waste matter which cannot be used is got rid of.

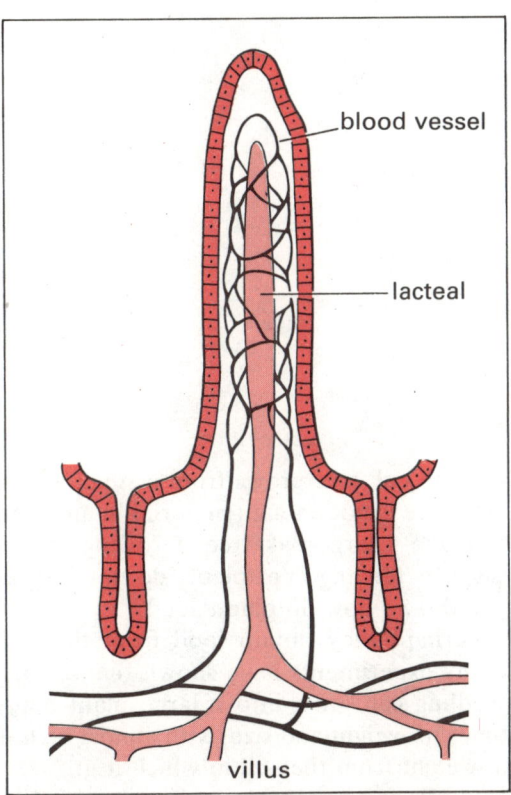

Structure of a villus from the intestine

1 Why is digestion necessary?
2 What is the meaning of 'hydrolysis'? Give an example.
3 Trace what happens to starch in food from the time it is taken into the mouth until the products are in the blood.

Table of digestive enzymes

Situation	Secretion	Enzyme	Chemical change
Mouth	saliva	amylase	starch to maltose
Stomach	gastric juice (acidic)	pepsin	proteins to peptones
		rennin	caseinogen to casein
Small intestine	pancreatic juice (alkaline) and bile from the gall-bladder	trypsin	proteins to peptones
		lipase	fats to fatty acids and glycerol
		amylase	starch to maltose
	secretion from walls of intestine	erepsin	peptones to amino-acids
		maltase	maltose to glucose
		lactase	lactose (milk sugar) to glucose and galactose.

Photosynthesis

Alcohol treatment
Boiling in ethanol removes the green pigment from the leaf

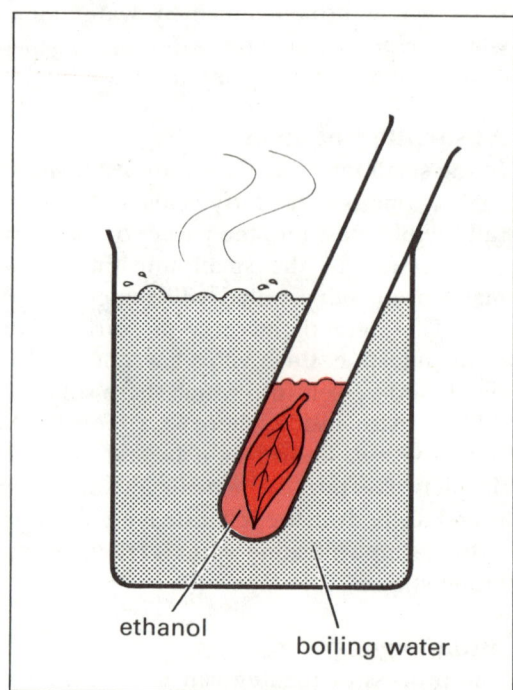

ethanol

boiling water

Warning
Do not boil ethanol with a naked flame; use boiling water

A large oak tree grows from a small acorn, but there is not enough material in an acorn to make a large oak tree! Growing in size suggests feeding. Yet plants do not feed as you do. So how do plants feed?

Perhaps they obtain food from the soil. Many experiments have shown that a small seedling can grow into a large plant many times its weight and size, with almost no loss in weight from the soil in which it grows.

In fact green plants in sunlight, make their food from carbon dioxide and water. This is called **photosynthesis**. One of the foods plants make is starch.

Test for starch in green plant tissues

Procedure

1 Dip the leaf in boiling water.

2 Boil in ethanol. For safety sake, put the test tube containing the ethanol in a beaker of boiling water. Ethanol boils at 78°C so the boiling water will keep the ethanol boiling without the use of a naked flame.

3 Take the leaf out and rinse it in water.

4 Place it in a dilute solution of iodine for a few minutes.

Reason

To kill and fix the tissues (i.e. as far as possible, prevent the contents of the leaf changing chemically while the test is done).

To remove the green colouring matter which is soluble in ethanol. It would be difficult to recognize the blue colour of the starch-iodide if the leaf was still green.

To soften the leaf after the ethanol has made it brittle.

If starch is present, a blue colour will be seen.

These tests show that starch is sometimes present in green leaves.

Conditions for starch formation

Sunlight. No starch is found in green leaves which have been shaded from sunlight for a few hours.

Chlorophyll. Tests carried out on a variegated leaf, which has been in sunlight for some time show that starch is formed only in the green parts, not in the colourless parts.

Carbon dioxide. No starch forms in a leaf if there is no carbon dioxide in the air around it.

Photosynthesis

Green plants can make sugars and starch from the raw materials, carbon dioxide and water. This building up (synthesis) can be done only with the help of energy in sunlight (radiant energy). Plants use only about 1 per cent of the sunlight falling on them.

The process is not possible without the green pigment, **chlorophyll** being present.

Photosynthesis can be summed up as:

$$\text{carbon dioxide} + \text{water} \xrightarrow{\text{sunlight}} \text{sugar} + \text{oxygen}$$

If more sugar is formed than the plant can use right away, it is changed into starch and stored in leaves, and other parts of the plant. Note that oxygen is given off as a by-product.

In darkness, sugars are no longer formed and the stored starch is changed back into sugars.

Chlorophyll

This pigment can be studied if a solution is made in acetone (see Topic 24).

The solution is fluorescent; it is green when seen with light passing through it, but dull red when seen by reflected light. This shows that the pigment absorbs light energy from sunlight.

Chlorophyll is really a mixture of pigments.

Proteins and fats

Some of the carbohydrates made by plants are used to build up proteins and fats. To change sugars into proteins, some nitrogen is needed and also sulphur and phosphorus. These are obtained from soil in the form of nitrates, sulphates and phosphates.

Proteins are made chiefly in the growing regions of root and shoot, and in fruits and buds.

Fats are found mostly in seeds, e.g. coconut, peanut.

Enzymes are found in plants, as might be expected since digestion of starch, proteins and fats takes place in the plant tissues.

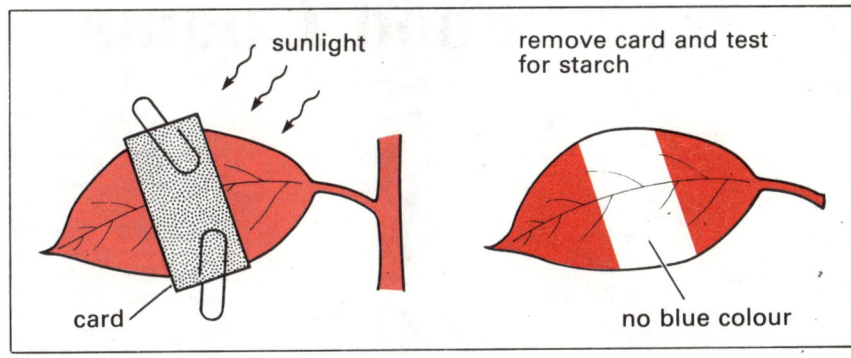

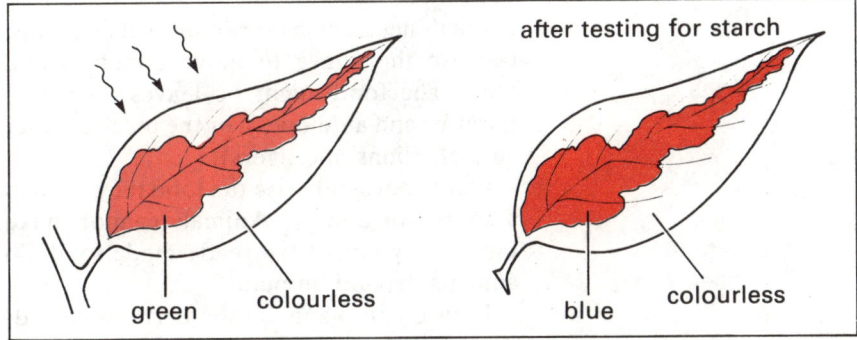

When a seed begins to germinate, an enzyme called diastase hydrolyses the starch in endosperm or cotyledon, to sugar. It is this sugar that gives energy for the young radicle and shoot to start growing.

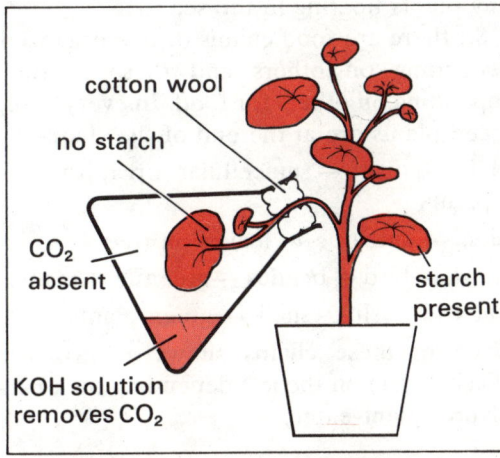

Conditions for starch formation
Testing the effect of sunlight (above)

Showing the effect of chlorophyll on starch production (below)

Showing the need for carbon dioxide for starch production

1 Name the three material substances needed for a plant to form carbohydrates. What is the source of the energy needed?
2 A green leaf, picked from the lower branches of a tree during daytime, gave a negative result when tested for starch. Suggest reasons to explain why starch was absent.
3 Can you explain why a seed does not germinate when it is kept dry?

Animals need much more energy than plants, because they have to move about to find food. The food might be leaves on a tree (giraffes and antelopes), or the meat of other animals (lions and tigers).

Plants can synthesise the food they need as a source of energy. Animals cannot make food. They must have ready-made food. So animals depend on plants.

If all green plants on the earth were suddenly killed by some disaster, man would not survive very long.

Think of some examples, You eat beef, but the cow feeds on grass which is a plant. Perhaps you prefer eggs, but the chicken that lays the eggs must feed on corn, which comes from a plant. Even if you decide to eat fish, the fish feed on smaller fish, but these feed on tiny plants floating in the sea.

So there are food chains of one organism depending on others and these in turn depending on others for food. In every case, green plants are at the end of the chain.

Man ← fish ← unicellular organisms ← plants

Lion ← antelope ← leaves on tree

Hawk ← bird ← beetles ← greenfly ← plants

Cat ← thrush ← snail ← green plants

Each of these chains shows a **carnivore** (flesh-eater) on the left depending on a **herbivore** (plant-eater).

Food chain
The cat eats the bird, which eats the snail, which eats the plants

Now it would take a long time for a hawk to collect enough greenflies to satisfy it. So the animals in between concentrate the food.

The number of smaller birds must be greater than the number of hawks and the number of greenflies must be even greater. So there is a kind of pyramid where the basic green plants support the herbivores, which in turn support a smaller number of carnivores at the top.

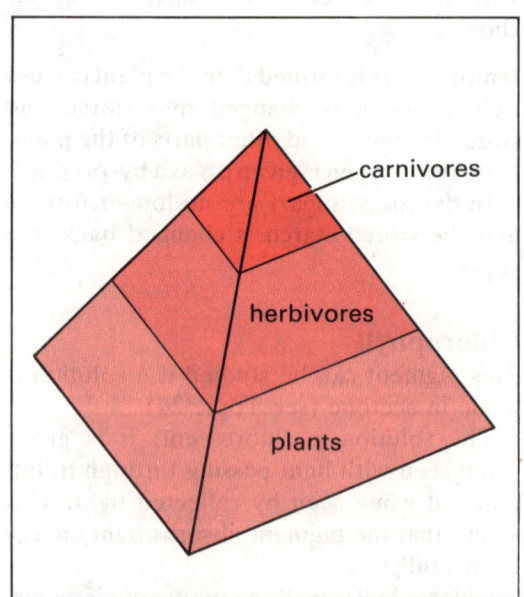

Food pyramid

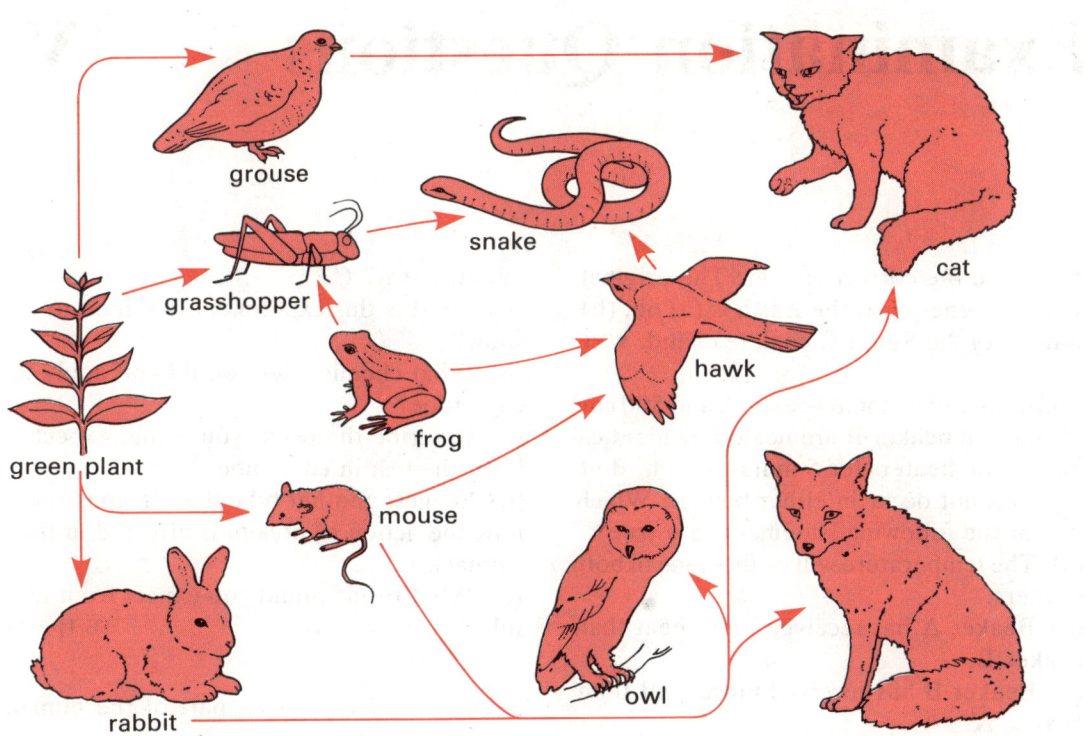

Food web

Food chains are just single strands of a more complex tapestry. Several food chains are involved in a food web of relationships.

Organisms such as green plants and some bacteria that can make their own food from simple chemicals, such as carbon dioxide and water, are called **autotrophs**.

Heterotrophs are those organisms, such as animals, that depend on getting ready-made organic substances by feeding on other organisms.

There are some plants that do not contain chlorophyll. Moulds, bacteria and fungi such as mushrooms are examples. A few flowering plants, such as bird's nest (*Monotropa hypopitys*) and bird's nest orchid (*Neottia nidus-avis*) have no chlorophyll and are heterotrophic.

Mucor

Mucor, or pin-mould, is an example of these unusual plants. If a piece of wet bread is kept moist and warm under a bell jar for a few days, a growth of white tufts like cotton wool will be seen. After some days black 'pin-heads' are seen at the ends of delicate colourless stalks.

These are spore-bearing organs concerned with reproduction (see Topic 95).

The body of the mould is a mass of branching thin threads called hyphae. This mass of threads, the mycelium, is found inside the bread. The material on which a mould grows is called the substrate.

The hyphae secrete enzymes and acids into the bread. With the help of these enzymes, the food material is broken down (digested) then it is taken into the hyphae. In other words, the fungus digests its food outside the body. Otherwise, its method of feeding is rather like our own.

Mucor

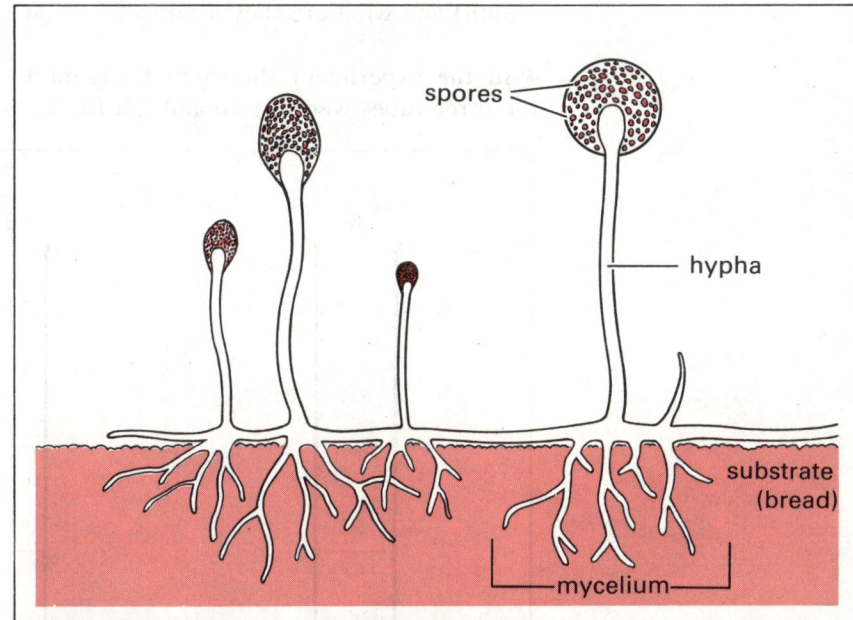

1 Name any three carnivorous animals.
2 Name any three herbivorous animals.
3 Name any three plants that do not contain green chlorophyll.

147

Examination Questions V

1 Choose the correct answer: The original source of energy on the Earth is (a) oil, (b) water, (c) the Sun, (d) coal, (e) wind. WYL

2 One litre of water in beaker A and 500 cm^3 of water in beaker B are heated by identical immersion heaters for 5 minutes each. Boiling does not occur in either beaker. Which two of the following statements are true?
(a) The temperature will be the same in both beakers.
(b) Beaker A has received more heat than beaker B.
(c) Beaker B has received more heat than beaker A.
(d) Both beakers have received the same amount of heat.
(e) The water in A is at a higher temperature than that in B.
(f) The water in B is at a higher temperature than that in A. EM

3 Mammals have four kinds of teeth. What are these four kinds of teeth called (starting from the back and working to the front of the mouth) and what are they used for? M

4 In the experiment shown in Diagram 1 the three tubes were set up and left for 15 minutes at 37°C.
(a) What is this experiment designed to show?
(b) What chemical test would you apply to each tube?
(c) Describe the result you would expect from the test in each tube.
(d) Suggest another tube that demonstrates how the action of ptyalin is affected in the stomach.
(e) What result would you expect from this tube? SE (part)

5 Diagram 2 represents part of the human digestive tract.
(a) Name the parts labelled A to F.
(b) What is the main function or purpose of the part labelled E?
(c) What is the main function or purpose of the part labelled D? ALS

6 Which of the following statements are true of enzymes?
(a) They act on any kind of substance.
(b) They work only in acid solutions.
(c) They are destroyed by heating them above 37°C.
(d) They change the speed of a reaction. WM

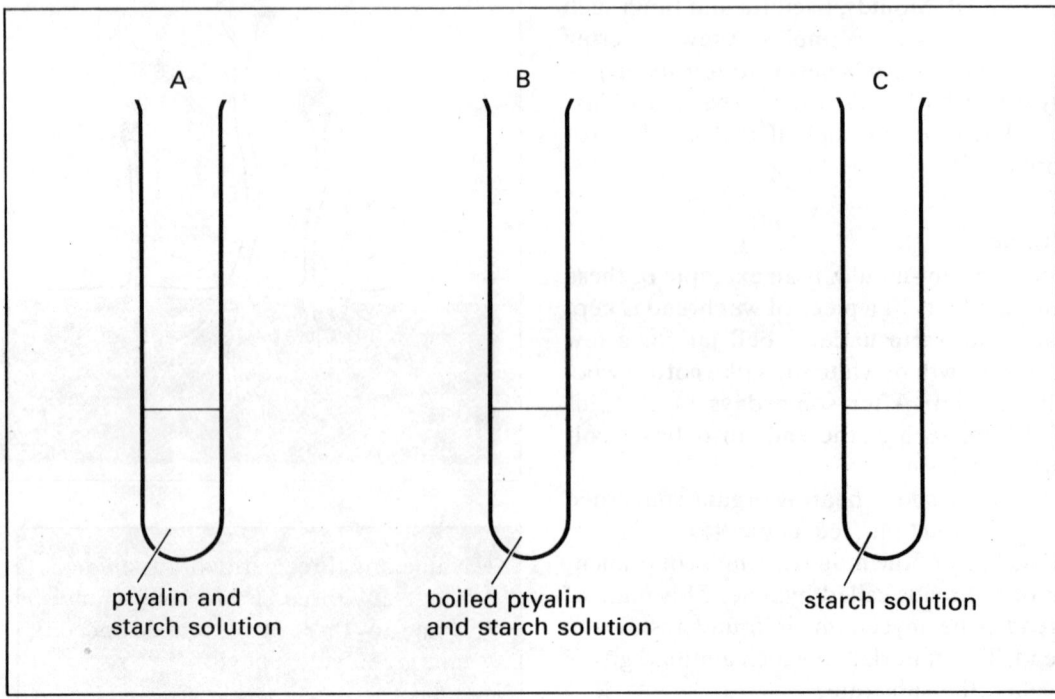

A — ptyalin and starch solution

B — boiled ptyalin and starch solution

C — starch solution

Diagram 1

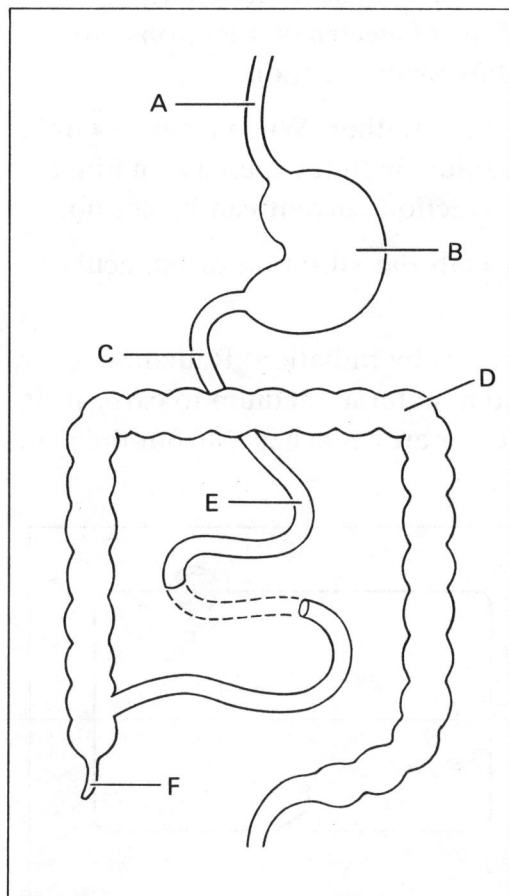

Diagram 2

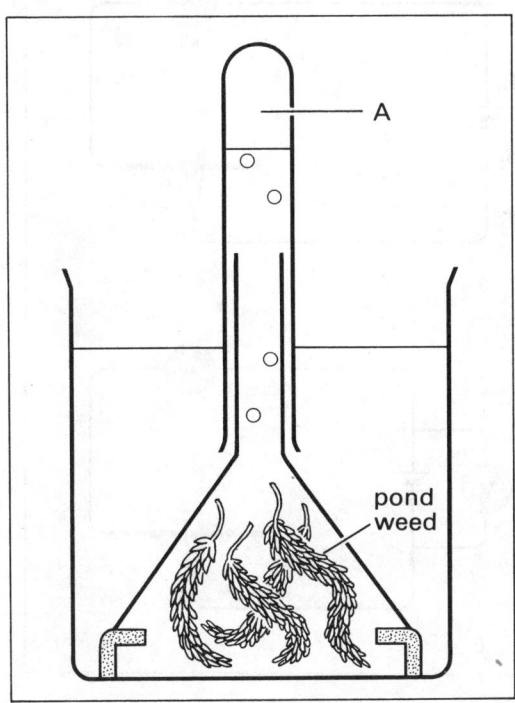

pond weed

Diagram 3

7 Pick out the correct answer: An athlete would obtain a rapid supply of energy by eating (a) butter, (b) milk (c) fruit juice, (d) beef, (e) glucose. EM

8 Benedict's solution, Fehling's solution or Clinistix may be used to test for the presence of: protein, vitamin C, starch, fat, reducing sugar. Which is the correct answer? ALS

9 Choose the correct answer: Plants receive the energy for photosynthesis from: (a) the sun; (b) the chlorophyll in the leaves; (c) carbon dioxide; (d) the mineral salts in the soil; (e) oxygen. Y

10 The experiment shown in Diagram **3** was set up and left in the sunlight. What gas collected at A? Explain briefly how you could test it. EM

11 The list below is of pieces of equipment which can bring about energy conversions: battery; steam engine; photo electric cell; solar cell; cross bow; gas light; electric lamp; nuclear reactor. Name the piece of equipment in each case which will bring about the following energy conversions: (a) Chemical energy to heat energy. (b) Light energy to electrical energy. (c) Electrical energy to light energy. (d) Chemical energy to electrical energy. EAN

12 The following organisms make up a food web: leaves; thrush; hawk; caterpillar; snail; earthworm. Arrange these organisms in a food web. EAN

13 Energy exists in many forms. Which one of the following is **not** a form or energy? (a) friction; (b) heat; (c) electricity; (d) motion. NW

ENERGY ON THE MOVE

Energy is not always needed at the place where it is found. An example is electricity. Although it is produced at the power station, it is used in home and factory for heating, lighting and driving machines. Electrical energy is transferred as electric current, which is a flow of **electrons**. Electrons are also involved in the passage of heat along metals by conduction.

Molecules can carry energy from one place to another. When a gas or liquid becomes hot the molecules vibrate more rapidly with the increase in kinetic energy. Since they are free to move, a convection current can be set up.

Sound travels from a vibrating source through the vibration of molecules, whether they be gas, liquid or solid.

Yet another method by which energy travels is by radiation. Radiant energy, such as light or radiant heat, does not need a material medium to carry it. It travels through space as, for example, when energy reaches us from the sun.

Electrical circuits

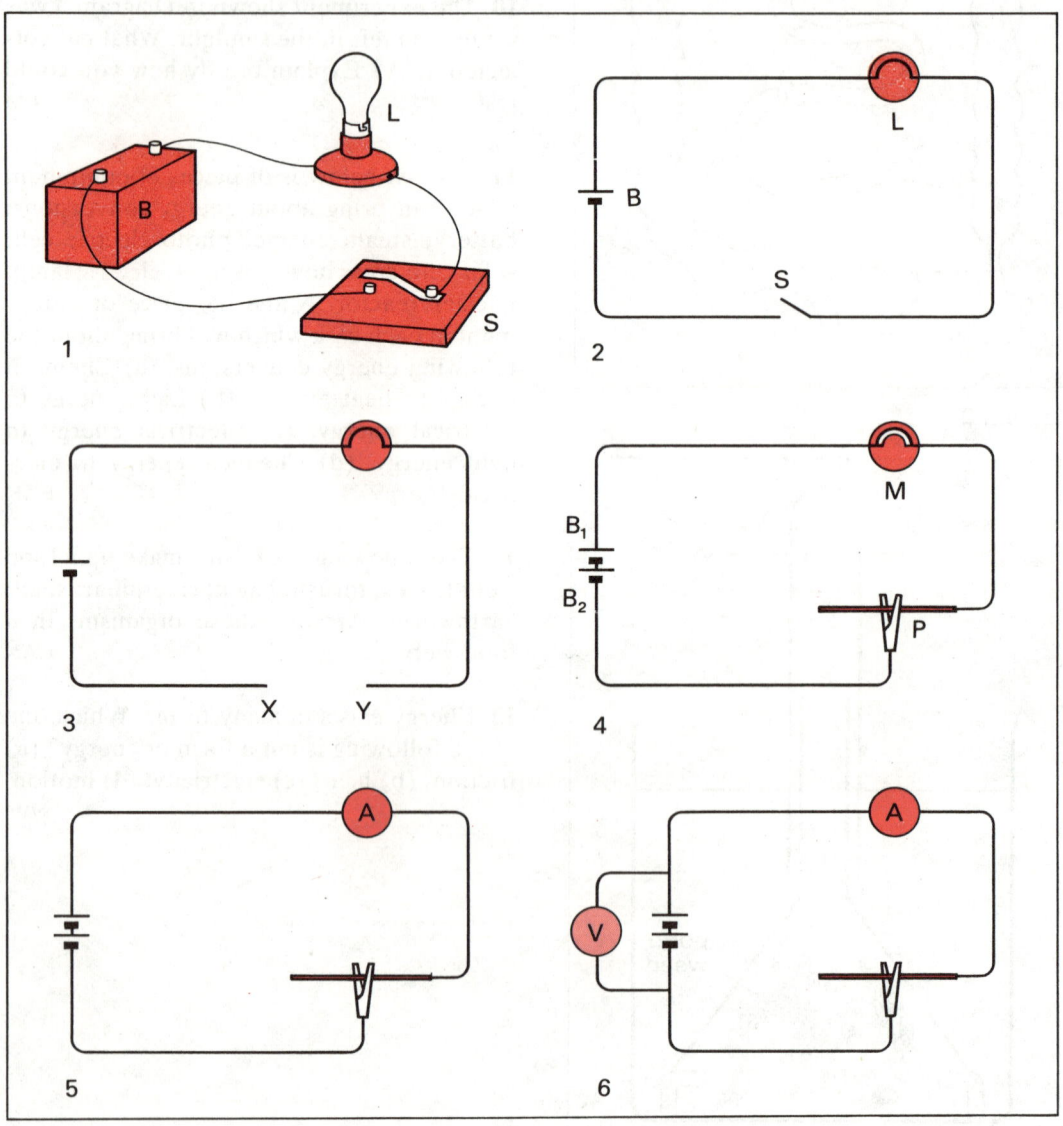

Electric Current

A number of experiments can be tried using a battery as the source of electric current. Some circuits are shown in the diagrams.

1 shows a battery B connected in series with a light bulb L and a switch S.

2 is the circuit diagram for the arrangement in **1**. When the switch is closed the lamp L lights. A complete path (or circuit) is needed for the electric current to flow.

3 A gap XY is left instead of the switch. Different things are used to complete the circuit across XY. Suitable things are copper wire, iron wire, a glass rod, a piece of paper, a pencil 'lead'. The lamp lights when metals are used to bridge the gap. Metals are electrical **conductors**. Paper and glass do not conduct electricity well; they are **insulators**.

4 Two batteries B₁ and B₂ are used in series, with a different lamp bulb M and a crocodile clip is used to tap off different lengths of the pencil 'lead' P. Pencil 'lead' is actually graphite – a form of carbon. As the length of carbon rod included increases, the **resistance** of the carbon to the flow of current makes the lamp become dimmer. The pencil 'lead' is acting as a **resistor**.

5 Instead of the lamp bulb, an ammeter A is used. The current flowing round the circuit can then be recorded, as different lengths (1 cm, 2 cm, 3 cm and so on) of carbon rod are included in the circuit. Current is measured in **amperes** (amps) (A).

There must be a kind of pressure pushing the electrons round the circuit. This pressure is measured in volts (V).

6 Read the battery voltage by connecting a voltmeter across the terminals.

Put a suitable resistor R in the circuit shown in **7**. Tap off one, two, three batteries in turn. Note the ammeter reading each time. Here are some typical results. 1.5 volt batteries were used.

No. of batteries	Voltage (V, volts)	Current (I, amps)	V/I
1	1.5	0.01	150
2	3.0	0.02	150
3	4.5	0.03	150
4	6.0	0.04	150

The value of V/I is always the same. The current flowing through the circuit is proportional to the voltage applied.

Ohm's Law states that the current flowing in a conductor is proportional to the potential difference across the ends of the conductor. **Potential difference** is what was called pressure. It is measured in volts.

A conductor that allows a current of 1 A to flow through it when a voltage of 1 V is applied at the ends has a resistance of 1 ohm (Ω).

$$\frac{\text{Potential difference } V \text{ (volts)}}{\text{current } I \text{ (amps)}} = \text{resistance } R \text{ (ohms)}$$

It follows that

$$V \text{ (volts)} = I \text{ (amps)} \times R \text{ (ohms)}$$

also

$$I = \frac{V}{R}$$

1 Name two substances that conduct electricity. Name two substances that act as insulators.

2 What names are given to the following: (a) a device that changes chemical energy into electrical energy; (b) an instrument that measures electric current in amperes; (c) an instrument that can be used to measure potential difference?

3 Copy the following table in your book and fill in the spaces (1 mA = 0.001A; 1 kΩ = 1000Ω).

V	I	R
100V	50A	*
*	2A	5Ω
12V	*	24Ω
*	5mA	10kΩ

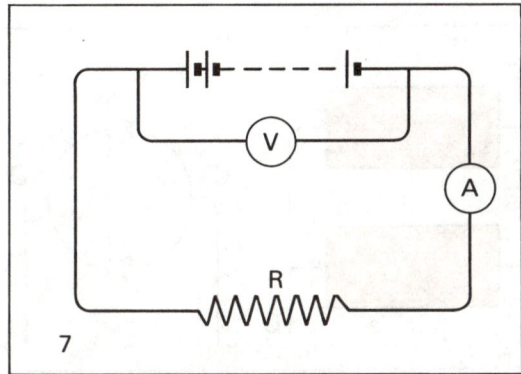

7

Cells are connected in **series** when they are joined 'head to tail' (**1**). They are in **parallel** when positive poles are connected together and negative poles are also joined (**2**).

The voltage across AB is 4 V, but the voltage across CD is 2 V. There would not seem to be any advantage in having two cells in parallel, but later you will see that there is.

Resistors can be linked in series (**3**) or in parallel (**4**). In order to find out what effect these combinations have, connect each to a battery and measure the current flowing in the main circuit. **5** shows a 2 V battery in series with resistors of 20 and 5 ohms in series. The current flowing is 0.08 A. **6** shows the same resistors in parallel joined to the battery. In this case the current is 0.5 A.

Compare the flow of current with the flow of water. Suppose water is flowing out of a tank (see Diagram **7**). A narrow tube that lets out the water corresponds with an electrical resistance. The narrower the tube the smaller the rate of flow (the greater is the resistance).

But if two narrow tubes are put side by side (in parallel) more water will flow (current is greater).

The current can be calculated in the two cases. When several resistors are in series, the total resistance is found by adding the separate resistances together:

$$R = r_1 + r_2 + r_3 + \dots$$

For the example in **5**

$$R = 20 + 5 = 25\Omega$$

So current $= \dfrac{2\ V}{25\Omega} = 0.08$ A

When they are in parallel, the single resistance R that can replace the separate ones is given by

$$\frac{1}{R} = \frac{1}{r_1} + \frac{1}{r_2} + \frac{1}{r_3} + \dots$$

In the example **6**

$$\frac{1}{R} = \frac{1}{20} + \frac{1}{5} = \frac{1 + 4}{20} = \frac{5}{20}$$

so $R = \dfrac{20}{5} = 4\Omega$ and the current

$= \dfrac{2\ V}{4\Omega} = 0.5$ A.

Cells in series and parallel

When cells are put in series, the voltages just add together. But what about cells in parallel? The voltage across the common positive and negative terminals still reads the same as for only one cell.

When the two cells in parallel are connected across a resistor and an ammeter (**8**), the current for the combination reads higher than for a single cell. This is because cells themselves offer a small resistance to the flow of electrons.

The internal resistance of the two cells in parallel is less than the internal resistance of the one cell. An accumulator (car battery) has a very small internal resistance. This is why it is possible to draw a very high current from it. If a 12 V car battery has an internal resistance of 0.04Ω, it could give a current of

$$\frac{12\ V}{0.04\Omega} = 300\ A.$$

1 What single resistance is equivalent to two resistors of 2 ohms and 8 ohms; (a) if they are in series, (b) if they are in parallel?
2 If the potential difference across RT in diagram 6 is 2 V, find the current that flows (a) through the 20 ohm resistor, (b) through the 5 ohm resistor.
3 What is the resistance of the element of an electric fire if a current of 3 A flows when it is connected to a 240 V supply?

Linking cells

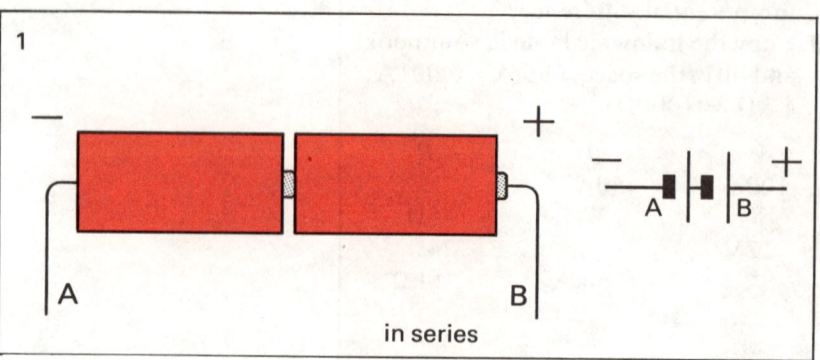

in series

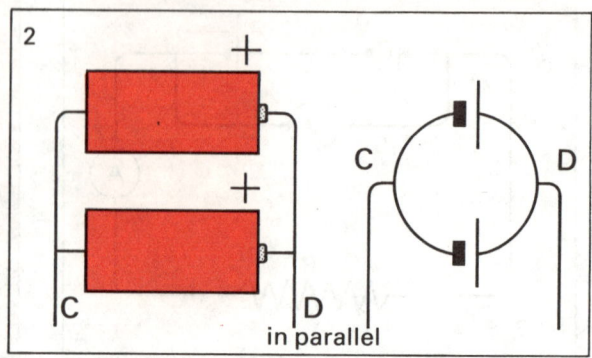

in parallel

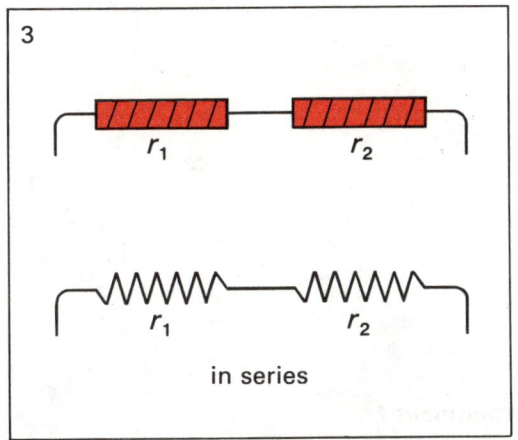

in series

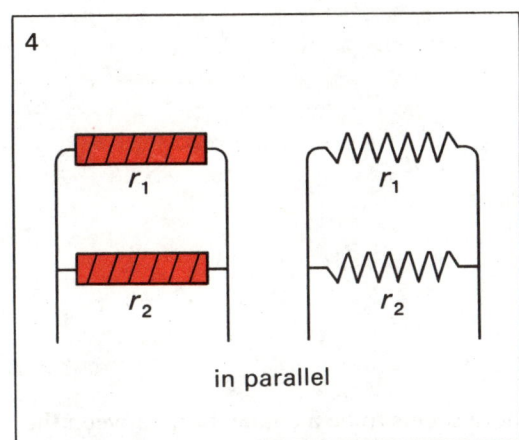

in parallel

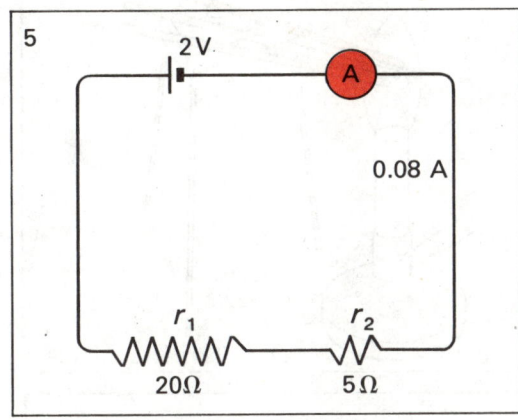

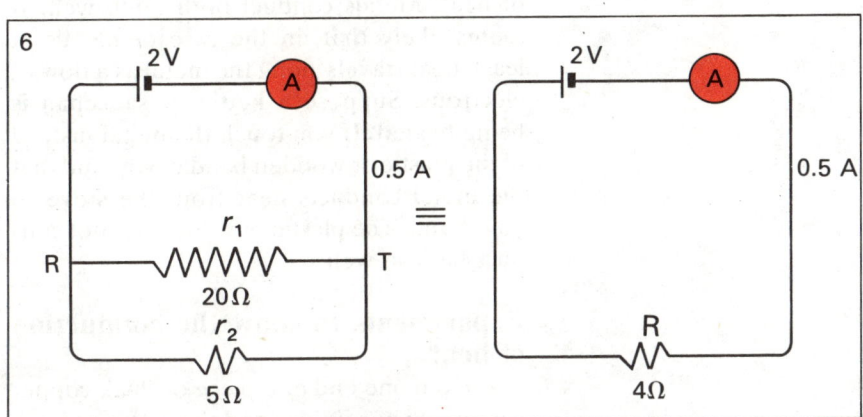

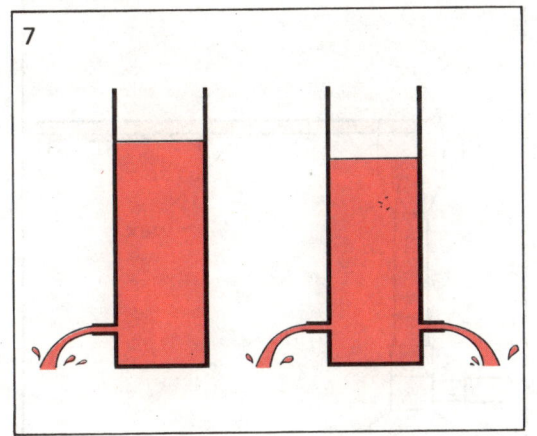

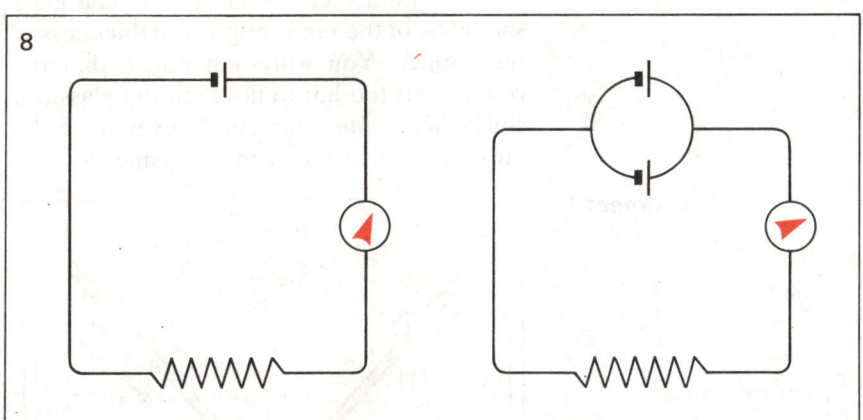

There seems to be a connection between the conduction of electricity and the conduction of heat. Metals conduct both quite well. It seems likely that, in the case of metals at least, heat travels along the metal as a flow of electrons. Suppose a kettle or saucepan is being heated. If you touch the metal instead of the plastic or wooden handle, you find that the metal conducts heat from the stove to your hand. The plastic or wood does not conduct heat so well.

Experiments to show the conduction of heat

1 Hold one end of a piece of thick copper wire with your fingers and the other end in a bunsen flame. At the same time, hold in the other hand, a piece of glass rod. The glass should be of the same length and thickness as the copper. You will soon notice that the copper gets too hot to hold but the glass can still be held quite comfortably, even when the far end begins to melt in the flame.

Experiment 1

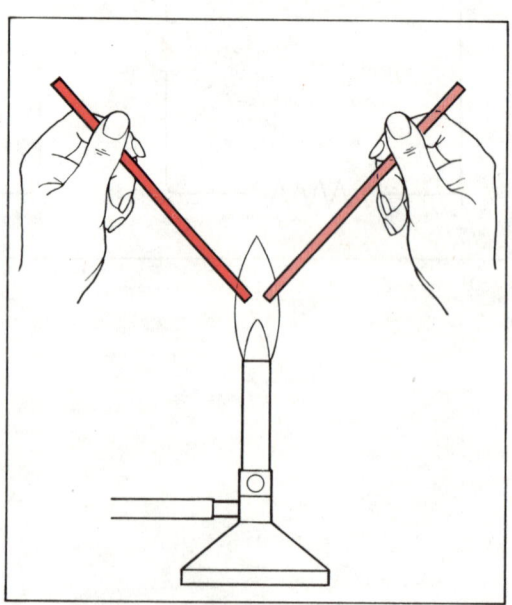

Experiment 2

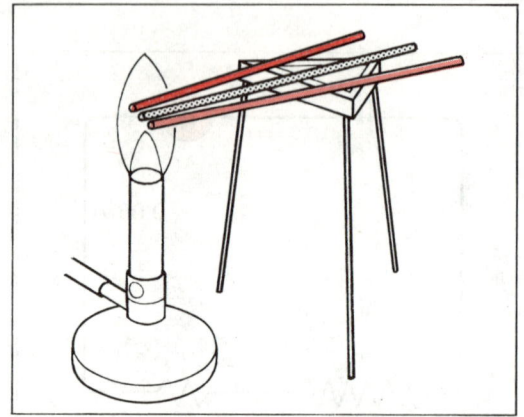

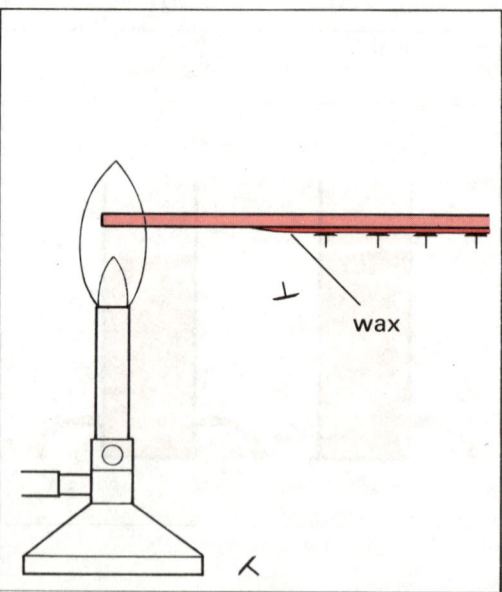

wax

2 To compare the rate at which heat passes along different rods, arrange the rods horizontally. Drawing pins or tacks are attached to the rods by paraffin wax. When the rods become hot enough the wax melts and the pins fall off. If rods of the same thickness and length are used, you find that copper conducts heat better than iron and iron better than glass or wood.

3 Put an unlit bunsen burner under a fine mesh copper wire gauze. Turn the gas on and light the gas **above** the gauze. The gas below the gauze does not get hot enough to burn for quite a while. The copper is conducting the heat away from the gas.

This is the principle of the Davy Miners' Safety Lamp. The gauze around the flame conducts the heat so that any inflammable gas in the air of the mine outside does not get hot enough to explode. A gas needs to be heated to its ignition point before it will burn.

4 Paper is wrapped tightly around a rod made of wood joined to copper. This compound bar is warmed evenly in a flame. You find that the paper close to the wood is charred, but not the part close to the metal. The metal has conducted the heat away, but the wood is a bad conductor of heat.

5 Boil some water near the top of a test tube. When the water at the top is boiling, that at the bottom of the test tube is still cold. This shows that water is a poor conductor of heat.

Metals are good conductors. Bad conductors or insulators include wood, cork, glass. Most liquids (except mercury) and gases are bad conductors.

The better conduction of metals is linked with the fact that the atoms making up the metal structure have loosely held electrons. These are free to move and carry energy from the hot part to the cooler part.

You will be able to find examples of bad conductors being used to protect your hands. For example, a kettle has a wooden or plastic handle. Water pipes are usually 'lagged' with felt or flannel to prevent heat passing to the air outside. This is particularly important during very cold weather.

Blankets are used on a bed to prevent your body losing much heat especially when you are sleeping and the body metabolism is low.

The reason why woollen materials, like blankets, are bad conductors is because they are fluffy and air is trapped among the strands of wool. Air is a bad conductor. So you keep warm if you wear wool or a string vest next to your skin. The layer of air trapped in the material prevents too much heat being conducted away from the body.

Air inside the cavity walls of a house, especially if filled with foam or fibre-glass, prevents undue loss of heat from the rooms inside. Double-glazed windows also keep the house warmer by trapping air between the two sheets of glass.

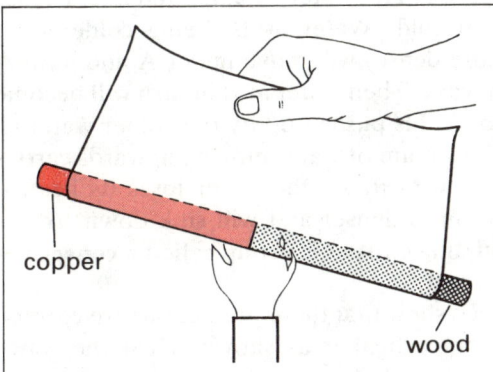

copper

wood

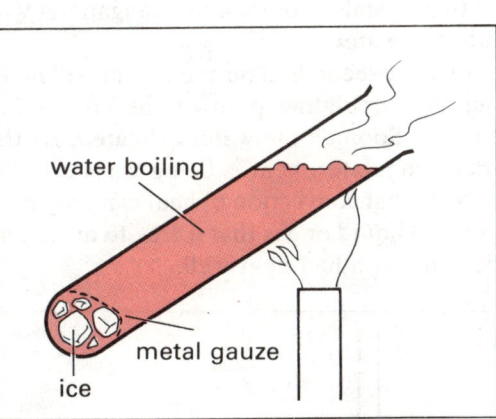

water boiling

metal gauze

ice

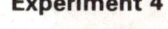

Experiment 3

Experiment 4

Experiment 5

1 Pick out the good conductors from the following list: copper; wood aluminium; water; ice; cork; tin; zinc; paper.
2 Why is it more difficult to solder a copper wire on to a large block of copper than on to a small piece of copper?
3 Explain why a metal fence can feel colder than a wooden one, although both are at the same temperature. (Remember that a feeling of coldness can be due to heat leaving your finger.)
4 Eskimos can keep warm when living inside an igloo made of ice. What does this tell you about the conducting property of ice?

Convection of Heat

70

Suppose a beaker of water is heated by a small flame. Neither glass nor water is a good conductor of heat, so the water at A will get hot first. Hot water expands and is less dense than cold. Water at B, being colder (and more dense) will move under A and push A upward. Then water at B in turn will become hot and is pushed up by the colder water C. The stream of water moving upwards carries heat with it. As the water loses its heat, it becomes denser and will sink down. A circulating current is set up called a **convection current**.

To show that these predictions are correct, clamp a beaker as shown. Heat the water with a small flame at one side. Then drop one or two crystals of potassium manganate(VII) into the water.

In a few seconds, a purple colour will mark out the circulating path of the convection current. Soon all the water is heated. As the water circulates, heat is being transferred.

Note that convection of heat can only happen in a liquid or gas that is free to move and does not conduct heat well.

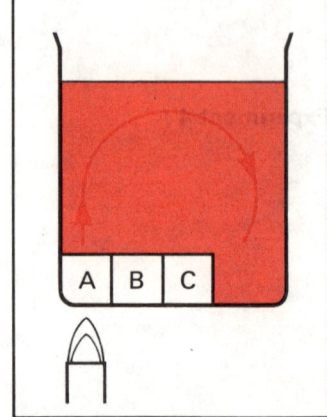

Convection currents
Theory (above) and practice (right)

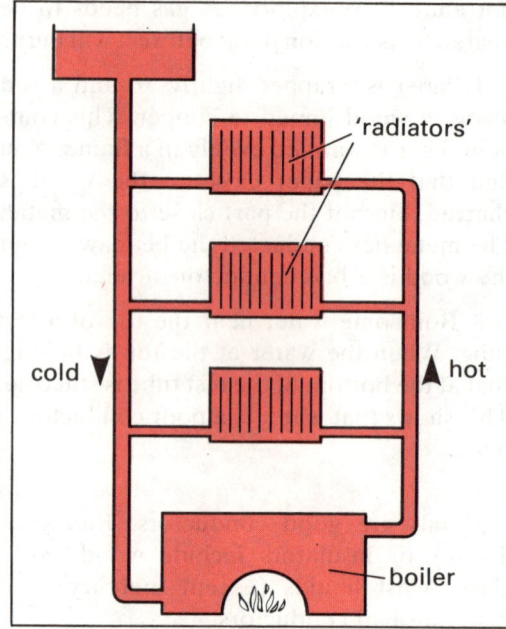

Central heating system

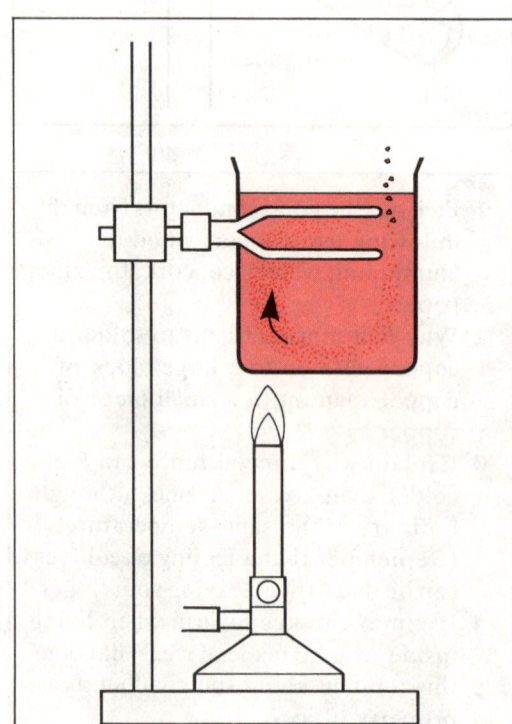

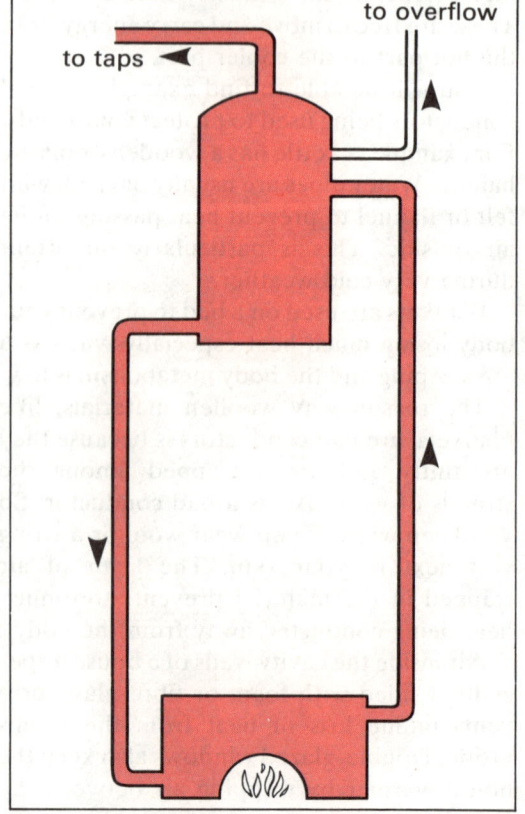

Hot water system

By looking back to Topic 48 you will see that the cooling of a motor car engine depends on heat being convected away from the engine block.

The central heating system of a house as well as the domestic hot water supply, which is often combined with the central heating system, depend on convection. In both the boiler and the storage tank, the hot water outlet is at the top and the cold water feed comes in at the bottom. Hot water moves up and cold water down.

Convection of heat also can happen in air. When a tall glass chimney is placed over a burning candle, the flame will go out. But if a metal dividing strip is placed in the chimney, the candle will continue to burn.

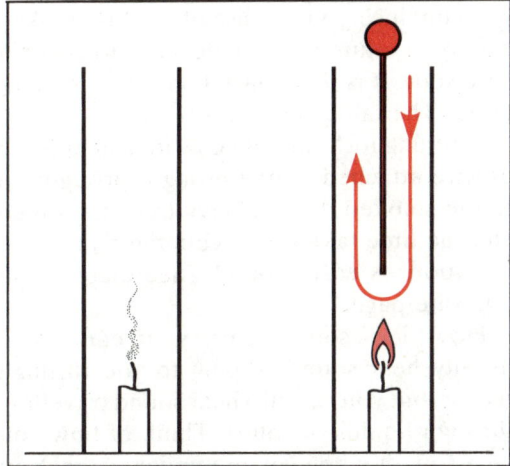

How to make a candle burn

You can see why this happens if a glass-fronted box with two chimneys is used. After burning a candle for a short time under one chimney, the smoke from a smouldering piece of paper, placed above the other chimney will show a convection current. This is a model of the ventilation system once used in mines. Provided no dangerous gas is present, a fire can be lit under one of the ventilation shafts.

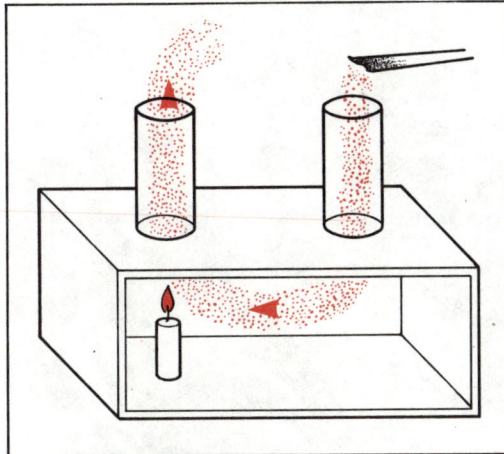

Convection currents in air

Large-scale convection

When the sun shines at the seaside, the land warms up more quickly than the sea. So during the day, heat is convected upwards above the land. This draws air in from the sea. At night the land becomes cool quicker than the sea and the current is reversed. Land breezes occur during night time.

Trade winds

Air at the equator is heated and expands. The hot air is lighter and rises. The surrounding air is drawn into this region. Since the world is rotating, these convection currents, known as trade winds in the northern hemisphere became north-easterly winds.

Right at the equator there is a region of still air called the doldrums.

Weather

Natural convection currents cause winds. But also forced convection currents can happen as a result of masses of cold and hot air being brought together by winds. These clashes of cold and hot air are responsible for some of the types of weather that occur.

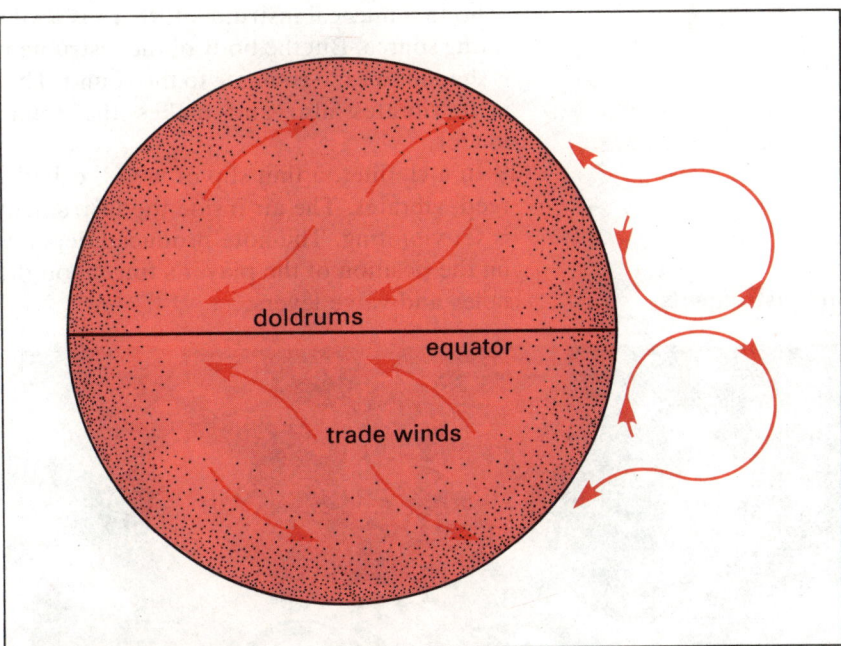

Trade winds

1 Can you explain why the ceiling above a 'radiator' or above a lamp, is often dirtier than elsewhere?
2 How does the convection of heat explain how a person sitting in the middle of a room, far from the 'radiators' is kept warm?
3 Explain why it is not possible for heat to be convected in mercury.

Sound

71

Sound is a form of energy. Molecules take part in the transmission of heat from place to place by convection. Molecules also take part in the movement of sound energy from one place to another.

Making sounds

Sounds are made when things vibrate. The sound may be a noise, as in the case of an explosion or when a stone falls on a sheet of metal or when a car engine is running.

If the vibrations are regular, a musical note may be produced. The thing vibrating may be a tightly-stretched skin, in the case of a drum, or a strip of metal as in a xylophone. It may be a string. In the violin the string is bowed. The strings of a guitar or a harp are plucked. In the piano the strings are set vibrating when struck by hammers.

Sometimes a column of air is set vibrating, as in wind instruments like the recorder or clarinet.

So in a musical instrument, there is a vibrating source. But the body of the instrument is shaped so as to **resonate** to the sound. That is it vibrates also and makes the sound louder.

In a clarinet, a thin slip of wood, called a reed, vibrates. The air inside the instrument is set vibrating. The note produced depends on the position of the player's fingers on the holes and valve levers.

In instruments like the bugle, trumpet and trombone the source of vibrations is the player's lips.

Frequency

A note may be high or low in pitch. If the string of a violin is shortened or if it is tightened, a higher note is produced. The pitch of a note corresponds with its **frequency**. For example, the note C in the middle of the piano range is caused by 256 vibrations per second, that is 256 hertz (Hz). If this is difficult to believe, the following experiment will convince you. A disc of card is smoked evenly in a luminous flame so as to cover it with soot. It is then placed on a record turntable. The table is set revolving.

A tuning fork (middle C) with a light piece of wire attached to one prong is brought up to the smoked disc. A wavy trace is formed and the time taken for a counted number of vibrations is easily found. (See diagram on opposite page.)

How does sound reach your ears? You usually hear sound coming to you through the air, but you can also hear sound travelling through liquids or solids. Think of how you hear sounds when you are under water. You probably know too that by putting your ear to the ground you may hear the sound of distant horse's hooves. The sound passes through the ground.

Wind instruments

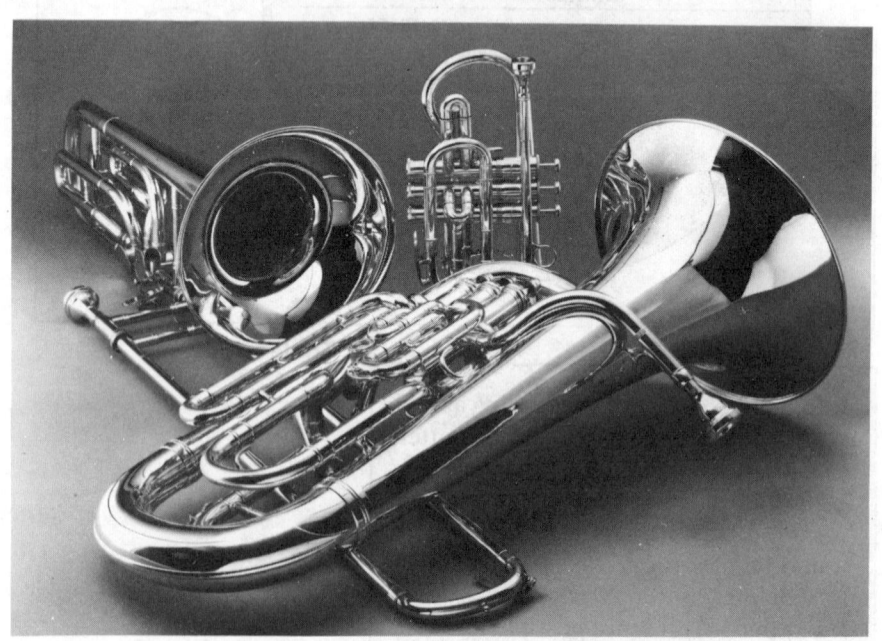

Sound needs a medium to carry it. Sound cannot travel through empty space, like heat and light. To show this an electric bell is hung inside a bell jar. Only two fine wires, for joining the bell to a battery, connect the bell with the outside. The air is pumped out of the bell jar. If the bell is now rung only a faint sound can be heard. When the air is let back into the jar, the ringing becomes clear.

Speed of sound

It is easy to measure how fast sound travels. For example, suppose you start a stop watch exactly as you see the flash of a gun fired at a distance of 1000 metres away. Then you stop the watch the instant you hear the sound. If the time between seeing and hearing is 3 seconds, the speed of sound is 1000/3 m/s = 333 m/s. We can neglect the time the light takes to cover the 1000 m because the light travels 300 million metres in 1 second.

Sound travels faster in water than in air, and quicker still through a solid.

Echoes

When you shout in a tunnel, or in a large empty hall, you notice the sound coming back to you. The sounds are reflected from the walls as echoes.

Depth sounding

Echoes can be useful. A ship at sea sends a sound signal down through the water. The echo from the sea bed is picked up by a hydrophone (an underwater microphone). The time taken for the sound to go down to the bottom and back gives the depth of the sea bed. An ultrasonic signal is used (about 50 000 Hz) which is too high a frequency to be heard by the human ear.

An echo device called ASDIC was used for detecting submarines during the Second World War.

How sound travels

An example may help you to understand what happens when sound travels through a medium.

Stand some dominoes on edge near to each other. When the end one is pushed down, a 'wave' travels along the row as the rest of the dominoes fall down.

Imagine that each domino is attached to the table by a spring, so that when it is pushed aside it comes back to its first position. If you keep pushing the end domino, like a vibrating source, then a series of pulses travels along the row.

This kind of to-and-fro movement passing through the medium is how the molecules of a medium are acting as a sound passes through the medium. This kind of motion is called **longitudinal wave** motion.

1 What is the vibrating source when a bugle is sounded?
2 The frequencies of four notes were found to be: doh 264, me 330, soh 396, upper doh 528 Hz. What connection can you find between these frequencies?
3 What difference in time is there between seeing and hearing a batsman strike a ball if the observer is at a distance of 264 m? Take the speed of sound as 330 m/s.

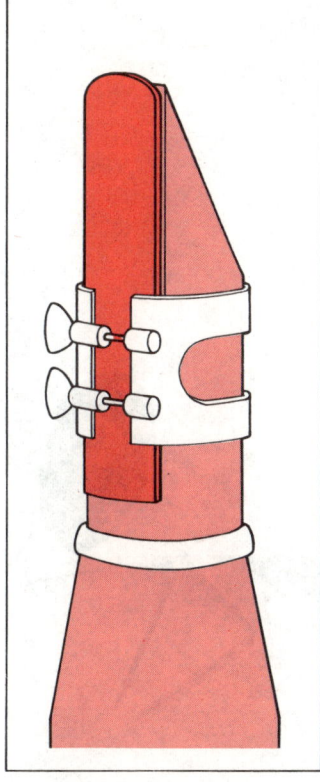

Clarinet mouthpiece with reed

How to find the pitch of a tuning fork

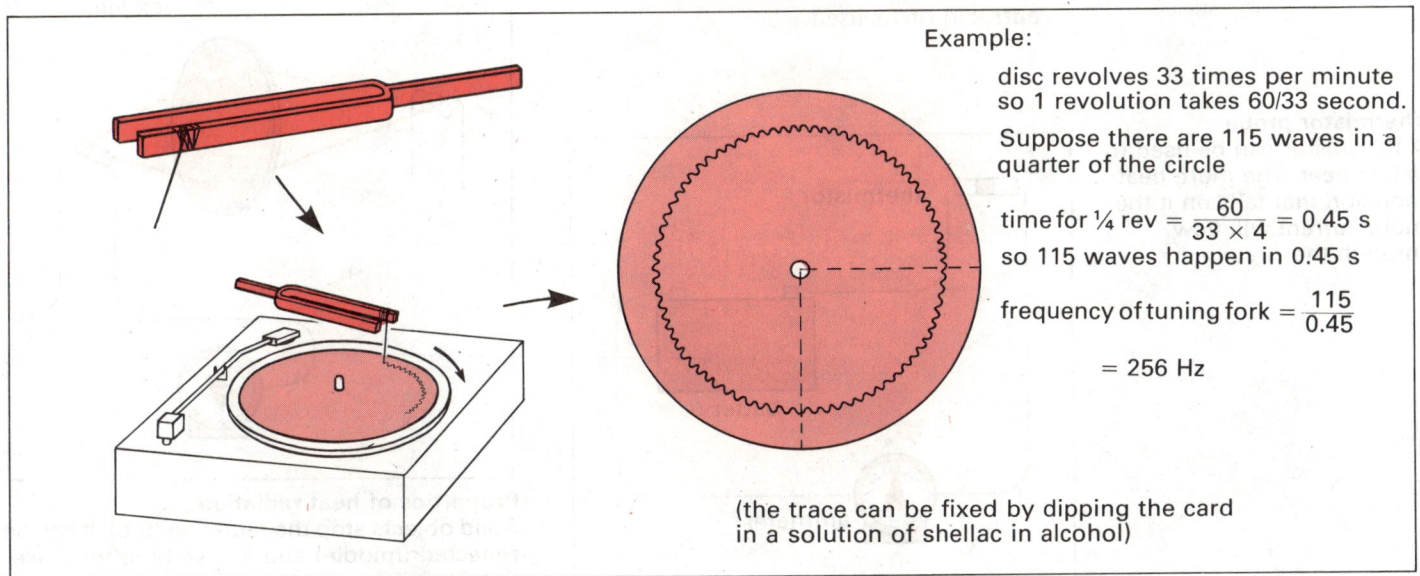

Example:

disc revolves 33 times per minute so 1 revolution takes 60/33 second.

Suppose there are 115 waves in a quarter of the circle

time for ¼ rev = $\frac{60}{33 \times 4}$ = 0.45 s

so 115 waves happen in 0.45 s

frequency of tuning fork = $\frac{115}{0.45}$

= 256 Hz

(the trace can be fixed by dipping the card in a solution of shellac in alcohol)

Radiation

72

The sun is 92 million miles from the earth. Except for a few miles of air round the earth, there is nothing between the sun and the earth to conduct or convect heat. But heat does reach us and in the sun's rays you feel warm and may get sunburnt. This heat reaches us by radiation. Some other forms of radiant energy are light, X-rays and radio waves.

Radiant heat can be felt in front of an electric fire. If a board or sheet of metal is put between the fire and your body, the heat does not reach you. If the screen only partly covers the fire, you feel the radiant heat only on that part of the body that is not in the shadow.

The properties of radiant heat can be shown experimentally. A suitable heat source can be an electric fire element or an infra-red lamp, or a cylinder of gauze made red hot by a bunsen burner flame inside it.

As a probe to detect heat, a thermistor is very useful. This is a special kind of transistor. Its electrical resistance becomes less as the temperature rises. So if it is connected in a circuit with a battery and an ammeter, the current shown on the ammeter will give a measure of how hot the thermistor is.

The diagrams show some experiments:

1 This shows that radiant heat casts a shadow; B is cold but A is hot. So heat travels in a straight line.

2 Radiant heat can be reflected; as with light, the angle of reflection is the same as the angle of incidence (see Topic 74).

3 Using a magnifying glass (convex lens) a point F can be found at which the heat after passing through the lens seems to be concentrated or focused.

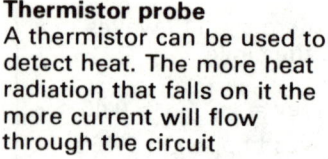

Infra-red lamp

Thermistor probe
A thermistor can be used to detect heat. The more heat radiation that falls on it the more current will flow through the circuit

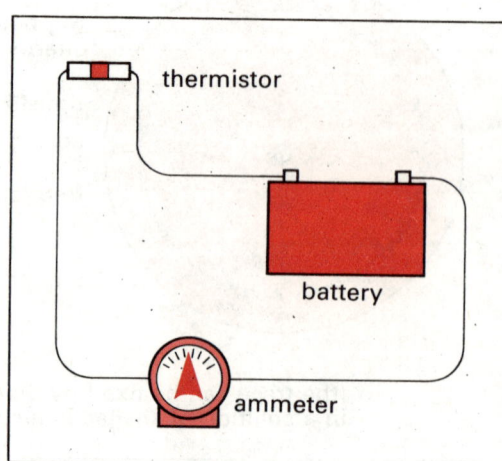

thermistor

battery

ammeter

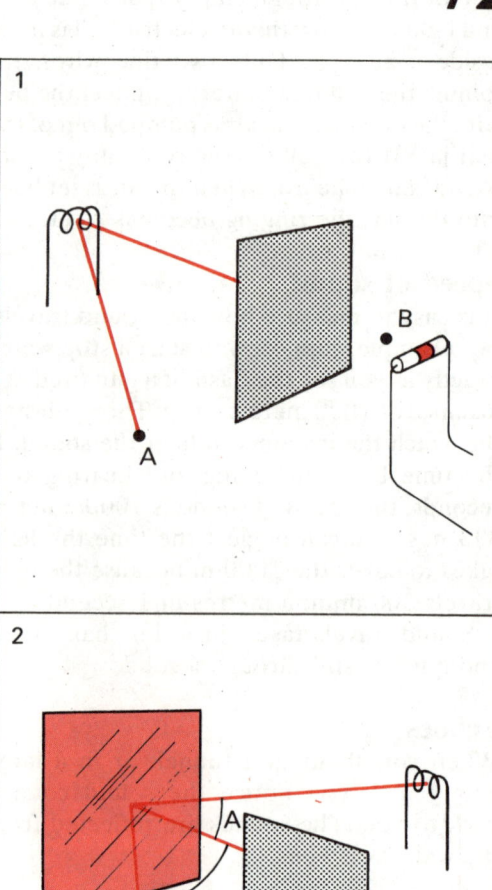

A

B

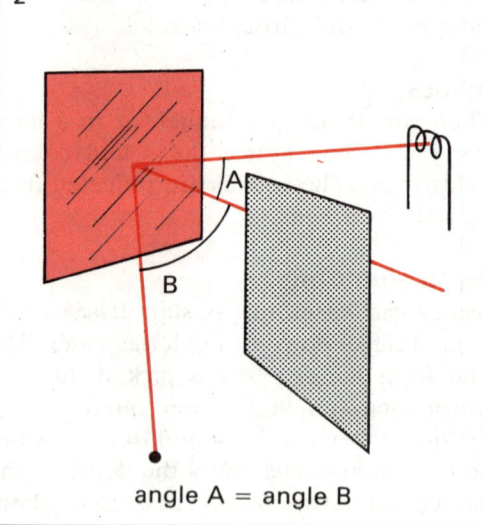

A

B

angle A = angle B

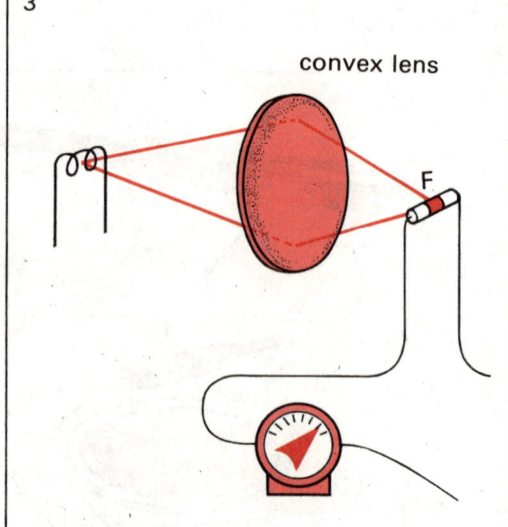

convex lens

F

Properties of heat radiation
Solid objects stop the radiation (top). It can be reflected (middle) and focused (bottom) like light

Passing through space

Radiation does not give up its energy while passing through space. Heat is found only when the radiant energy falls on an object. In fact, the upper atmosphere on which the heat first falls is very cold, probably because the air is so 'thin'. When radiant heat reaches the earth, the heat is absorbed.

Absorption of heat

The amount of heat energy taken up depends on the kind of surface the object has.

Two similar squares of tin are supported vertically at the same distance from the heat source: **A** is shiny; **B** is blackened. Drawing pins are fixed on the back of the squares with candle wax.

Or the heat can be picked up by two cans of water (same amount in each can) with the can **A** shiny, but **B** blackened. In both cases the blackened surface absorbs heat more quickly than the polished one.

Giving out heat

You might try to find out how the nature of the surface affects the giving out of heat. But cans of hot water radiate heat only very slowly. So any difference noticed would not give conclusive results.

If the hot object is really very hot, it can be shown that it gives out heat more quickly if it is dull than if it is polished.

So-called 'radiators' (which convect heat more than they radiate) give off more radiant heat when painted dull black.

Heat is absorbed less by light-coloured than by dark-coloured surfaces. Light coloured clothing is worn in summer and in hot countries. Sometimes buildings are whitewashed to keep them cool.

The vacuum flask shows very well the various ways in which heat travels. The flask is designed to avoid loss of heat from a hot liquid put inside. It can also keep a cold liquid from getting hot.

Frost on a clear cloudless night

It may seem strange that frost is more likely to form on a clear night than on a cloudy one. The earth receives radiation from the sun in daytime, but at night it radiates heat itself. If there are no clouds to reflect back this radiant heat, the earth's surface becomes cold enough for water vapour to condense and form dew or frost.

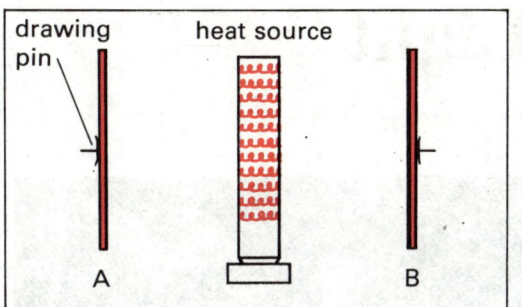

Absorption of heat
The black surface absorbs more heat than the polished one

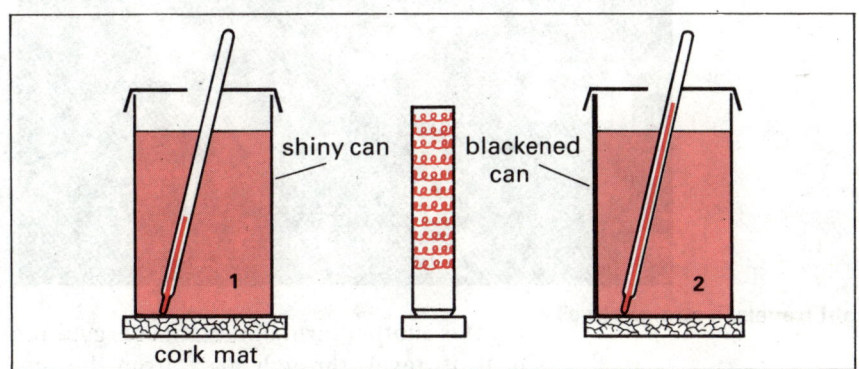

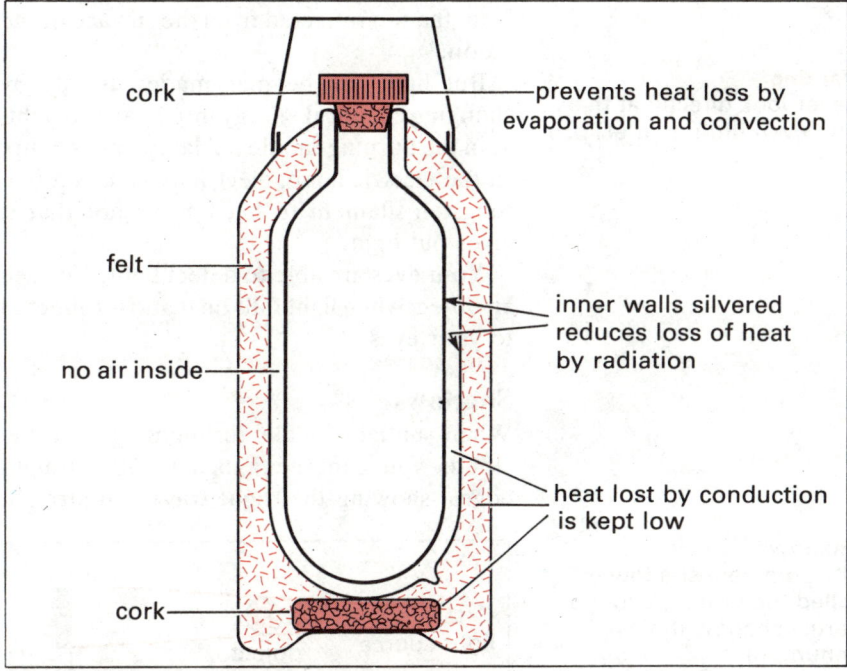

cork — prevents heat loss by evaporation and convection

felt

no air inside

inner walls silvered reduces loss of heat by radiation

heat lost by conduction is kept low

cork

Vacuum flask

1 In the absorption of heat experiment do you think it unfair that can **2** is painted, but can **1** is not? What would happen if can **1** were painted with white paint?
2 When finding out how the cans of water absorb heat why are the cans given cardboard lids and why do they stand on cork mats.
3 A can of hot water put on a table loses heat in four different ways. What are these ways?

Light

73

Light travels in straight lines

Warning
Never look directly at the sun, even during an eclipse

Light is another form of radiant energy. Like heat, it travels through space from the sun, reaching us as daylight. Moonlight is the light from the sun reflected from the surface of the moon.

But light can be man-made; usually by changing chemical energy into heat and light, as in the burning candle, oil lamp or gas lamp. In the electric lamp, electricity causes a fine tungsten filament to become so hot that it gives out light.

Your eyes are able to detect light. You see an object when light falls on it and is reflected to your eyes.

Shadows

When sunlight breaks through a gap in the clouds you can see a beam with straight edges, showing that light travels in straight lines. Light casts shadows with sharp edges provided the source of light is small or far away.

Light from a lamp is not always from a concentrated point, as in a projector. If the light is from a larger lamp the shadow cast by an object is dark at the centre (**umbra**), but only partly dark (**penumbra**) outside this central region. The kind of shadow formed on a screen depends on the size of the lamp and the size of the object.

Eclipses

When the moon moves into a position directly in line with the earth and the sun, there is an eclipse of the sun (solar eclipse). If you were at point A on the earth's surface, the sun would be completely blotted out. This is a total eclipse. At point B you would see only a partial eclipse. The next total eclipse for viewers in Britain will be in 1999; it will be total for people in Cornwall.

It sometimes happens that the moon is further than usual from the earth and its umbral shadow does not reach the earth. If you were standing on the earth at C you would see a dark moon surrounded by sun. This is an annular eclipse.

An eclipse of the moon (lunar eclipse) happens when the moon passes into the shadow of the earth.

Shadows
The complete shadow is called the umbra, and the partial shadow the penumbra

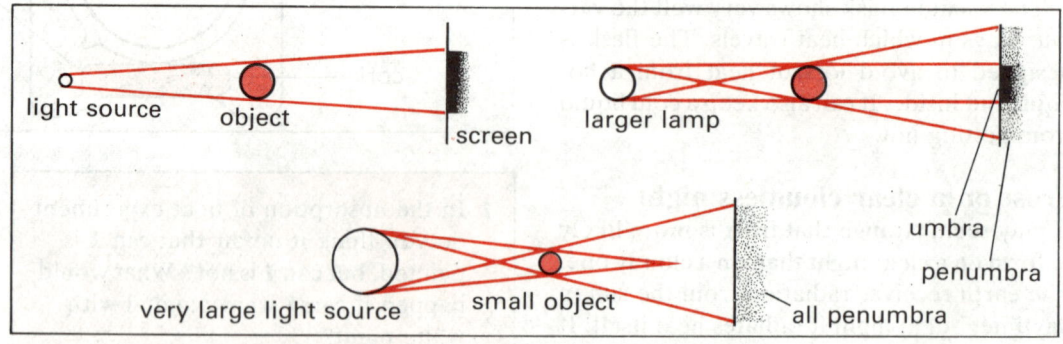

Solar eclipse

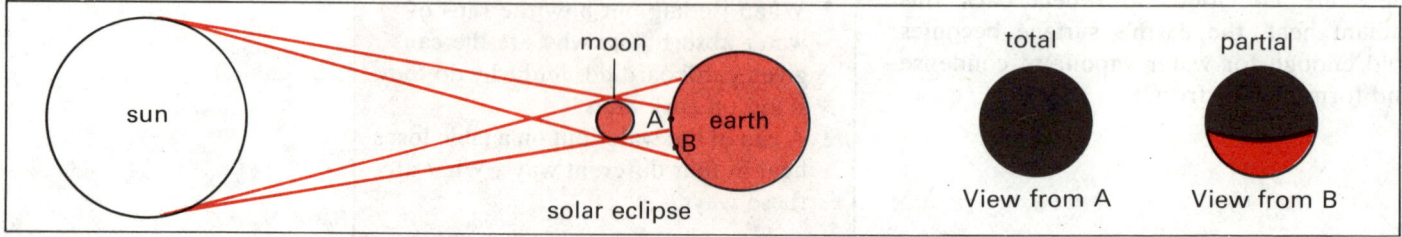

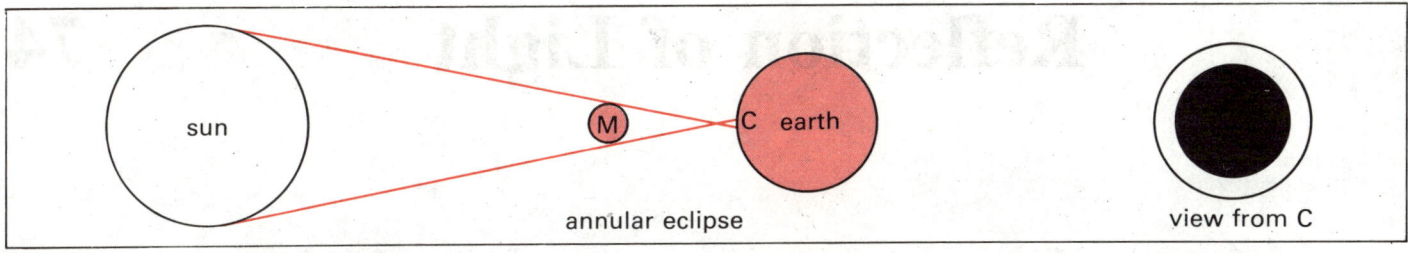

annular eclipse

view from C

Annular eclipse

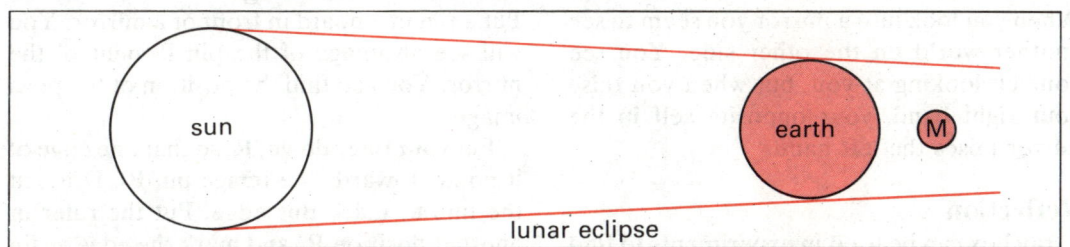

lunar eclipse

Lunar eclipse

Pin-hole camera

A simple camera can be made from a box with a pin-hole at one end. The far end is cut away and a greaseproof paper screen put in its place. Light from different points of a distant object passes through the pin-hole and forms a small inverted picture (image) on the paper.

A better camera can be made from two cardboard tubes one sliding in the other. A more elaborate one could be a wooden box painted dull black inside.

To take a photograph on a photographic plate, the plate would have to be put in place of the paper screen. Several minutes exposure would be needed even in bright daylight.

Speed of light

The speed of light is about 186 000 miles/second or 300 000 kilometres/second! That means it is instantaneous as far as all practical purposes are concerned, although it becomes important in these days of space flights. The light from Jupiter, for example, takes 43 minutes to reach the earth. Light from the nearest star after the sun takes 4.3 years to reach us.

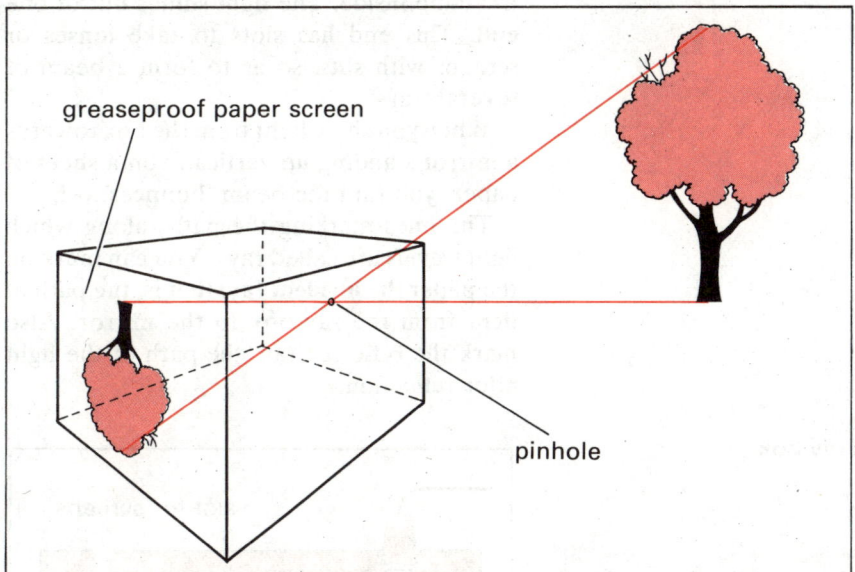

greaseproof paper screen

pinhole

Pin hole cameras

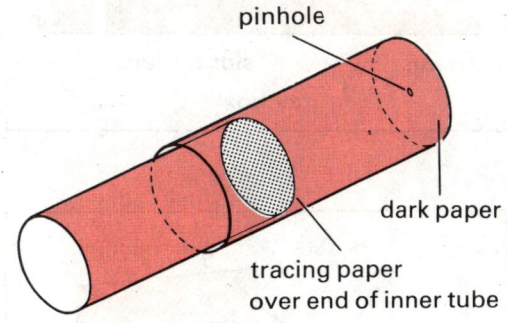

pinhole

dark paper

tracing paper over end of inner tube

1 In what ways is light different from sound?
2 In the pin-hole camera what will be the effect on (i) the amount of light entering the camera and (ii) the sharpness of the picture formed, if the pin-hole is made larger?

3 The shadow of a boy, 2.5 m tall, cast by a street lamp is 5 m long. The lamp is 5 m above ground level. Draw a diagram and find out how far the boy is from the foot of the lamp-post.

Mirrors

When you look into a mirror you seem to see another world on the other side. You see yourself looking at you, but when you raise your right hand, your opposite self in the mirror raises the left hand.

Reflection

A ray box can be used in experiments to find out about reflection. It is a box with an electric lamp inside. The light shines out of one end. This end has slots to take lenses or screens with slits, so as to form a beam of several 'rays'.

When you shine light from the box towards a mirror standing up vertically on a sheet of paper, you find the beam 'bounces' off.

The lines marking the paths along which light travels are called rays. You can mark on the paper the **incident** ray, that is, the path of light from the ray box to the mirror. Also mark the **reflected** ray, the path of the light after reflection.

A ray box

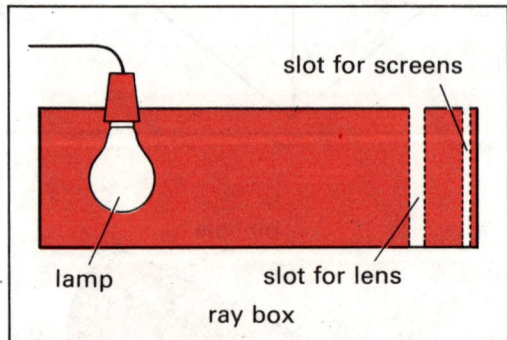

ray box

slot for screens

lamp

slot for lens

Reflection
A ray of light bounces off the mirror at the same angle as it arrived

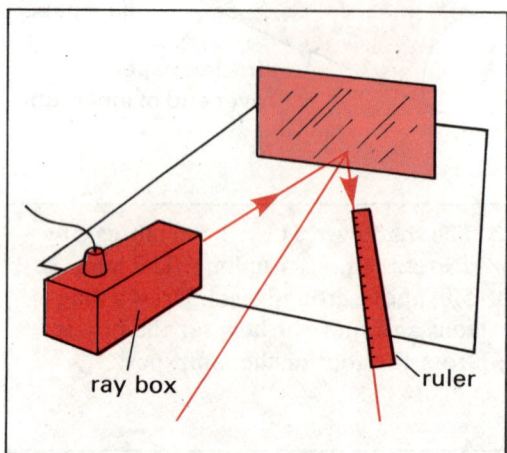

ray box

ruler

Position of the image

Put a pin in a board in front of a mirror. You will see an image of the pin P 'behind' the mirror. You can find the position of the pin's image.

Put your ruler down, R, so that one edge of it points towards the image pin P'. Draw in the line to mark this edge. Put the ruler in another position R' and mark the edge again when pointing at the image. Take away the mirror and continue the two lines you have drawn until they meet. This will be the position of the image pin.

You can check that the image of the pin is as far behind the mirror as the real pin is in front. Also if you join the position of the pin to the position of the image, the line makes a right angle with the mirror.

Draw in a line perpendicular to the mirror; this is called the **normal**. The angle between the normal and the reflected ray, is called the **angle of reflection**. The angle between the normal and the incident ray is called the **angle of incidence**. Measurement shows that the angle of reflection is equal to the angle of incidence. This is true whatever the angle of incidence. It is the first law of reflection.

The second law of reflection states that the incident ray, the normal and the reflected ray all lie in the same plane. After all, you drew them all on the flat piece of paper.

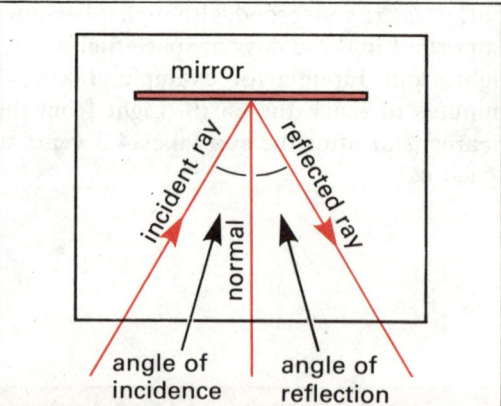

mirror

incident ray

reflected ray

normal

angle of incidence

angle of reflection

Normal
The normal is the line between the incident and reflected ray perpendicular to the mirror

1 If you move a mirror through 30°, through what angle does the reflected ray from an object move?

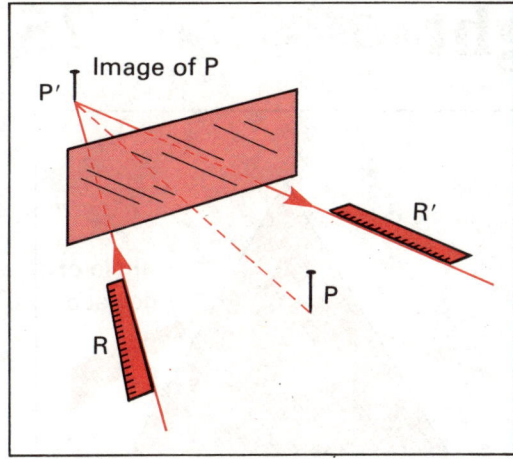

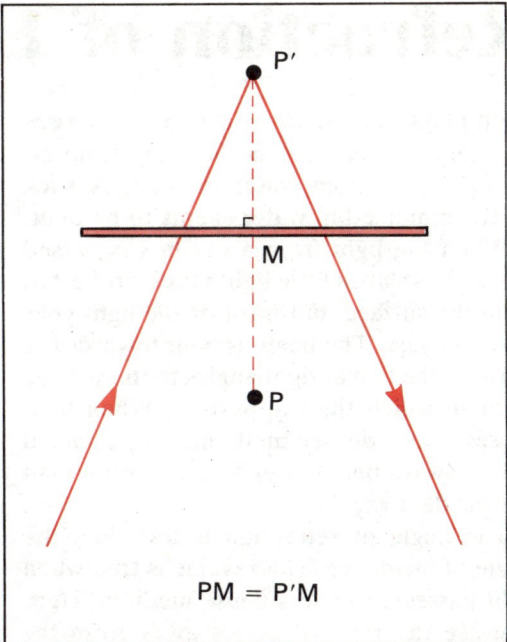

Image
The image appears to be behind the mirror. You will find it is as far behind the mirror as the real pin is in front

$$PM = P'M$$

Diffuse reflection

Light is reflected from all ordinary things, not only mirrors and bright metal surfaces. You may have noticed that when you are reading a book in bright sunlight, there is a position of the book that is most uncomfortable because light is reflected so strongly that you are dazzled. But light is also reflected from the book in other positions. You see objects because light is reflected by them. You cannot see anything in a completely dark room.

Periscope

You may like to make a periscope. You could use a cardboard tube. Set two mirrors, as shown, at 45° to the sides of the tube.

This is a useful device to let you see over the heads of people in front of you, when you are at a football match or a golf match.

Curved mirrors

In and around most homes you can find curved mirrors. A convex mirror (which bulges out) is often used as a driving mirror. It gives a small picture of a wide field of view. A concave mirror (which 'caves in') is used as a shaving mirror. It gives an enlarged image of a chin placed close to it.

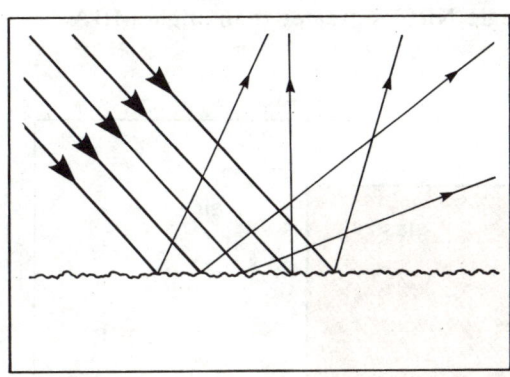

Diffuse reflection

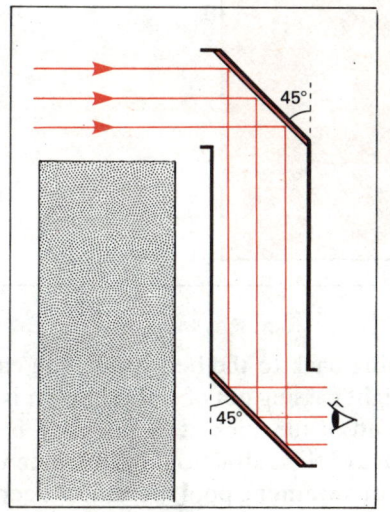

Periscope

2 A man, 6 ft tall, looks into a long mirror. What is the smallest length that the mirror can be for him to see the top of his head and his feet while standing still in front of the mirror?

3 What kind of mirror would be best (a) for a dentist to use when looking behind the teeth in your mouth; (b) in a large store to enable the store detective to watch customers in all parts of the store; (c) as a reflector behind the electric bulb in a torch or a car head-lamp?

Refraction of Light

75

Light plays some funny tricks when it passes from one medium to another, e.g. from air into glass, or from water into air. A stick partly immersed in water seems to be bent.

When the light from a ray-box is passed into a glass slab, a little light may be reflected from the surface, but most of the light goes into the glass. The beam is bent towards the normal (the line at right angles to the surface through which the ray passes). When light passes into a denser medium, the refracted ray is always bent closer to the normal than the incident ray.

The angle of refraction is less than the angle of incidence. The reverse is true when light passes into a less dense medium. Here you see the ray AB bends away from the normal BN, when it goes out of glass into air. Angle NBC is greater than angle MBA.

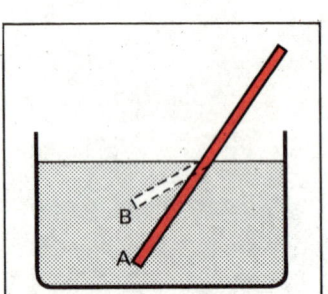

A stick half immersed in water appears to be bent

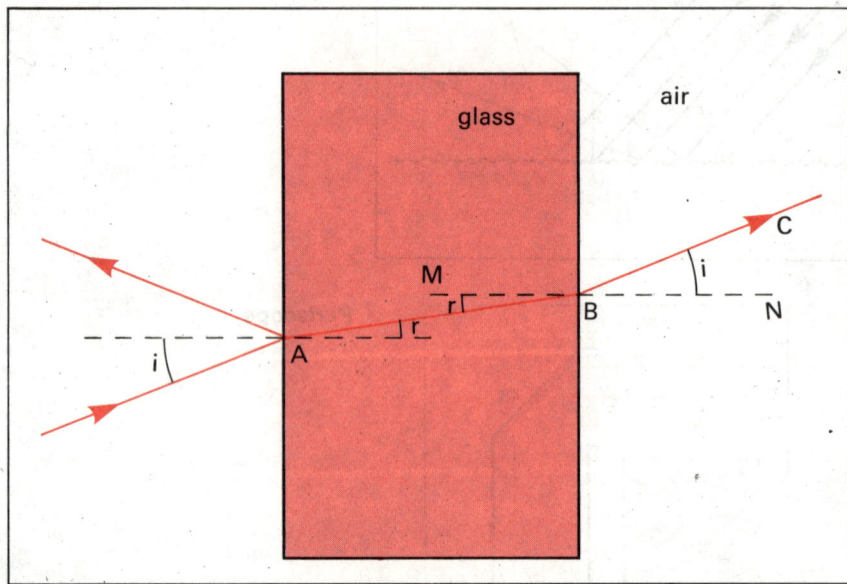

Refraction
A ray of light passing into glass is bent towards the normal

Going back to the bent stick, you can see that light passing out of water into air is bent. The end of the stick at A seems to be at B because of this refraction. In the same way, a pond or swimming pool seems less deep than it actually is.

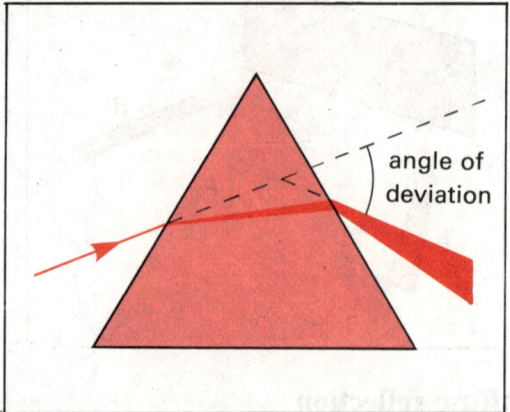

Prism
A prism bends light through the angle of deviation

Passing through a prism

In a prism, the opposite sides of the glass are not parallel. Light passing through is turned through an angle. This is because of two refractions. Coloured edges can be noticed on the beam that comes out of the prism (see Topic 76). The beam inside the prism is also faintly coloured.

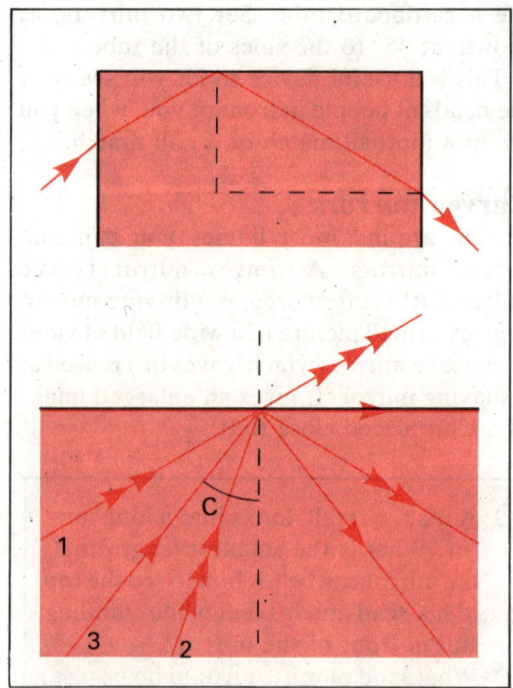

Internal reflection
If the light ray is at too steep an angle to glass it will all be reflected and not refracted. At smaller angles only some light is reflected

Internal reflection

You may have noticed, when shining a light beam through the glass block, that you can have a beam not passing out of the glass. Instead, the beam is reflected back into the glass. The light is **internally reflected** at the glass-air surface.

Critical angle. Rays of light pass through glass towards the air outside the glass. Ray 1 is internally reflected. No light passes out. For ray 2 the incident angle is so small that most of the light passes out of the glass. There is an angle C for which ray 3 will neither pass out nor be reflected internally. The angle C is called the **critical angle**. For glass to air, the angle is about 39°.

Prisms and mirrors

A prism with angles of 90°, 45° and 45° can be used to reflect light. Light can be reflected through a right angle (see diagram). This is because the beam falling on the hypotenuse side is incident at 45°. But 45° is greater than the critical angle, 39°, so the light is all internally reflected. Light can be turned back through 180° (see diagram). In this case, the object viewed is seen upside-down.

Prisms are better for reflection than the usual mirrors. You may have noticed when using a mirror that although some light is reflected at the surface of the glass, most is reflected at the silvered surface at the back only after it has been refracted through the thickness of glass. When an object, such as a candle flame, is viewed at a glancing angle in a thick glass mirror, several images can be seen.

Mirages

On a hot day, the air close to a hot road becomes hotter and less dense than that above it. So light from the distance is bent more and more as it comes near the ground and meets hotter air.

An observer near the ground receives light which seems to come from the ground, X. But the light comes from a point in the sky at Y. So the sky appears on the surface of the road ahead. Since he does not expect sky on road, he thinks of the bright patch as being water. No doubt you have seen this 'water' on the road ahead of you on a hot day. It is a mirage. This is how a traveller in the desert sometimes thinks he can see an oasis in the distance.

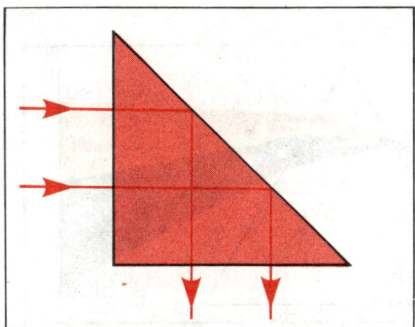
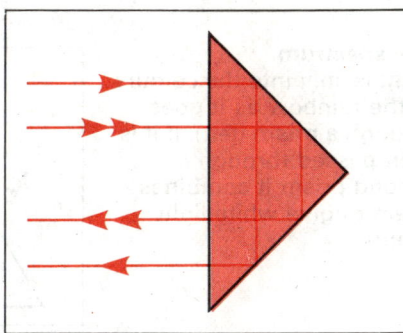

Prism
Light rays bent through 90° (left) and 180° (right)

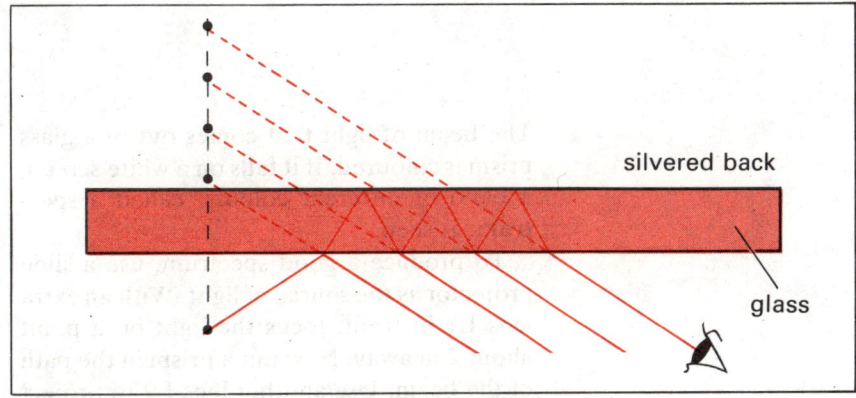

silvered back

glass

Multiple image
A glass mirror has the rear of the glass silvered. Because some light is also reflected from the front of the glass faint multiple images are seen

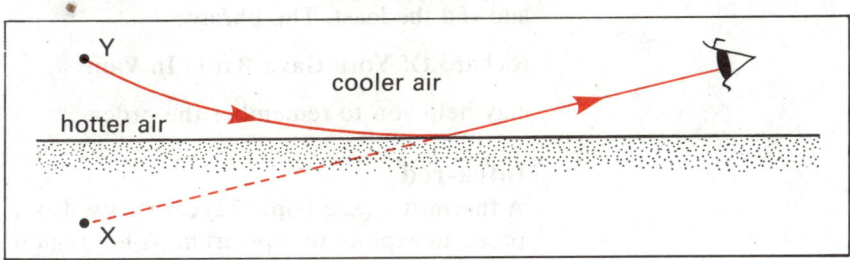

Y

cooler air

hotter air

X

Mirage
On a hot day rays of light from the sky are bent and appear to be coming from the ground

1 In the diagram, the eye at E cannot see the coin C. What will be the effect of pouring water into the vessel? Draw a diagram to show your answer.
2 When I look out of the window of my room I seem to see a television set in the branches of the tree outside. Can you explain how this can happen?
3 Look at the illustration of multiple images, and try to decide why the second image is the brightest, not the first.

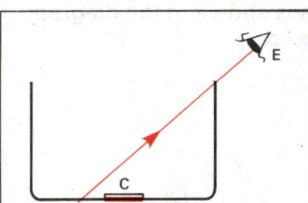

Colour

76

The spectrum
Light is split into the colours of the rainbow as it goes through a prism (left). If it is then passed through a second prism it combines again to give white light (right)

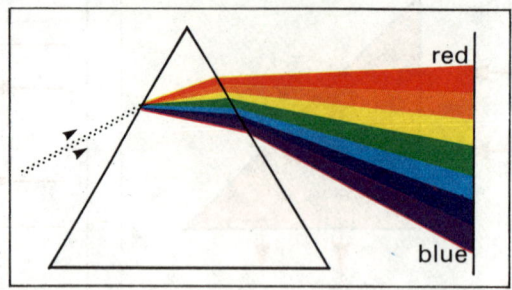

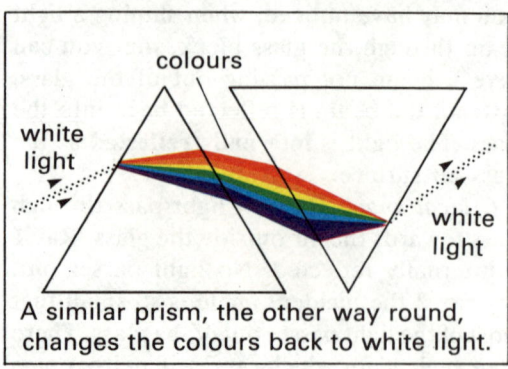

A similar prism, the other way round, changes the colours back to white light.

The beam of light that comes out of a glass prism is coloured. If it falls on a white screen, a band of different colours, called a **spectrum**, is seen.

To produce a good spectrum, use a slide projector as the source of light. With an extra lens L1 in front, focus the light on a point about 2 m away. Now put a prism in the path of the beam. Use another lens L2 to project the colours on to a white screen.

The colours are red, orange, yellow, green, blue, indigo, violet. Violet is refracted most and red the least. The phrase:

Richard Of York Gave Battle In Vain

may help you to remember the order.

Infra-red

A thermister (see Topic 72) can be used as a probe to explore the spectrum. A hot region is found just beyond the red. This is due to invisible infra-red radiation reaching this region.

Ultra-violet

Just beyond the violet end is another region of invisible radiation. This is ultra-violet. Although we cannot see it, a photographic film is very sensitive to ultra-violet radiation.

Coloured lights: additive mixing

When beams of red, green and blue lights overlap other colours are produced. These colours can be seen at the start of some television programmes. Since red, green and blue all mixed together make white light, these colours are called primary colours.

The colour of an object depends on what kind of light falls on it and what happens to this light. Some things are transparent and allow light to pass through. A piece of red glass allows red light only to pass through. All other colours are absorbed by the glass.

A red book appears red in red light, but looks black if only blue light falls on it. The blue light is absorbed. The red book only reflects red light (and perhaps some orange).

The spectrum
The best spectrum of colours is seen if the arrangement of lenses shown in the diagram is used

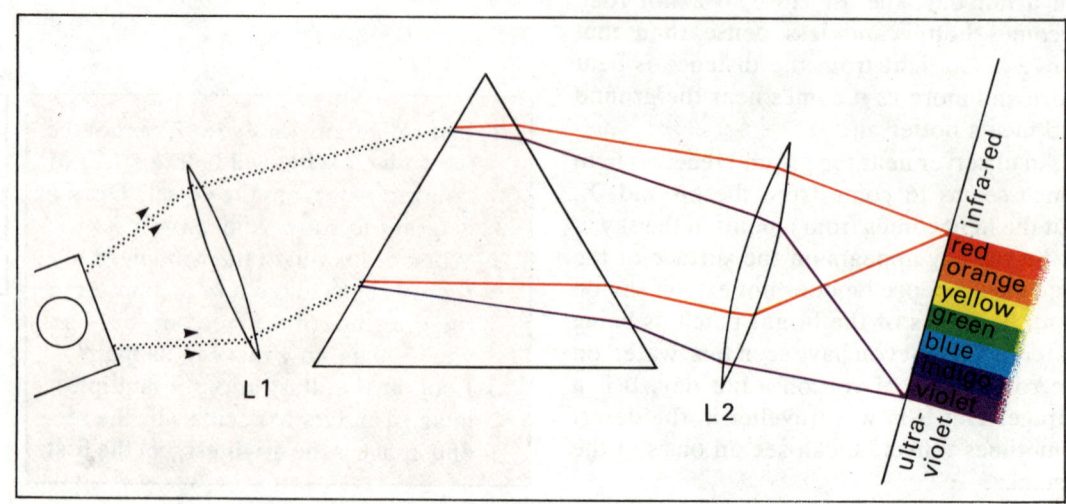

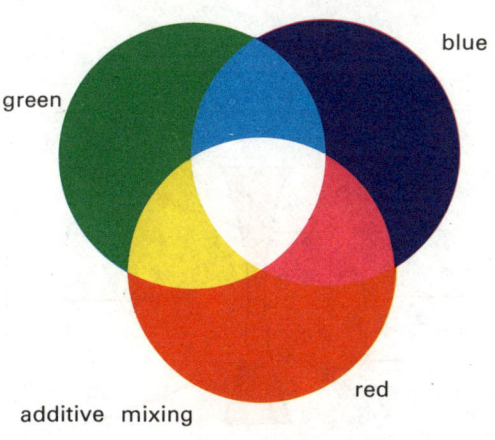

green · blue · additive mixing · red

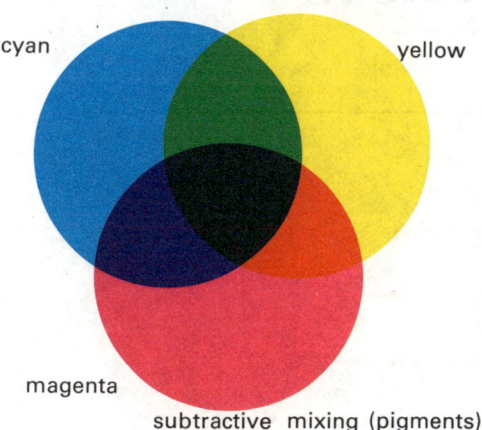

cyan · yellow · magenta · subtractive mixing (pigments)

Pigments and paints

A red pigment seen in white light absorbs blue, green and yellow. It reflects red and to some extent, orange.

A yellow pigment absorbs blue, green and red. It reflects yellow with a little green and a little orange.

When red and yellow pigments are mixed, they will, between them, absorb all colours except orange. So mixing pigments is a case of **subtractive** mixing, whereas mixing coloured lights is **additive** mixing.

Note that mixing several different pigments produces black not white.

Fluorescent lighting

When electricity at a high voltage is passed through a tube in which there is a gas under low pressure, a bright glow is produced. If the gas inside is mercury vapour, the glow is rich in ultra-violet radiation. Mercury lamps are sometimes used in street lighting. (It is worth noticing that in this bluish light, red colours, such as that of lipstick, seem quite black!)

Fluorescent lighting

Some substances, called phosphors, glow brightly when put in ultra-violet radiation. These substances change ultra-violet radiation into visible light. You can see this effect if you have an ultra-violet lamp, by putting vaseline, or crystals of quinine sulphate under the lamp.

A small amount of a phosphor is sometimes added to detergent powders. The clothes washed in the detergent, appear quite bright when held in daylight, which contains enough ultra-violet radiation to cause the fluorescence.

The familiar long fluorescent lamps used in shops and homes, are discharge lamps in which ultra-violet radiation falls on a mixture of phosphors lining the tube. The substances used can be chosen to produce either light similar to daylight or slightly tinted light.

Warning
Ultra-violet can be dangerous to the eyes

1 Write down the colours of the spectrum in order, starting with the colour that is refracted most. Where would you put infra-red in your sequence?

2 A pop singer has a red shirt and blue trousers. How will he appear if the stage on which he is standing is lit by (i) red light only; (ii) blue light only?

3 What colours are produced when (a) red and blue lights are mixed, (b) yellow and blue paints are mixed.

4 When you next have the chance of seeing a rainbow, find out if the colours in it are the same as those in the spectrum. Also what colour is lowest in the bow? If a secondary rainbow is visible, find out if the colours are the same as in the primary bow and in the same order.

Lenses

77

Convex lens (left)

Concave lens (right)

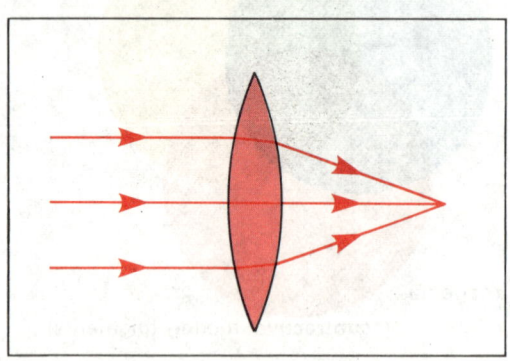

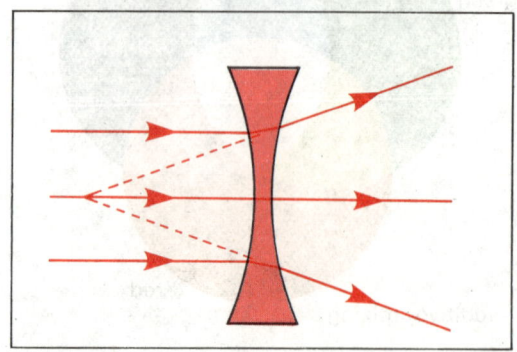

Some interesting results are found when light undergoes refraction by passing through a lens. A ray box can be used to study this.

A convex lens is one that bulges in the middle. It makes rays of light converge towards a point known as the focus.

A concave lens is one that 'caves in' in the middle. It makes rays of light spread out or diverge.

But what happens when you look through these lenses?

Convex lens

Hold a magnifying glass (convex lens) at arm's length and look through it at something in the distance: houses, trees or clouds. The scene will seem to be small and upside down.

You can form this small inverted picture (image) on a piece of paper. Hold the paper upright on the other side of the lens and move the lens back and forth until the picture is quite clear.

If the object viewed is a great distance away, the distance between lens and paper will be the focal length of the lens. This is a

good way of finding the focal length of a convex lens.

This may remind you of showing your friends the slides of your holidays, using a projector. In this case you put a small photograph upside down in the projector and a larger picture the right way up forms on the screen, some distance away. But a convex lens can be used in a different way. If it is brought near some print it acts as a magnifying glass. A large upright image of the printing is seen.

You cannot form this kind of image on a screen. It is just an image that you seem to see. It is a **virtual image**. The image that you see in a mirror is also a virtual image.

When the image can be formed on a screen as in the case of the projected picture, it is said to be a **real image**.

Concave lens

When you look through a concave lens, the view you see is always upright and smaller than the object. It is a virtual image.

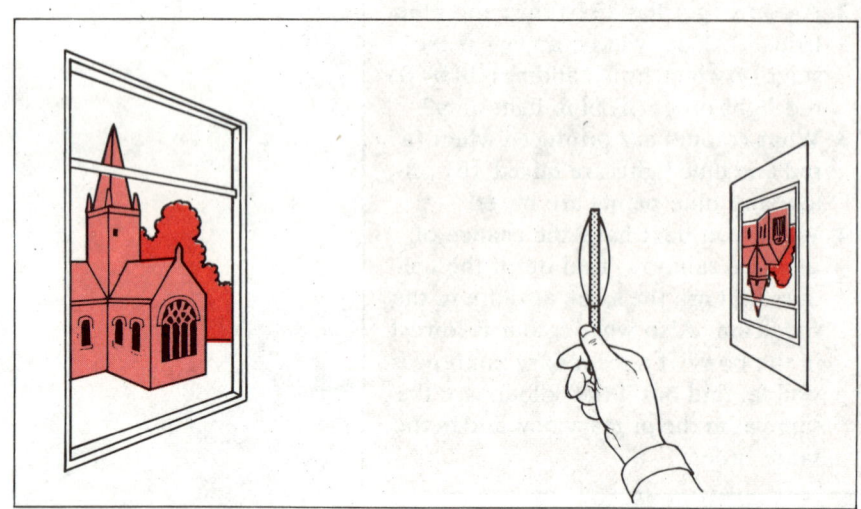

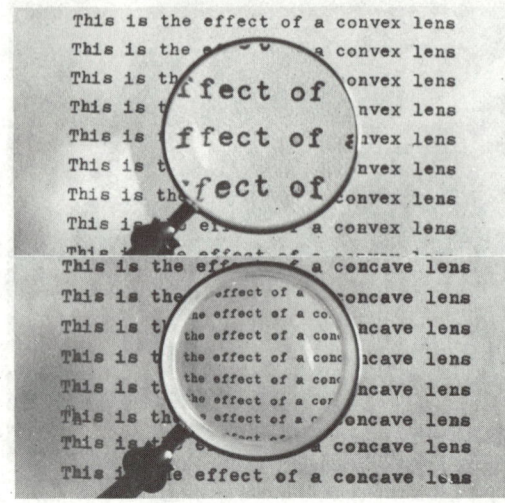

Ray diagrams

You can draw diagrams of the paths of light (rays) through lenses if you remember two important facts:

1 Light which is at first travelling parallel to the axis of the lens goes through the focus after passing through the lens, e.g. ray 1 shown in Diagram **1**.
2 Any ray of light passing through the centre of the lens passes straight on, e.g. ray 2.

In Diagram **1** the object (OB) is a distance further than the focal length away from the convex lens. It forms a real inverted image (IM). If the object is closer than the focal length to the same lens a virtual image is formed (Diagram **2**).

In Diagram **3** the concave lens forms a smaller upright virtual image of the object closer than the focal length.

Curved mirrors

Diagrams can also be drawn for reflection from curved mirrors. Here again, two rays are important:

1 A ray at first parallel to the axis, either passes through, or appears to come from the focus after reflection.
2 A ray passing through or coming from the centre of curvature C is reflected back along the same path.

Summary

There are similarities between lenses and curved mirrors.

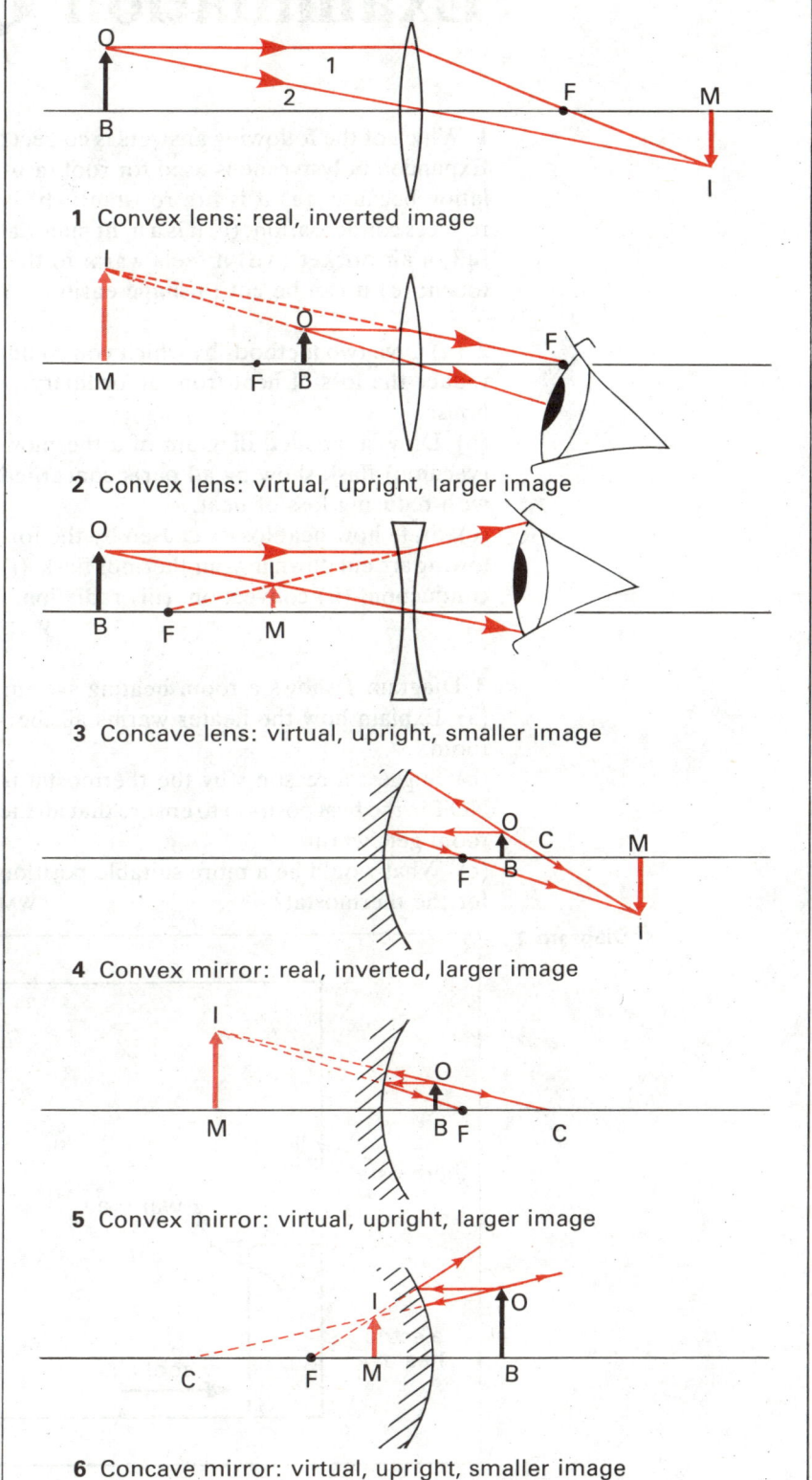

1 Convex lens: real, inverted image

2 Convex lens: virtual, upright, larger image

3 Concave lens: virtual, upright, smaller image

4 Convex mirror: real, inverted, larger image

5 Convex mirror: virtual, upright, larger image

6 Concave mirror: virtual, upright, smaller image

1 In which of the following instruments would you find a convex lens: microscope, barometer, telescope, cineprojector, field glasses, stethoscope, camera?
2 The image formed on a screen after light from an object has passed through a convex lens is sometimes fringed with red or blue at the edges. What would cause this effect?
3 Explain, with the aid of diagrams, how a convex lens can be used to form either a magnified picture or a smaller image of an object on a screen.

	Concave lens	**Convex mirror**
	Both give upright, virtual images, smaller than the object	
	Convex lens	**Concave mirror**
Object close	Both give upright, virtual enlarged images	
Object further away	Both give inverted, real images, larger or smaller depending on distance of object.	

171

Examination Questions VI

1 Which of the following answers is correct? Expanded polystyrene is used for roof insulation because: (a) it is fire resistant; (b) it reduces condensation; (c) it is a light material full of air pockets; (d) it feels warm to the touch; (e) it can be cut to shape easily. Y

2 (a) Give two methods by which you could reduce the loss of heat from an ordinary house.
(b) Draw a labelled diagram of a thermos (vacuum) flask showing all parts concerned with reducing loss of heat.
(c) State how heat losses caused by the following are cut down in your thermos flask. (i) conduction, (ii) convection, (iii) radiation. WYL

3 Diagram 1 shows a room heating system.
(a) Explain how the heater warms all the room.
(b) Suggest a reason why the thermostat is NOT in the best position to ensure that all the room gets warm.
(c) What would be a more suitable position for the thermostat? WM

5 Pick out the correct answer. A total eclipse of the Sun occurs on Earth when:
(a) it is the Vernal equinox;
(b) the Moon is directly between the Earth and the Sun;
(c) the Earth is directly between the Moon and the Sun;
(d) whenever there is a new Moon on the longest day;
(e) whenever there is a new Moon on the shortest day. WYL

6 A white screen is illuminated by beams of primary red, primary blue and primary green light of the same intensity. Where the beams overlap the resulting colour is: (a) magenta; (b) orange; (c) black; (d) peacock blue; (e) white. Which answer is correct? Y

7 Choose the correct answer and write down the sentence.
The ray of light striking the surface of a plane mirror at right angles is called the: (a) normal ray; (b) incident ray; (c) reflected ray; (d) refracted ray; (e) critical ray. EM

Diagram 1

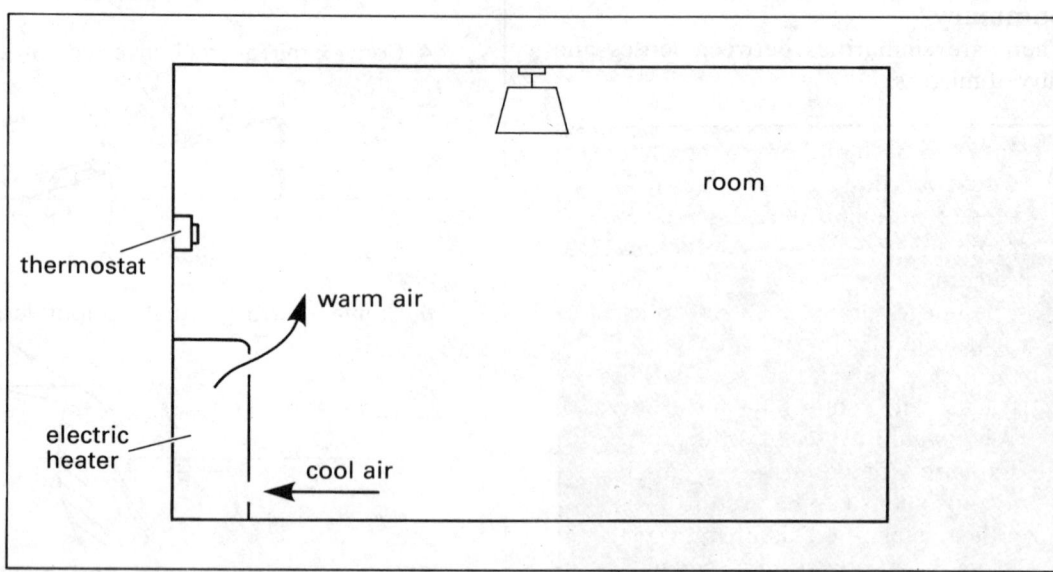

4 Which TWO of the following answers are correct? A concave lens (a) is a diverging lens; (b) is a converging lens; (c) may throw an image onto a screen; (d) may be used as a magnifying glass; (e) is used in microscopes; (f) cannot throw an image on a screen. EM

8 Choose the correct answer and write out the sentence.
Electric current is measured in: volts; amps; ohms; watts; joules. M

9 Diagram 2 shows the side view of a pin-hole camera.
(a) Draw TWO lines representing rays of light to show how an image of the tree could be formed on the screen.
(b) Draw a square about 2 cm edge. In it make a sketch to show what the tree would look like on the screen.
(c) What is the common fault of a home-made pinhole camera which causes the image to be blurred?   ALS

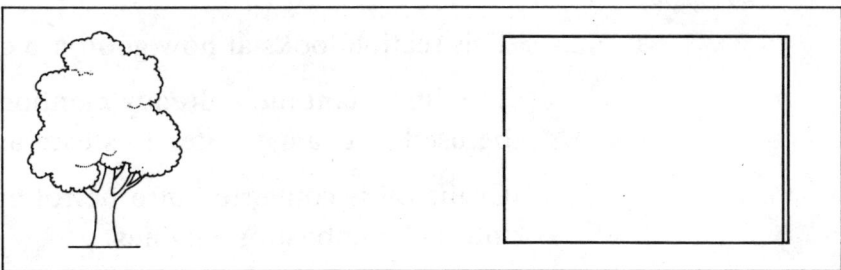

Diagram 2

10 A boy standing 1220 metres from a cliff fires a starting pistol. He hears the echo 8 second later. What is the speed of sound in air?   EM

11 In the circuit shown in Diagram 3, which instrument is represented by the symbol A? Copy the drawing, and: (i) mark with an X the position of a switch which would control both lamps C and B; (ii) mark with a Y the position of a switch which would control lamp B only.   Y

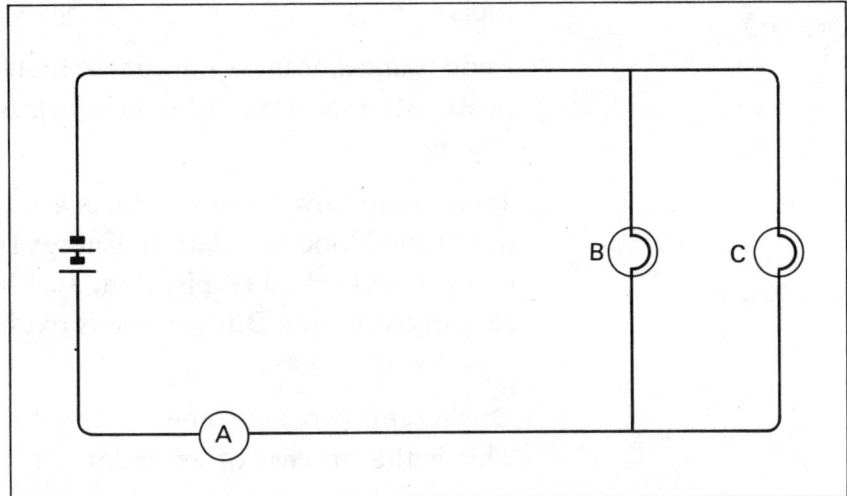

Diagram 3

12 Which of the following describes the sort of image you would see if you were using a convex mirror?
(a) It was the right way up but blurred in places. (b) It was the right way up but looked bigger. (c) It was upside down but blurred in places. (d) It was the right way up but looked smaller. (e) It was upside down and clear. EM

13 Complete the rays in Diagram 4 to show the paths of the rays of light.   EAS

Diagram 4

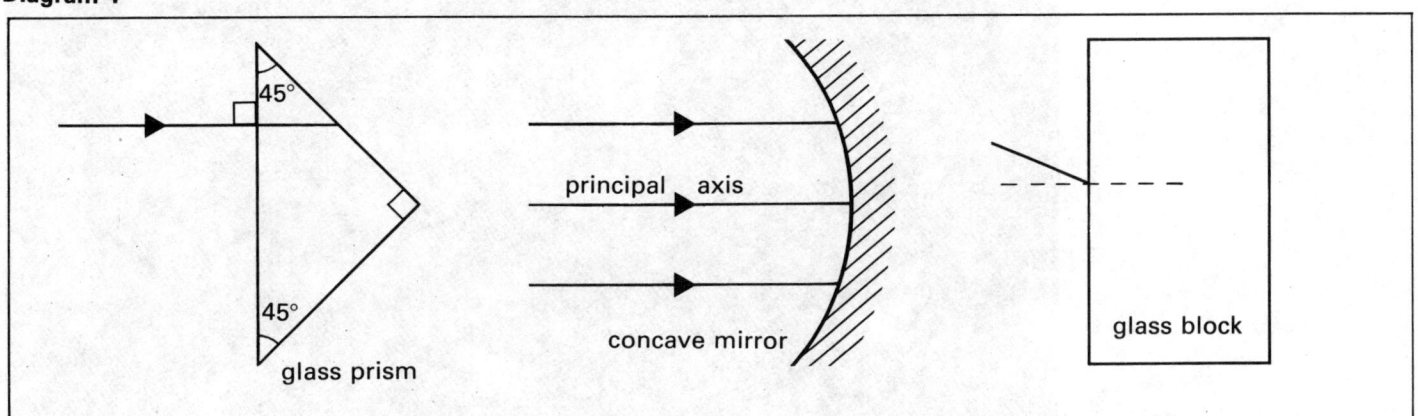

RELEASING ENERGY

This section looks at how good use can be made of the fuels available to us.

Two important fuels already mentioned in Topic 60 are coal and oil. Coal can be used to change water to steam and steam can drive steam engines.

Crude oil is converted into petrol and diesel oil which are used to drive internal combustion engines.

Coal gas and natural gas are handy fuels, because they can be taken in pipes, without any mess, to where they are needed. The laboratory burner uses these fuels.

Food is another most important fuel. As noted in Topics 63 and 64, food must be digested so that soluble products, such as glucose, are taken into the blood system.

To find out how these products are used in the body it is necessary to study blood and blood circulation. Energy from glucose is used in the cells of the body in a process called **respiration.** Just like the burning of petrol, respiration is a case of oxidation. But glucose is oxidized in the tissues of the body in a much less dramatic way.

During this process some waste products are formed and must be got rid of. This is the process of **excretion.**

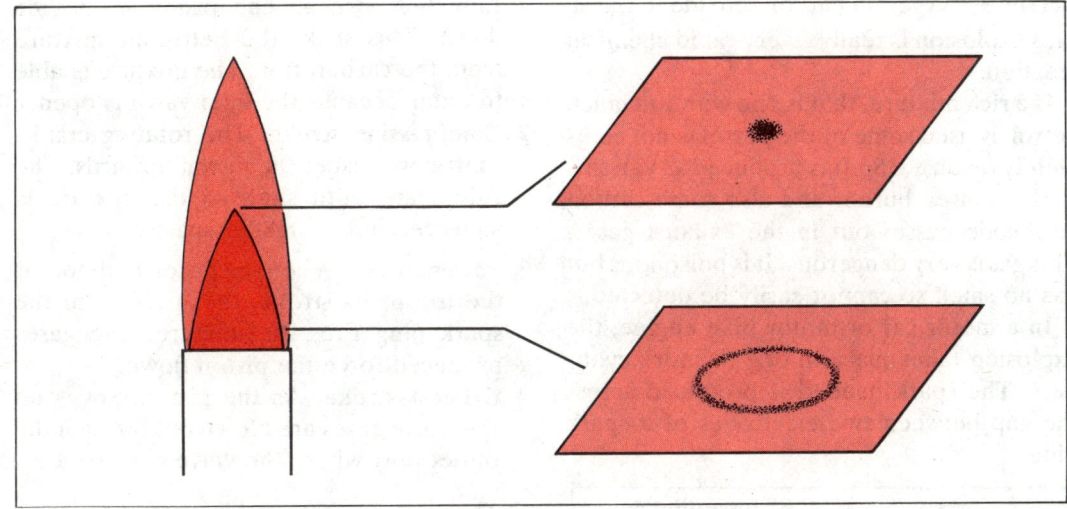

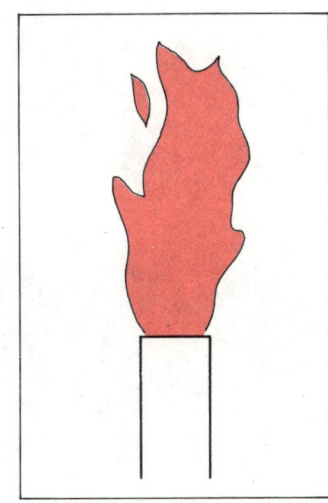

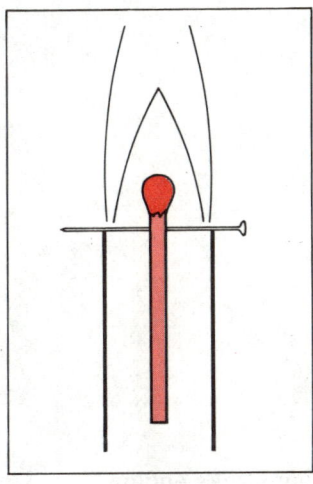

The bunsen burner

Methane (CH_4) is the main constituent in North Sea gas. When it burns it is oxidised:

$$CH_4 + 2O_2 \rightarrow CO_2 + 2H_2O$$

methane   oxygen   carbon dioxide   water

A gas or vapour will start burning when its temperature reaches its ignition point. Even a spark is sufficient to make methane burn. Carbon disulphide has such a low ignition point that even a hot glass rod put in the vapour will set it alight.

If you open the air inlet on the bunsen burner and light the gas, you see a hot flame with a distinct blue cone. Hold a piece of card horizontally in the flame for two seconds only. The charred marks show where the hot parts of the flame are. The hottest region is just above the tip of the blue cone.

Put a live match, with a pin pushed through it, at the top of the chimney of an unlit bunsen burner. Then light the gas. It is found that the match does not catch fire because the dark region inside the blue cone is unburnt gas.

When the air hole is closed, the flame is luminous, less vigorous and not so hot. If a cold piece of porcelain is put in this flame it is quickly covered with a deposit of fine soot.

The methane gas does not have enough air mixed with it to burn completely. Some of the carbon in the methane is left on the porcelain without being burnt.

Steam engines

If you think of railways and ships you can easily understand that steam engines have had a noble and romantic history. In a steam engine water is heated and changes to steam. The steam is under high pressure and it is used to force a piston backwards and forwards. This to-and-fro movement is changed into a rotary movement by the connecting rod and crankshaft.

It is possible to make a model that uses steam to drive a wheel. The rotor can be made from pieces of tin fixed into slots in a cork. The cork has a piece of glass tubing through it which fits over a wire axle. A jet of steam from a heated tin can with a small hole in the lid is directed at the rotor. This is a model of a steam **turbine**.

Bunsen burner
Finding the hottest area in a bunsen flame (above left). The match will not light because it is an area of unburnt gas (below). This flame is yellow and sooty because air has not been mixed with the gas (above)

1 Trace the sequence of energy changes that take place after coal is put into the fire-box of a steam railway engine.
2 A flame is luminous when tiny particles of unburnt carbon become white hot. Why is there a luminous flame when the air inlet of a bunsen burner is closed, but not when the air inlet is open?
3 Make a list of the fuels that are used in your own home.

Petrol is a mixture of hydrocarbons; one example is **octane**. A mixture of petrol and air explodes when ignited by a spark:

octane + oxygen → carbon dioxide + water

This explosion is really a very rapid chemical reaction.

If a rich mixture, that is one with too much petrol, is used some of the petrol is not completely oxidized. Soot is produced as you saw in the bunsen burner and also some carbon monoxide passes out in the exhaust gases. This gas is very dangerous. It is poisonous but has no smell so cannot easily be detected.

In a motor car or motor bike engine, the explosion takes place in one or more cylinders. The spark needed is produced across the gap between two electrodes of a spark plug.

Spark plug

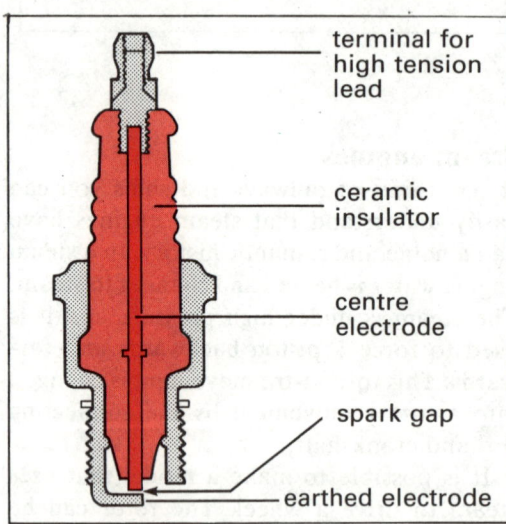

terminal for high tension lead

ceramic insulator

centre electrode

spark gap

earthed electrode

Four-stroke engine

What happens can be seen in four stages:

1 **Induction stroke.** The piston is moving down. This sucks the petrol-air mixture from the carburettor. The mixture is able to enter because the inlet valve is open.
2 **Compression stroke.** The rotating crankshaft now pushes the piston upwards. The valves are both shut, so the mixture is squeezed into a smaller space.
3 **Power stroke.** When the piston is almost at the top of its stroke, the spark from the spark plug fires the mixture. The gases produced force the piston down.
4 **Exhaust stroke.** As the piston moves up the waste gases are blown out through the outlet port where the valve has opened.

This is the way that a four-stroke engine works. The stages could be remembered as:

Suck in — Squeeze — Fire — Blow out

Most engines have more than one cylinder. This makes for a smooth continuous movement. The common crankshaft determines the order of firing. The valves open and close as required. This is done by a camshaft which links with the crankshaft.

In an air-cooled engine, there are fins on the outside of the cylinder to encourage loss of heat. In the usual car engine, water circulates around the engine block (see Topic 48). Heat is convected upwards and passes into the radiator where a fan assists the cooling process.

Four-stroke engine

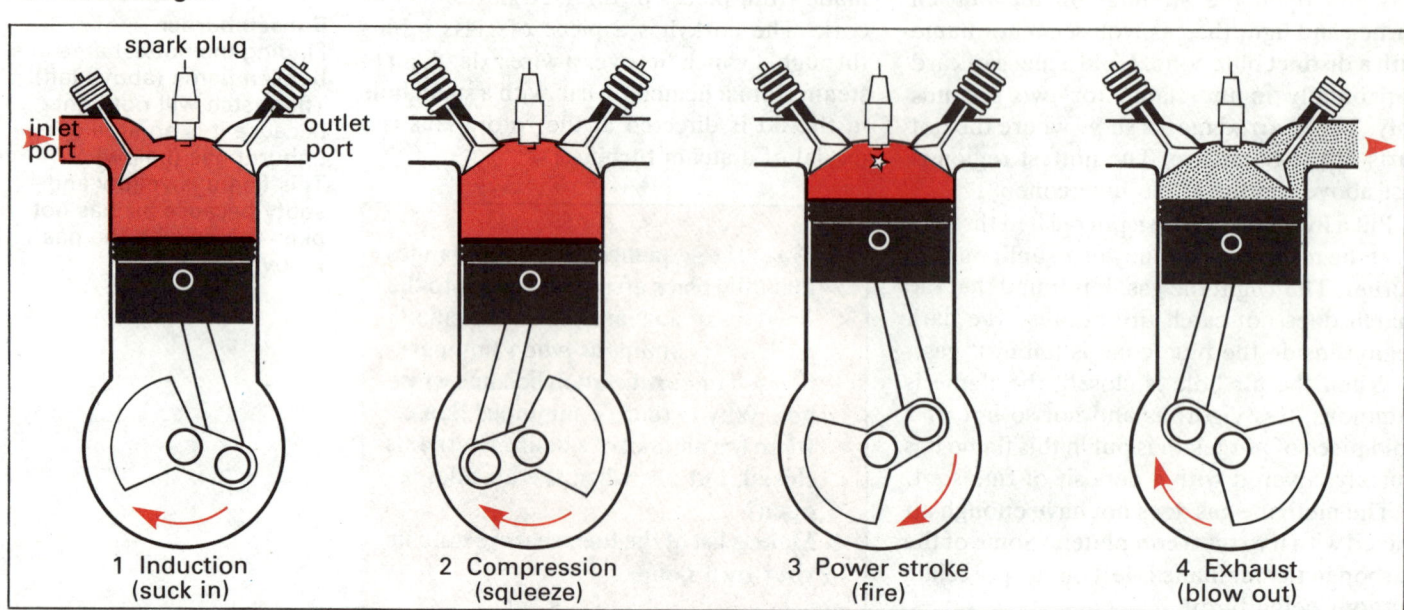

spark plug

inlet port

outlet port

1 Induction (suck in)

2 Compression (squeeze)

3 Power stroke (fire)

4 Exhaust (blow out)

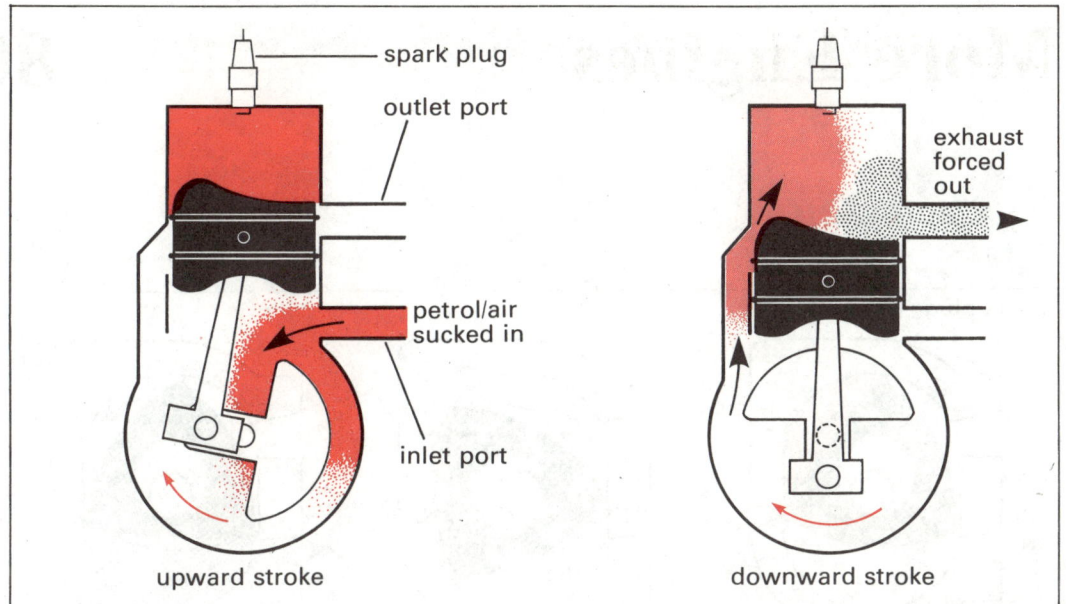

spark plug

outlet port

exhaust forced out

petrol/air sucked in

inlet port

upward stroke

downward stroke

Two-stroke engine

In this case there are only two strokes of the piston.

1 The upward stroke compresses the mixture and allows fresh petrol-air mixture to come in.

2 The downward power stroke pushes the spent gases out and fills the cylinder with fresh mixture.

The two-stroke engine is not as efficient as the four-stroke one. Small lawn-mowers and motor cycle engines are usually two-stroke.

Diesel engine

This engine uses a heavier oil in place of petrol. No spark plug or ignition system is used. Air is very much compressed (to about one-sixteenth of its volume) in the compression stroke. Because of this compression the air becomes very hot. When fuel is injected into the cylinder it explodes and pushes the piston down.

Diesel engines have an efficiency of 35–40 per cent compared with 25–30 per cent for four-stroke engines. But they need to be very strong and heavy, and so tend to be used on large vehicles, ships and railway engines. Recently it has become possible to make them much lighter and now they can be used in cars as well.

1 Draw a diagram of a four-stroke engine and use it to explain what happens on each upward stroke.

2 What happens in a two-stroke engine on the upward stroke?

3 Why is it dangerous to stay in a closed garage when a car engine is running?

Diesel engine

More Engines

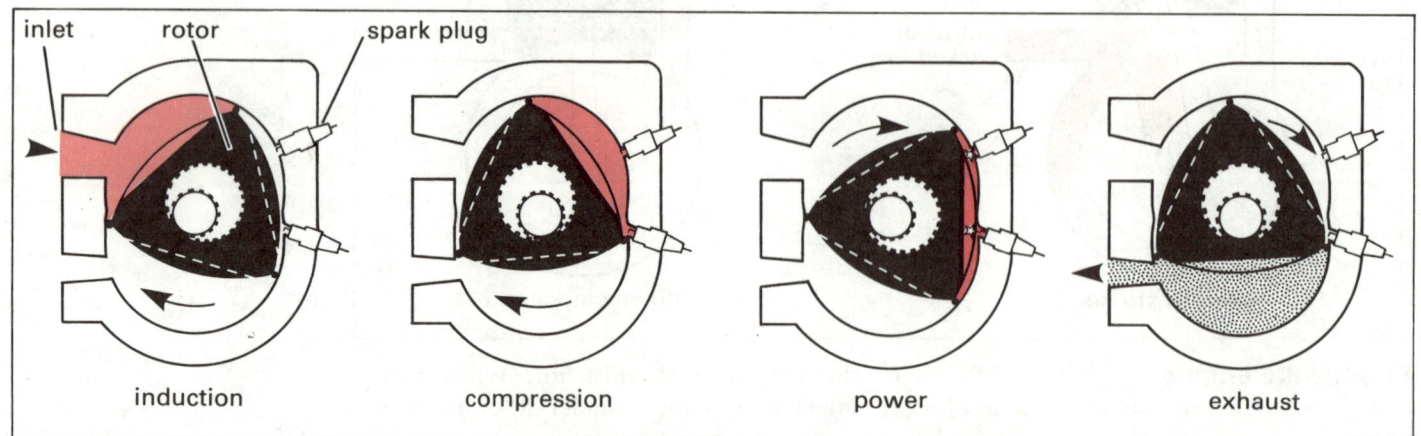

induction   compression   power   exhaust

Wankel engine
Engine cycle (above);
shape of rotor (right)

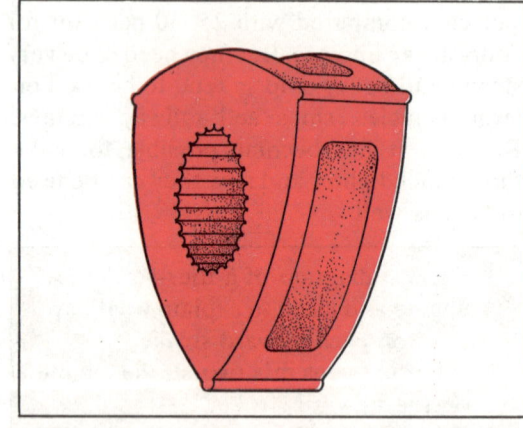

Turbine blades
This arrangement of fixed
and moving blades is used
in, for example, the turbojet
engine

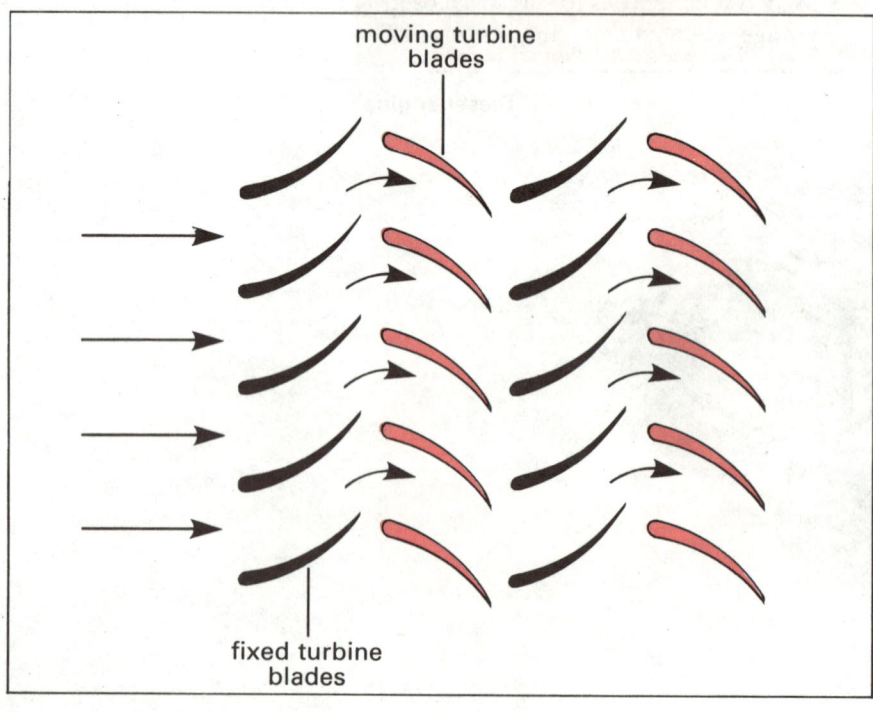

moving turbine
blades

fixed turbine
blades

Rotary engine

The Wankel rotary engine is used on some
cars. It has no piston. It might replace the
piston engine in the future.

The sequence of stages is much the same.
But in this engine the casing is oval-shaped,
with a narrow waist. The rotor is roughly
triangular; the points of the triangle touch
the engine casing all the time. This triangular
rotor moves on a toothed gear wheel and
makes an unusual wavy movement inside the
casing.

Each rotor face is hollowed out and a
petrol-air mixture is drawn in as it passes the
inlet port. By the time the rotor moves to the
spark plug region, the mixture has been compressed. The mixture is fired and space is
allowed for expansion at the lower end. The
exhaust gases are pushed out.

Since there are three faces on the rotor,
there are three explosions during one turn of
the rotor.

The advantages of this engine are that it is
simple and has fewer moving parts, nothing
goes up and down – only round and round.

Turbines

A jet of steam or water can be made to rotate
a paddle of blades. This type of device is
called a **turbine**.

In hydro-electric power stations the rotating turbine shaft drives a generator which
produces electricity.

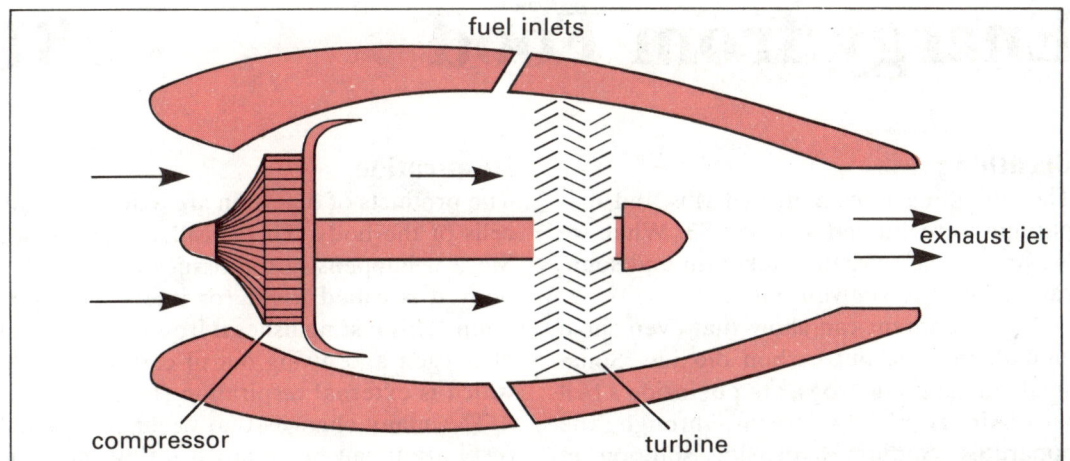

fuel inlets

exhaust jet

compressor

turbine

Turbojet engine

Another example of a turbine is the turbojet engine, used in some aeroplanes.

As the plane moves forward, air rushes into the compressor at the front of the engine. Here the air is so compressed that it becomes very hot. When it reaches the combustion chamber, fuel is injected. This burns in the air, to produce the thrust of expanding gases. These not only drive the plane forward, but pass through a turbine on the way. The turbine operates the compressor.

Jet action

But what makes the aeroplane move? To answer this you need to remember that every action has an equal and opposite reaction. (This is called Newton's third law of motion.)

No doubt you have, at some time or other, blown up a balloon, then let it go. Air rushes out of the balloon. The stretched fabric provides the force and pushes the air out. But the air itself pushes on the balloon and this reaction sends the balloon backwards. The recoil of a gun when it is fired is a similar reaction.

So the jet engine of the aeroplane forces exhaust gases out at great velocity. The reaction is that the plane is pushed forward in the opposite direction.

Ram-jet engine

Jet action does not depend on air in the atmosphere resisting the exhaust jet. In fact, planes like Concorde can fly faster at a great height where there is less air.

When a plane travels well above the speed of sound, the engine can do without a compressor. The air enters with great force. What is known as the 'ram' effect comes into play to compress the air so much that its temperature can rise from below 0°C to over 700°C.

The efficiency of jet engines increases from about 20 per cent at low speeds to over 30 per cent at higher speeds.

Rockets

Rockets are used to launch space craft. Many different materials are used as fuel. The fuel is carried in tanks together with tanks of an oxidising agent such as liquid oxygen. An inert gas such as nitrogen is carried in liquid form and is used to force the fuel and oxidant into the combustion chamber.

The rocket moves upwards as the reaction to the tremendous force of the gaseous products leaving the exhaust.

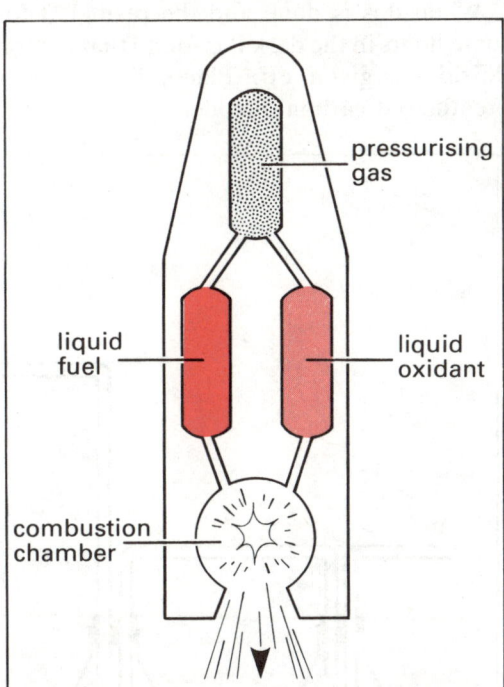

pressurising gas

liquid fuel

liquid oxidant

combustion chamber

1 What is the meaning of the 'efficiency' of an engine?
2 What features of a motor car are introduced to keep waste energy to a minimum?

Energy from Food

81

Breathing

The breathing movements of ribs and diaphragm were studied in Topic 33. When we breathe, some oxygen is taken in and some carbon dioxide is given out.

An experiment can show that even small animals breathe out carbon dioxide. Some small animals, e.g. frogs, are put inside a bell jar. Air is slowly drawn through the apparatus. Sodium hydroxide solution in flask A takes all carbon dioxide out of the air, so the lime water in flask B stays clear. But slowly a precipitate forms in flask C. This shows that carbon dioxide is given off by small animals when they breathe.

Do plants breathe? Using the same apparatus, a plant can be put under the bell jar – a potted geranium (*Pelargonium*) is suitable. The pot is wrapped carefully in a plastic or rubber cover, so that there is no chance of carbon dioxide coming from the soil. Also a dark cloth or light-proof cover is put over the bell jar, so that the plant is in the dark. In this way the plant cannot photosynthesise.

When this is done and the plant left for some hours in the dark it is found that carbon dioxide is given off. Plants like animals breathe out carbon dioxide.

Respiration

The products of digestion are oxidised in the cells of the body. This is called **respiration.** Since it happens in the tissues of the living body it is called tissue (or internal) respiration. This distinguishes it from the taking in of oxygen and giving out of carbon dioxide which is external respiration (breathing).

The many changes that occur in internal respiration can be summed up by the equation:

$$\text{sugar} + \text{oxygen} \rightarrow \text{carbon dioxide} + \text{water} + \text{energy}$$

But the process is not as simple as this equation might suggest. There are more than twenty reactions making up the overall process. Each of these gives out a small amount of energy, with the help of many enzymes.

Respiration and photosynthesis

All living things respire. Most take in oxygen and give up carbon dioxide. Respiration in plants is on a much smaller scale than in animals. In sunlight, it is quite masked by the process of photosynthesis. Only green plants are able to make carbohydrates from carbon dioxide and water by photosynthesis.

Respiration takes place all the time, but photosynthesis only happens in light.

Breathing
This apparatus shows that small animals produce carbon dioxide when they breathe

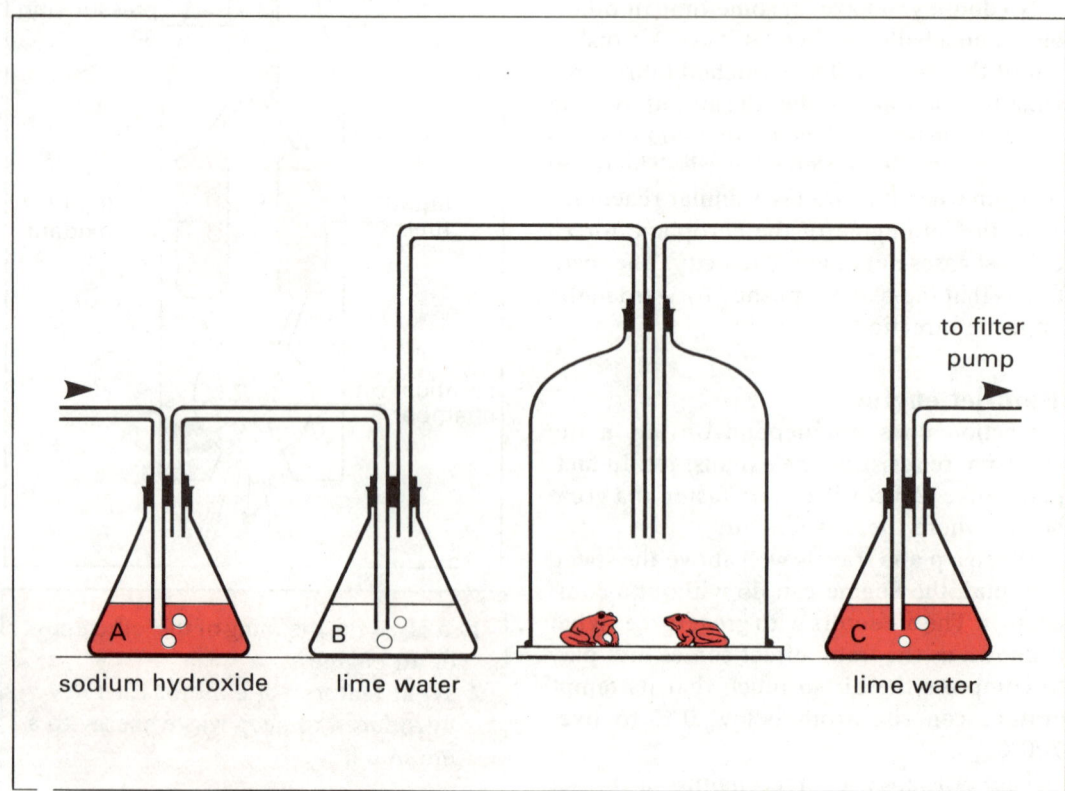

sodium hydroxide   lime water   to filter pump   lime water

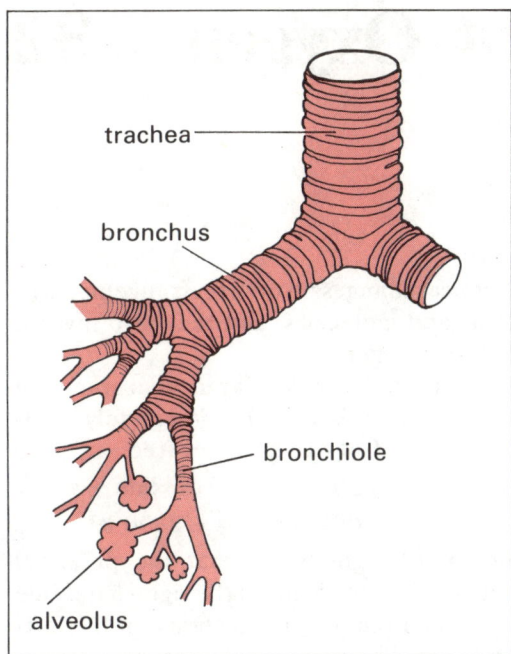

Structure of the lungs

trachea

bronchus

bronchiole

alveolus

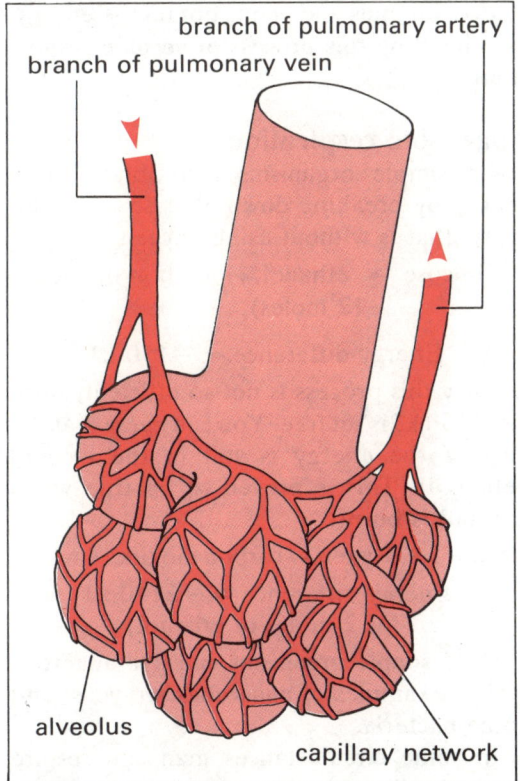

branch of pulmonary artery

branch of pulmonary vein

alveolus

capillary network

Alveolus
The blood is brought into close contact with the air so that the oxygen can be dissolved in the blood

1 Where, in the body, is glucose oxidized to carbon dioxide and water? Where does the carbon dioxide leave the body?
2 Is respiration the same as breathing?
3 What is the precipitate formed in the lime water test for carbon dioxide?

In darkness:

$$O_2 \to \text{plants} \to CO_2$$
$$\quad\,\downarrow \text{animals} \,\downarrow$$

In light:

$$O_2 \updownarrow \to \text{plants} \updownarrow \to CO_2$$
$$\quad\,\downarrow \text{animals} \,\downarrow$$

It is important to know the difference between the two processes:

$$\text{sugar + oxygen} \underset{\text{photosynthesis}}{\overset{\text{respiration}}{\rightleftarrows}} \text{carbon dioxide + water}$$

Respiration
CO_2 is set free
O_2 is used
Energy is set free
Always in plants and animals
Many enzymes needed
Greater in animals

Photosynthesis
CO_2 is used
O_2 is set free
Energy of sunlight is needed
In green plants only
Chlorophyll as well as enzymes needed
Masks respiration, except in dark.

Respiratory surfaces
There must be a place where oxygen passes into the body and carbon dioxide passes out. In simple organisms like amoeba and *Hydra* the gases can pass in and out by diffusion alone.

In more complex animals there is a large respiratory surface. In man this is the lungs. Here the air comes into close contact with tiny blood vessels, and the gases in solution can pass across the boundary.

Frogs breathe partly by lungs, but also through their skin. Fish have gills and most insects have tracheal tubes as respiratory surfaces.

Lungs
From the back of the mouth a tube leads downwards. It is called the **trachea** or windpipe. This tube divides into two branches called **bronchi**. These divide again and again (**bronchioles**) until they end in a frothy cluster of tiny compartments called **alveoli**. It is here that the air in the alveoli comes close to the blood capillaries. A large amount of air comes near a large amount of blood and the exchange of gases takes place.

Brewers' yeast
Magnification × 1000

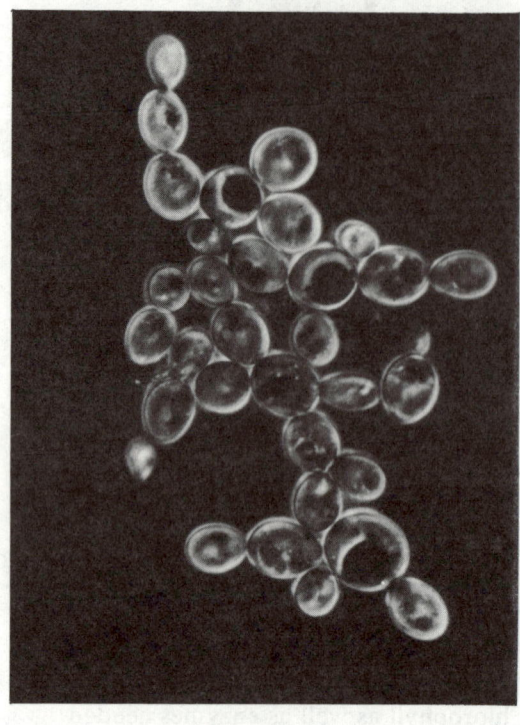

Chemical changes involve rearrangements of atoms and molecules. But they also involve changes in energy.

When a mole (molecular mass in grams) of glucose (which is 180 g) is completely oxidised, 2890 kJ of energy is set free.

glucose + oxygen → carbon dioxide + water

Energy difference = −2890 kJ

The minus sign means that the potential energy of the products on the right-hand side is less than that of the reactants on the left-hand side. The quantity of energy set free is proportional to the mass of glucose oxidised.

Most animals and plants obtain the energy they need by this process of **aerobic respiration.**

Anaerobic respiration

Some simple organisms can obtain their energy by breaking down glucose **anaerobically,** that is without using oxygen.

Glucose → ethanol + carbon dioxide
(2 moles)

Energy difference = −150 kJ

Clearly this process is not so efficient, since only 150 kJ is set free. You can guess that the rest of the energy is still in the alcohol (ethanol). This is proved to be true when ethanol is burnt:

ethanol + oxygen → carbon dioxide + water

Energy difference = −2740 kJ
(for 2 moles of ethanol)

Only simple organisms can live anaerobically. Examples of **anaerobes** are yeast and some bacteria.

In some circumstances man can respire anaerobically.

Oxygen debt

During vigorous exercise, energy is needed faster than oxygen can be supplied, in spite of more rapid breathing. So an 'oxygen debt' is set up. Lactic acid forms instead of carbon dioxide.

When exercise ends and the oxygen debt is repaid, the lactic acid is dealt with. Some of it is broken down to carbon dioxide. The rest is changed back to glycogen and stored in the liver.

Preparation of ethanol
The air lock keeps the air out but allows the carbon dioxide to escape

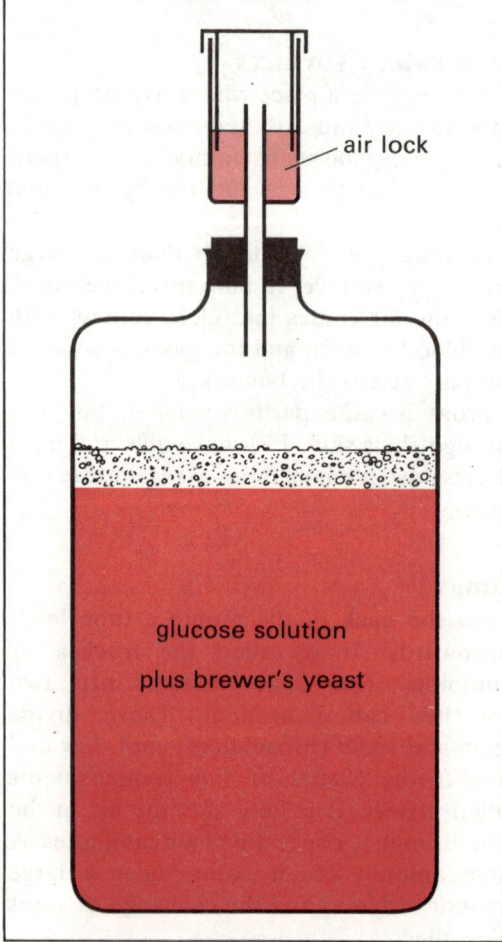

air lock

glucose solution

plus brewer's yeast

Whisky distillery

Preparation of ethanol (ethyl alcohol)

A solution of glucose is warmed to about 27°C. Some brewer's yeast is added. The solution is kept warm for a few days. An air-lock is fitted on the bottle so that although gases from inside can escape, no air or dust can get inside the bottle.

A little nutrient, such as ammonium phosphate, helps the process. Frothing is seen as carbon dioxide escapes and an alcoholic smell can be noticed.

This is the basis of wine and beer making. The reaction takes place because there is an enzyme, called zymase, present in yeast.

In industry, fermentation of a malt extract gives beer which contains 3 – 7 per cent ethanol. Fermentation of grape juice gives wines of about 12 per cent ethanol. Port and sherry may contain as much as 20 per cent ethanol, because they are 'fortified' by addition of spirit.

Distillation

If a stronger solution of ethanol is required, the liquid obtained above by fermentation is concentrated by fractional distillation. Compare this with oil refining (Topic 60). The liquid that is produced contains about 95 per cent ethanol.

When carried out under licence in distilleries, whisky, brandy and other spirits are made. These contain about 40 per cent ethanol or more.

Methylated spirit

This contains 80–90 per cent ethanol. Methanol (which is poisonous) and some other substances and colouring matter are added to the ethanol. In this way it is hoped that people will be dissuaded from trying to drink it.

---

1 Write down the names of two enzymes involved in digestion of food and one enzyme met with in fermentation.

2 What are enzymes? State their more important properties (You could look at Topic 64, if necessary).

3 Which of the following answers are correct? Ethanol is formed from glucose in the process of (a) photosynthesis, (b) fermentation, (c) aerobic respiration, (d) anaerobic respiration, (e) fractional distillation.

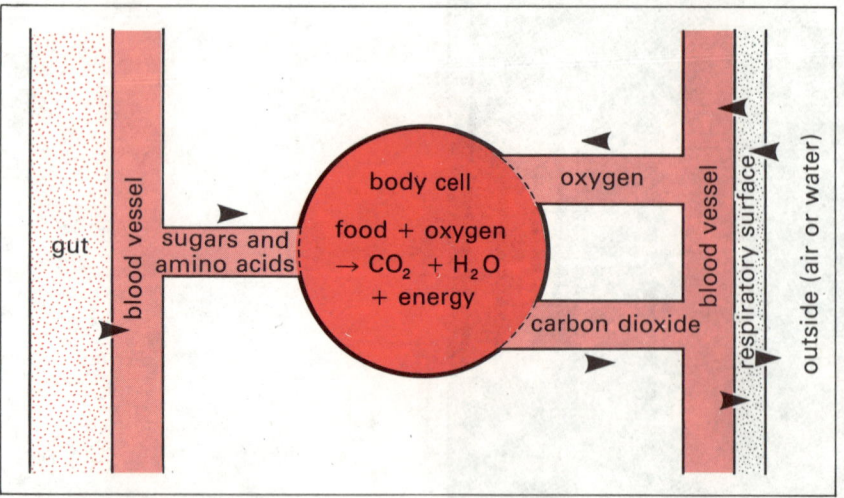

Energy transfer in man
Food from the gut provides sugar and amino acids to the blood stream. They are transported to the body cells that need them for growth and movement. The lungs provide oxygen to the blood stream, which also removes the carbon dioxide

The whole collection of chemical changes needed in the body to provide energy and get rid of waste substances is called **metabolism.** The blood system in man is the 'conveyor belt'.

Blood system
All animals that are made up of many cells need a liquid flowing round the body so that materials can reach all the cells. In many animals, but not all, this liquid is blood. The blood is forced round the body either in blood vessels or in wide spaces called sinuses bathing the tissues. Fish and insects have sinuses.

The pulse
Hold out the left hand palm upwards. With the finger tips of the right hand find the point on the wrist at the base of the thumb where the pulse can be felt. This is a **pressure point**, so called because near the surface there is a blood vessel with a bone behind it. If you press at this point, you can stop the blood flowing into your hand. By touching lightly you can feel this flow.

Measure your pulse rate by counting the number of beats in one minute, at this radial artery pressure point.

Working with a partner, compare your rate of heart beat with your rate of breathing in and out. Your partner can measure your pulse rate while you count the number of in-and-out breaths in a minute.

Now go for a quick, brisk run. Come in and immediately take the same measurements again, your partner on the pulse and you on the breath rate. What differences do you find?

You could also take readings whilst lying down quietly.

During exercise more energy is needed. More oxygen must be taken into the body cells.

Pressure points
There are several pressure points in the body. It is wise to learn how to find them. You may then be able to save life in an accident by knowing how to stop the flow of blood from a wound.

Here is a simple experiment to show how pressure points control blood flow. First find the ulnar pressure point on the left hand corresponding with the radial point but on the other side of the wrist. Hold your left hand high above your head so that blood flows downwards.

With the first two fingers of the right hand, press firmly on both radial and ulnar pressure points. Keeping this pressure on firmly, bring your hands down. Note that the palm of the hand looks pale.

Now release the pressure. You will see that the hand quickly becomes red as the blood flows back into it.

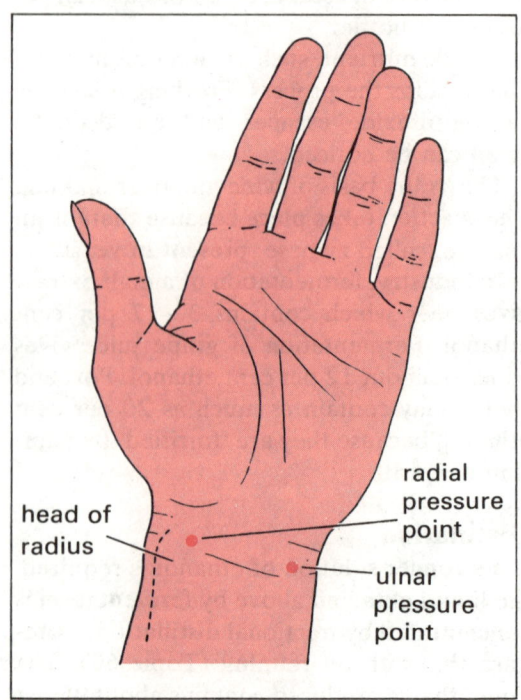

Pressure points

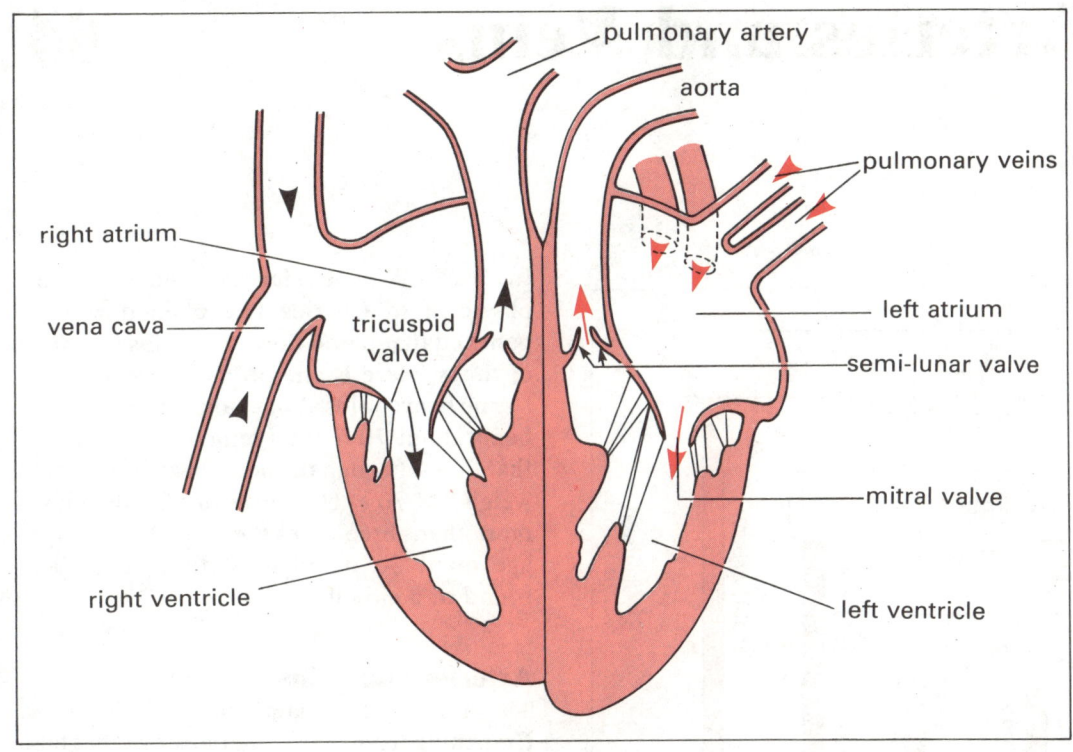

pulmonary artery
aorta
pulmonary veins
right atrium
vena cava
tricuspid valve
left atrium
semi-lunar valve
mitral valve
right ventricle
left ventricle

The heart
The actual structure is shown on the left. A diagrammatic representation is given below

The heart

When cut open the heart is seen to have four compartments. The **atria** (singular: atrium) are thin-walled, on both the right and the left. The **ventricles** have thick muscular walls. There is no direct passage connecting left and right sides of the heart.

Blood from all parts of the body flows into the right atrium. As the atrium fills, nerve action makes the muscular wall contract and force the blood into the right ventricle. It cannot go back into the vena cava, because valves prevent this.

In the right ventricle, flaps of tissue forming the tricuspid valve, come together and prevent the blood from going back into the atrium.

When the ventricular wall contracts, the blood must pass through the pulmonary artery into the lungs. Here the oxygen and carbon dioxide exchange takes place.

The blood returns in the pulmonary vein to the left atrium. Muscle action again forces the blood into the left ventricle.

When the left ventricle contracts, blood passes into the aorta. This is the main artery and leads to all parts of the body except the lungs. Valves at the entrances to both pulmonary artery and aorta prevent the blood from 'bouncing back' into the heart after it has been forced out. The walls of these arteries are very elastic. The atria and ventricles contract because a stimulus comes from a nerve centre. This is at the base of the right atrium and sends impulses along nerves in the dividing wall of the heart. You cannot consciously affect your heart rate, but some influences, such as emotion like fright and certain drugs, can alter its beat.

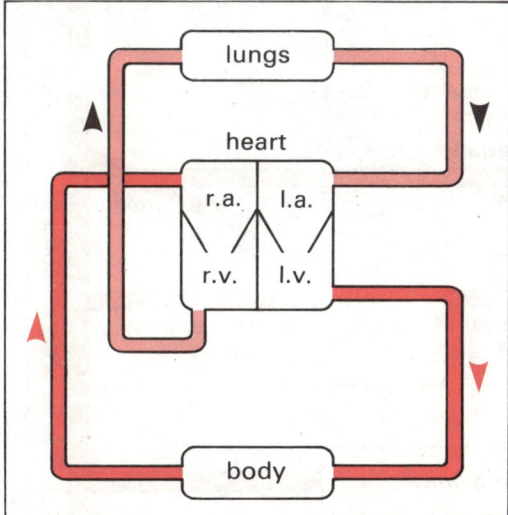

lungs
heart
r.a. | l.a.
r.v. | l.v.
body

1 Trace the path of blood from the right atrium till it reaches the left atrium.
2 In what ways is blood in the left atrium different from blood in the right ventricle?
3 Where is blood when it has just left (a) the left atrium, (b) the left ventricle, (c) the vena cava?

Blood system
The arteries are the dark colour and the veins the light colour

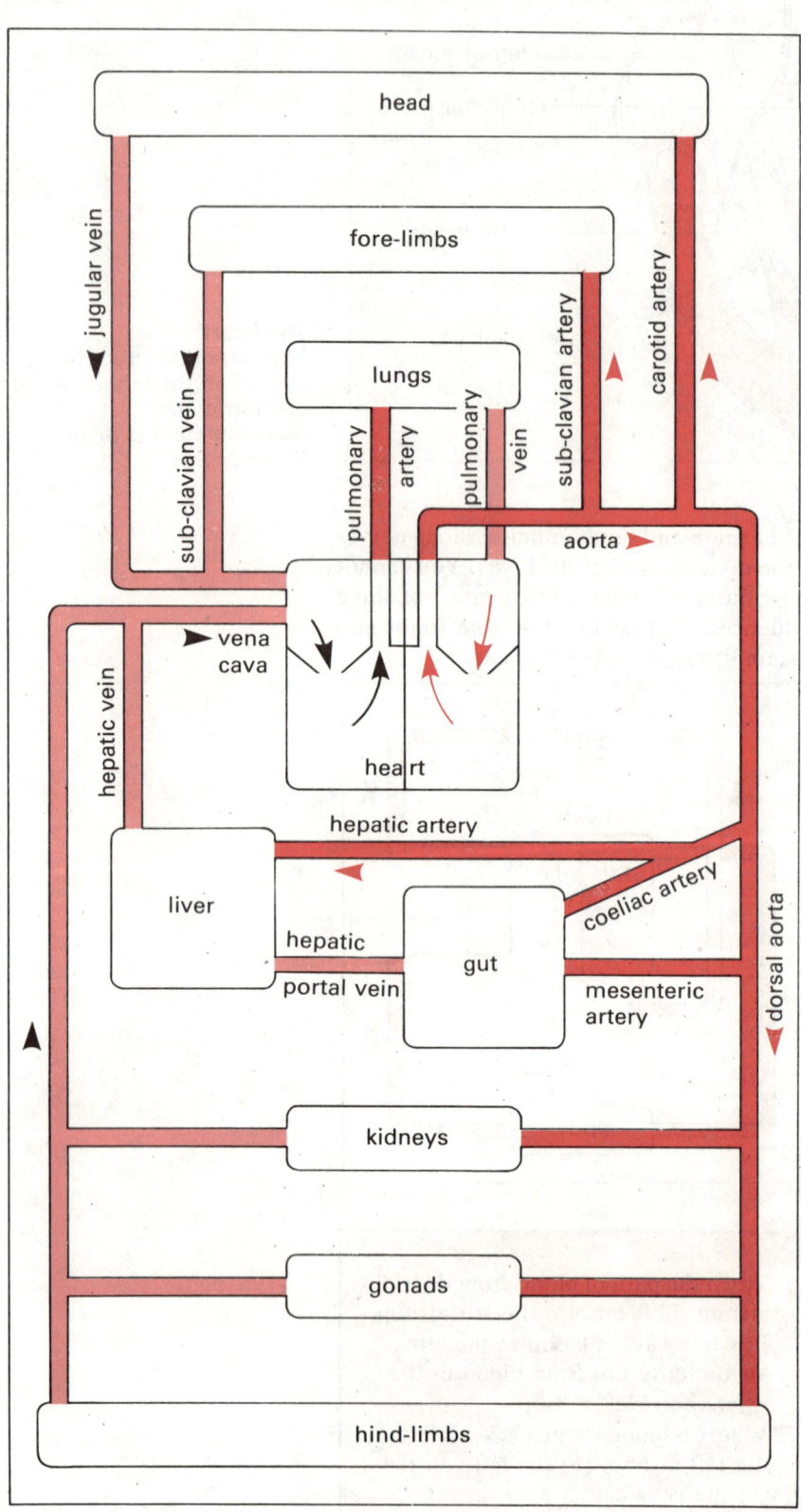

In 1628, William Harvey, who was court physician to Charles I, explained why he believed that blood flowed in a closed system of tubes. Here is one of his arguments.

The heart holds about 2 ounces of blood. It beats about 70 times a minute. So the blood that flows through the heart in an hour would weigh $2 \times 70 \times 60$ ounces or 525 lbs. This is more than three times the weight of an average man! So blood must flow round and round in a circuit.

Arteries and veins

The aorta is the main artery of the body. Branching from it are other arteries that take the blood to the liver, the digestive system, the kidneys, the sex organs and the muscles of all parts of the body. An **artery** is a vessel in which blood flows away from the heart. A **vein** is a vessel in which blood flows towards the heart.

The large collecting veins leading into the heart itself are the upper and lower venae cavae.

The blood in an artery almost always has more oxygen and is brighter red in colour than blood in a vein. The only exception is that the **pulmonary artery** takes deoxygenated blood (dark in colour) to the lungs and the **pulmonary vein** brings bright red, oxygenated blood to the heart.

The blood in arteries is forced along by great pressure in a jerky manner. After being pushed through the capillaries, blood passes along the veins in a sluggish manner. The pressure is low in the veins. In the legs blood has to be pushed upwards against gravity. But small 'pocket valves' help to prevent the blood going backwards. In varicose veins these valves no longer function so the weight of blood bearing down makes the veins larger.

Artery
   thick-walled
   elastic
   has a narrow lumen (channel)

Vein
   thinner walled
   not so elastic
   wider lumen

Arterial blood
  bright red
  contains more oxygen
  flows away from the heart
  high pressure
  flows in jerks

Venous blood
  dull red or purple
  less oxygen
  flows towards the heart
  low pressure
  steady flow

Blood pressure

The pressure of the blood in the arteries can be measured by using a special manometer, called a sphygmomanometer. A closed rubber tube is wrapped round the upper arm. The inside of this tube connects with the manometer. A small hand pump is used to increase the pressure inside the tube. The pressure is noted when this pumping causes the pulse to stop.

This (systolic) pressure is normally 100–120 mmHg.

Then the pressure is reduced by allowing air to escape slowly from the tube until the pulse is again felt. This gives the diastolic pressure, which is normally about 70–90 mmHg.

An experienced doctor can find these pressures by listening to the noises heard in a stethoscope placed at the brachial artery pressure point.

Recently it has become more common for a pressure gauge to be used rather than a manometer (see photograph).

Capillaries

Arteries branch to form narrower vessels, arterioles. These branch again and again until the vessels become so fine (capillaries) that a blood corpuscle can hardly pass. Yet a corpuscle is only a few thousandths of a millimetre in diameter!

Transport in plants

Nothing quite like a blood system is found in green plants. There is a system of tubes or vessels passing up the root and stem to the leaves. These are often called 'veins'.

Water and mineral salts absorbed in the root pass up through xylem vessels. The carbohydrates and other products made in the green leaves by photosynthesis pass to all parts of the plant through phloem vessels. This process is called **translocation**.

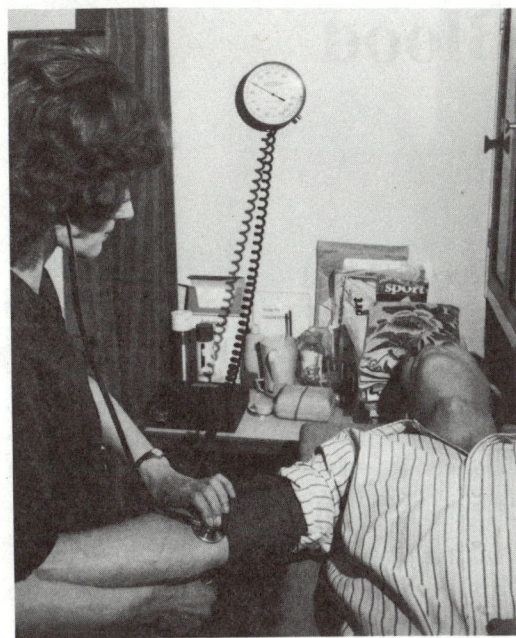

A doctor measures a patient's blood pressure

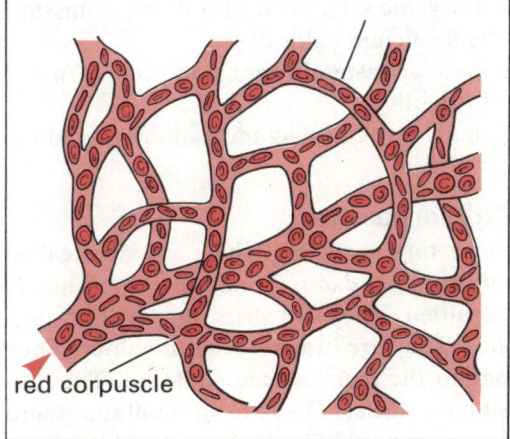

Capillaries
These allow the blood to come into close contact with the muscles that require oxygen

red corpuscle

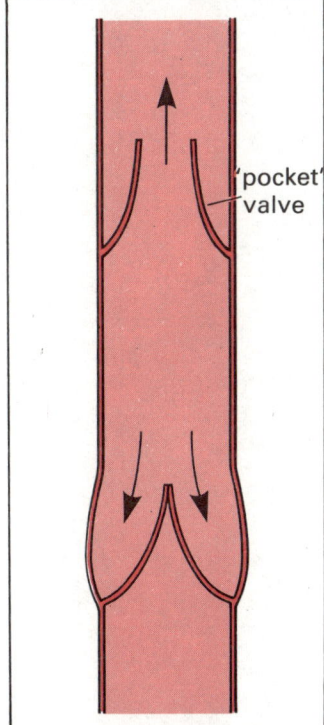

'pocket' valve

Veins
Veins have valves to stop the blood flowing backwards

1 Name the blood vessels that take blood away from: (a) the left ventricle; (b) the liver; (c) the kidneys; (d) the lungs.
2 How is the blood in the pulmonary vein different from the blood in other veins?
3 Explain why blood from a cut vein just ooozes out but blood from a cut artery spurts out.

Blood

85

Nature of blood

Blood consists of a pale yellow liquid, called plasma (55% of the volume), which contains particles called corpuscles. There are two kinds of corpuscle: red and white. Some tiny fragments of cells are also present. These are platelets. It is the red blood corpuscles that makes blood look red.

Plasma

This liquid contains several substances in solution:

1 digestion products, such as sugars and amino-acids
2 waste carbon dioxide, chiefly combined as sodium hydrogencarbonate
3 inorganic salts, such as sodium, potassium and calcium chlorides
4 urea, a waste nitrogen compound formed in the liver
5 many proteins, including albumin and fibrinogen

Red corpuscles

These tiny particles (which are also called erythrocytes) can only be seen when highly magnified. They are discs shaped like pastilles. They are orange–pink in colour. They contain the pigment haemoglobin. They do not have nuclei. They are so small that there are about 5 000 000 in 1 mm³ of blood.

Red corpuscles are produced from cells in the bone marrow. They last on average about 8 weeks only. Since the volume of blood in the body is about 6 litres this means that as red corpuscles are broken down in the liver, they are being replaced at a rate of about 5 million per second!

Red corpuscles are important because they carry oxygen from the lungs to the tissues of the body. Haemoglobin combines with oxygen to form oxyhaemoglobin.

When this reaches the tissues, it loses oxygen to the cells:

$$\text{haemoglobin} + \text{oxygen} \underset{\text{in tissues}}{\overset{\text{in lungs}}{\rightleftarrows}} \text{oxyhaemoglobin}$$

(venous blood) (arterial blood)

The carbon dioxide released during respiration is carried chiefly as sodium hydrogencarbonate, partly in the red corpuscles, partly in the plasma.

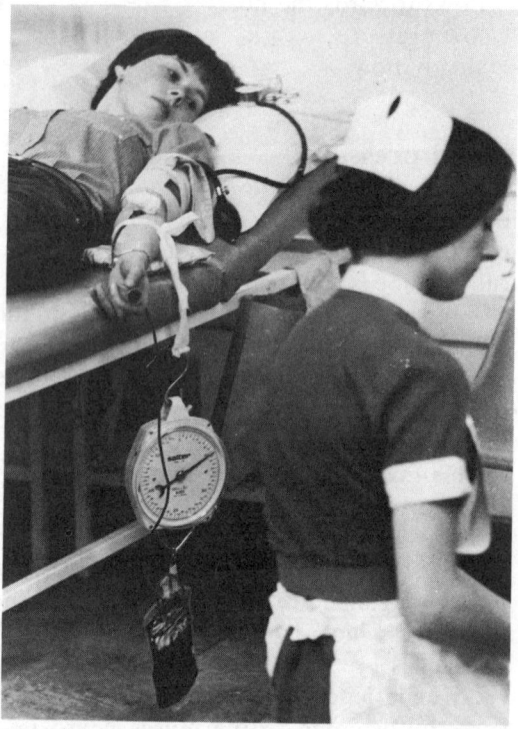

Blood donor

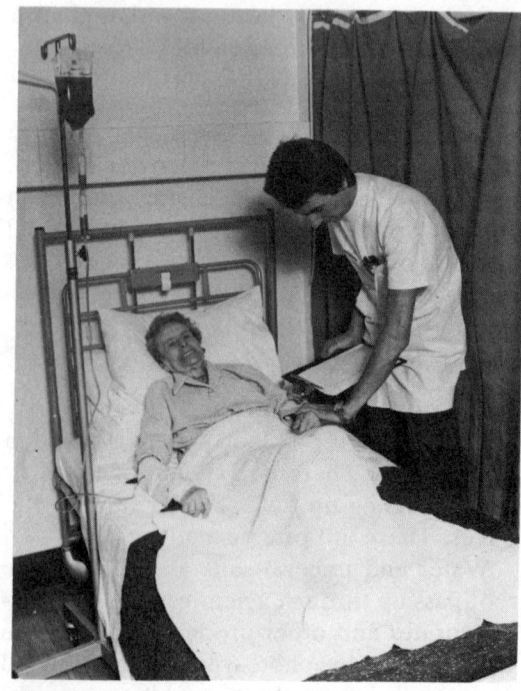

Blood transfusion

White corpuscles

These are not so regular in shape and size as the red corpuscles, which are all about the same size, about 7 μm (millionths of a metre) across. White corpuscles have nuclei. There is one white cell for about 500–600 red cells.

They are formed in the lymph system but are able to move in and out of blood vessels. They act as scavengers by removing waste materials and engulfing bacteria. The pus sometimes found in a wound contains white corpuscles that have been killed by the poisonous products of bacteria.

Most proteins foreign to the blood can act as **antigens** if they enter the blood. They cause the blood to form **antibodies.** These react with the antigens and make them less harmful. The poisons formed by bacteria also cause the blood to form antibodies or anti-toxins.

Platelets

The smallest bodies found in the blood are platelets. These have no nuclei. They seem to be small pieces of large cells found in the tissue spaces. Platelets help in the clotting of blood.

Clotting

When a blood vessel is cut the platelets and the damaged tissue together form a substance called **thromboplastin.** A succession of reactions changes fibrinogen into fibrin. The **fibrin** forms a network of strands which trap the escaping corpuscles. So a clot is formed and this contracts and seals the cut.

In contracting it squeezes out a clear liquid called **serum.** Serum is like plasma, but contains no fibrinogen.

Lymph system

The blood is not the only transporting liquid found in the human body. There are also lymph vessels, through which a colourless liquid, called lymph, slowly flows. It flows through lymph nodes (such as tonsils) on its way to the veins in the neck. Here it joins the blood system.

The flow is maintained by muscle movement. There are no valves in the vessels to stop backward flow. Lymph is similar in composition to blood plasma. White cells are present, but not red cells.

Blood transfusion

Some accidents and surgical operations can prove fatal because of loss of blood. Many lives are saved by giving the patient a blood transfusion. Blood is slowly dripped into the patient's vein.

Care must be taken in selecting the blood used. Some types of blood make other types of blood clot when they are mixed.

A person's blood belongs to one of four groups, A, AB, B and O. In the table an X shows the cases in which the donor's blood cannot be mixed with the patient's:

Donor	Patient			
	A	B	AB	O
O	–	–	–	–
A	–	X	–	X
B	X	–	–	X
AB	X	X	–	X

Group O subjects are universal donors. Group AB subjects are universal recipients.

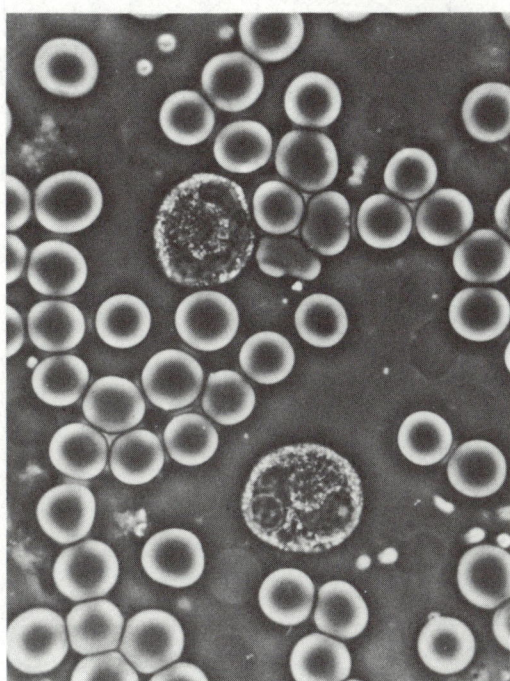

Microscopic photo of blood
The small particles are red blood cells, the large particles are white blood cells

1 Why do we need white corpuscles in the blood?
2 Write down three differences (other than colour) between red and white corpuscles.
3 Which of the following substances would you find dissolved in blood plasma? (a) sodium chloride; (b) starch; (c) urea; (d) glucose; (e) nitrogen; (f) sodium hydrogencarbonate; (g) haemoglobin.
4 What percentage volume of blood is taken up by the corpuscles?

Excretion

86

Structure of the kidney

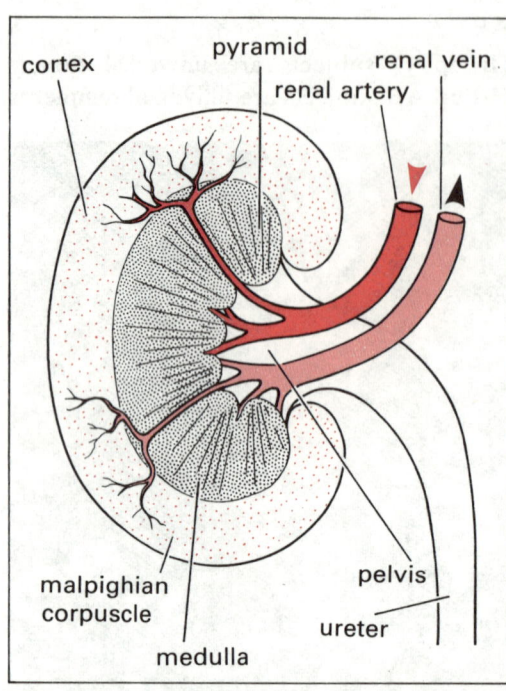

Detailed structure of the Malpighian corpuscle

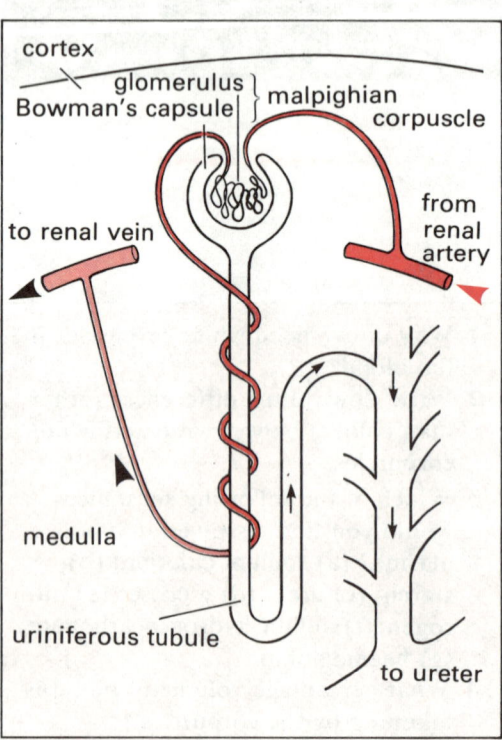

Not all the substances formed during the chemical reactions in the body (metabolism) can be used. Some are waste substances and need to be removed. They would be harmful if they stayed in the tissues.

The process of removing the unwanted products of metabolism, either by passing them out of the body or by changing them into harmless insoluble substances, is called **excretion.**

Getting rid of waste solid matter as **faeces** is not really an example of excretion. It is called elimination. The discharged matter has not been involved in the tissues of the body.

Mammals possess the following four main excretory organs:

Excretory organ	Excretory products
Lungs	Carbon dioxide, water
Liver	Bile pigments formed from haemoglobin; urea formed from waste ammonia
Kidneys	Urea filtered off from blood
Skin	Traces of sodium chloride, urea and water

Liver

This important organ controls the amount of glucose circulating round the body. The blood carries the glucose and amino acids which have been absorbed in the intestines. They go straight to the liver by the hepatic portal vein. The liver changes excess glucose into insoluble glycogen, which it stores. This can be changed back to glucose as needed.

Unwanted amino acids are broken down to form ammonia. This substance would be harmful if left in the body. In the liver it is changed into urea. This urea then passes in the blood stream to the kidneys where it is excreted.

The liver also changes some poisons and drugs into harmless substances to be excreted.

Kidneys

The pair of kidneys is in the hollow of the back, behind the intestines. The left one is slightly higher than the right.

When a kidney is cut across, two regions

can be seen. The streaky **medulla** is made up of several conical patches of tissue called pyramids. The streaks are really tiny tubes that lead out to the funnel-shaped pelvis at the top of the ureter.

In the darker **cortex** there are a number of specks. Under the microscope each of these bodies (Malpighian corpuscles) is seen to have a network of blood capillaries, called a glomerulus. Each of the glomeruli is surrounded by a thin-walled funnel called a **Bowman's capsule.**

The blood enters the kidney by the renal artery, which divides into branches. When the blood reaches a glomerulus, liquid is squeezed out of the blood into the capsule. The walls of the blood vessel and the capsule are very thin.

Urea is thus filtered into the capsule. But glucose, amino acids and salts also pass out of the blood. Proteins and blood corpuscles do not filter off.

Glucose, amino acids and soluble salts that have been squeezed out are required in the blood. They are taken back as the blood vessel passes close to the uriniferous tubules carrying these substances.

The remaining waste substances pass, as urine, into the pelvis of the kidney and then through the ureters to the bladder. From time to time the urine is passed out of the bladder through the urethra. This action is controlled by both voluntary and involuntary muscles.

There are approximately one million Malpighian corpuscles in a kidney. The total filtering surface is about 1.5 square metres. Each day about 150 to 200 litres of filtrate pass into the tubules. Of this only about 1.5 litres of urine is passed out of the body. Much of the water is taken back from the tubules into the blood.

One substance that must not be lost from the blood is sugar. In the disease diabetes mellitus, sugar passes out of the body. This can be spotted by testing for glucose in the urine.

Skin

The skin acts as an excretory organ by getting rid of very small amounts of solutes in sweat.

Sweat is being lost from the body in slight amounts all the time. It contains only about 0.5% of dissolved solids, mostly salt. Roughly the same amount of urea as there is in blood is also excreted. Traces of sulphur compounds are lost from the body in the form of keratin. This is present in the dead

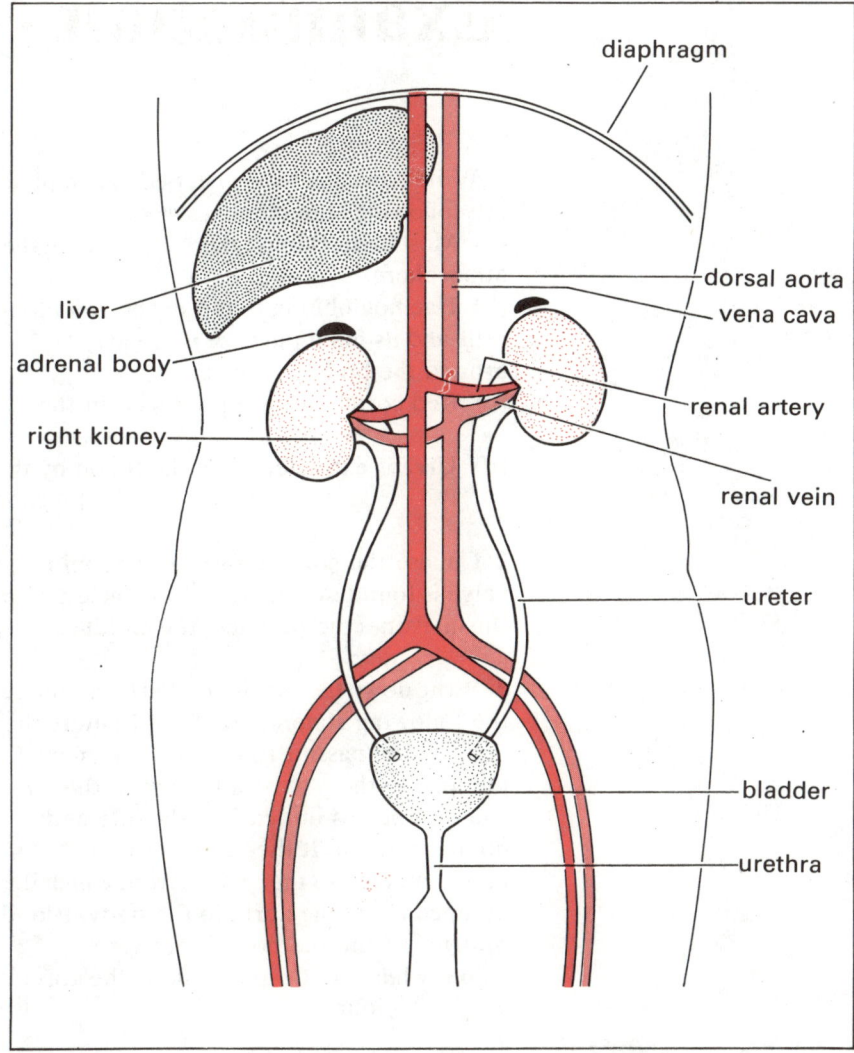

Arrangement of kidneys in the body

cells that flake off from the outer layers of the skin.

Excretion in plants

Plants build up carbohydrates and proteins from simple substances such as carbon dioxide, water and nitrates. Few waste substances are formed, however some are deposited in the cells of the plant as crystals. An example is calcium oxalate, but this is insoluble and does not interfere with the activities of the cell. Waste carbon dioxide and water vapour from respiration, and waste oxygen from photosynthesis diffuse out through the stomata (see Topic 34).

1 Which of the following are excretory organs? heart; liver; kidneys; pancreas; lungs.
2 In what part of the body is urea separated from the blood?
3 What is the difference between excretion and elimination?

Examination Questions VII

1 Write down the words needed to complete the following statements:
(a) Most living things take * out of the atmosphere.
(b) Haemoglobin is found in the red blood cells and its main purpose is to carry * around the body to the cells.
(c) Red corpuscles are produced in the *
(d) Urea is extracted from the blood by the *
WYL

2 Choose the correct answer: The mitral valve is found in: the heart; a bicycle pump; the intestine; the stomach; the middle ear. M

3 Write down the words needed to complete the following statements: Blood enters the * after passing round the body; next it is pumped to the * and then to the lungs where it passes out carbon dioxide and obtains oxygen. It then passes to the * and is pumped to the * from which it is pumped along the aorta to the body. Blood returning from the body is always a * colour while blood passing along the aorta is a * colour. EM

4 (a) Diagram 1, represents the upper part of one cylinder of a four-stroke petrol engine:
(i) Which stroke of the cycle is indicated in the diagram?
(ii) In which direction would the piston be moving?
(iii) How does the pressure in the cylinder vary during this stroke?

(iv) What is the purpose of this pressure variation?
(b) (i) Why does the diesel engine not need a spark plug?
(ii) How does the efficiency of a diesel engine compare with that of a petrol engine? Y

5 Choose the correct answer. Organisms which feed on sugar to give alcohol are: antibodies; viruses; yeasts; amoebae; spirogyrae. M

6 Which answers, if any, are correct? The products of respiration are: oxygen; water vapour; heat; carbon dioxide. WM

7 What are the parts labelled A and B in Diagram 2 (the heart) and A to D in Diagram 3 (excretory system).

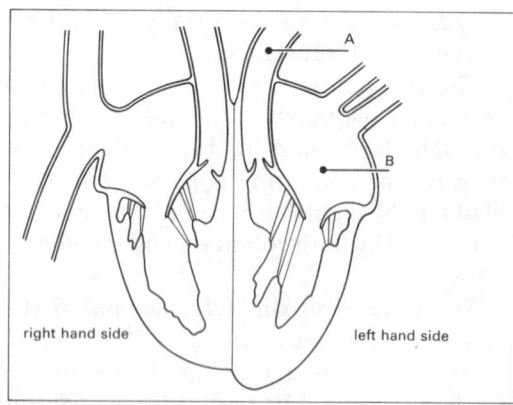

right hand side   left hand side

Diagram 2

Diagram 1

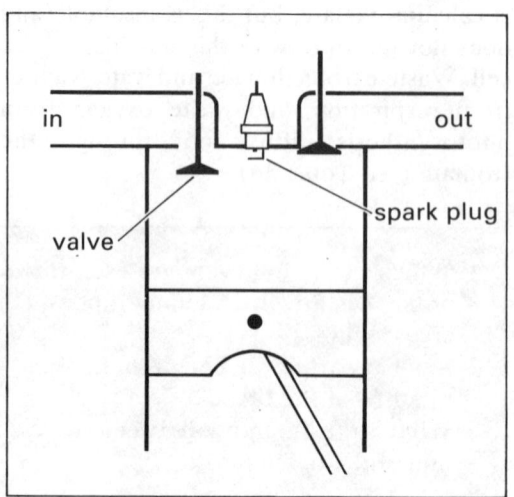

in   out

spark plug

valve

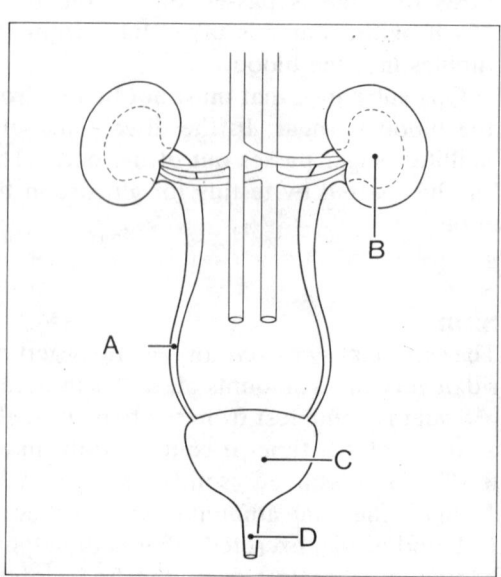

A

B

C

D

Diagram 3

8 A person had his pulse rate and body temperature checked and the results obtained are shown in the table below:

Activity	Result	Pulse rate (beats/minute)	Result	Body temperature (in °C)
Resting	A	66	F	37.0
Walking	B	85	G	37.1
Running	C	165	H	37.3
Sleeping	D	96	I	36.9
Writing	E	70	J	32.0

Which two results would you expect to be wrong in this table? EM

9 Which two statements are NOT true of veins?

(a) they have thin, supple walls;
(b) they have valves at regular intervals;
(c) they often pass near to the surface of the body;
(d) they carry blood away from the heart;
(e) they carry blood towards the heart;
(f) they have no valves. EM

10 The table below shows the blood groups which may be mixed safely.

Donor group	Recipient group A	B	AB	O
A	✓	X	✓	X
B	X	✓	✓	X
AB	X	X	✓	X
O	✓	✓	✓	✓

Key: ✓ Safe; X Unsafe.

(a) In an emergency, which blood group would be best suited for a recipient whose blood group is not known?
(b) Which answer is correct in the following, if incompatible blood groups are mixed together:
(i) the red cells will clump together;
(ii) the white cells will be destroyed;
(iii) more plasma will be made;
(iv) the platelets will dissolve in the plasma. WM

11 (a) Name **one** waste product removed by the kidneys.
(b) Name **two other** parts of the body which remove waste products. W

12 (a) What is the chemical action which produces heat in the internal combustion engine? (b) Give another way in which heat is produced in the internal combustion engine. (c) Name **two** waste products added to the atmosphere during the operation of an internal combustion engine. (d) Give **three** reasons for the inefficiency of an internal combustion engine. (e) Why is an internal combustion engine unsuitable for space travel? SW

13 Which **two** of the following are organs of excretion? (a) penis, (b) lungs, (c) kidney, (d) anus, (e) mouth. ALS

14 Choose the correct answer and write out the complete sentence.
Blood entering the left atrium (auricle) has returned to the heart from the: kidney; aorta; lungs; right ventricle; stomach. ALS

15 Choose the correct answer: Bile is produced by (a) the liver, (b) the stomach, (c) the kidney, (d) the colon or (e) the pancreas. WYL

FIELDS OF FORCE

The word 'field' has been used already. It means a region in which a force can be detected. There is a magnetic force acting on a piece of iron put near a magnet. In this **magnetic field** the iron itself gets magnetised and is pulled towards the magnet.

A stone thrown into the air soon falls downwards. In the **earth's gravitational field** a force pulls the stone down towards the earth. This same gravitational force acts on a satellite and stops it from moving off into space.

A compass needle is put near a wire. When an electric current is passed through the wire, the compass needle is deflected. This shows that a magnetic field is set up around the wire.

It can also be shown that an electric current can be set up when a magnet moves.

This link between magnetic and electric fields is very important. Many modern devices, such as electric motors, transformers and microphones, depend on it.

Jodrell Bank radio telescope

Magnetism and Gravity

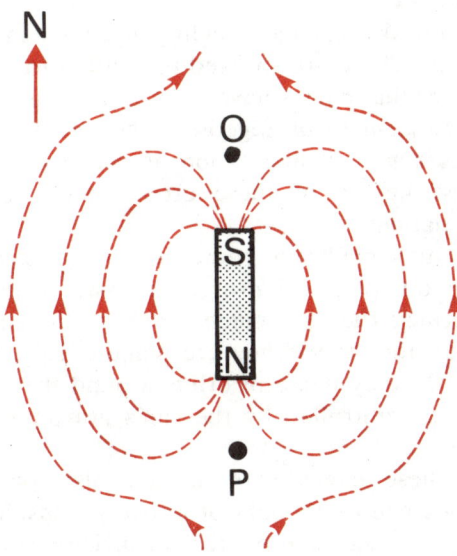

87

The greater the masses, the larger the attractive force between them. Also the greater the distance between them, the smaller this force is.

Somewhere between the earth and the moon is a 'neutral point' where the gravitational pull on a man due to the moon exactly balances the earth's pull on him. The man has no weight at this point. An astronaut can seem to be weightless in a satellite. But in this case the earth's gravitational pull is balanced by the centrifugal force due to the satellite moving in its orbit round the earth.

Magnetic fields

You can study the field near a magnet. Place a sheet of paper over the magnet. Sprinkle iron filings on the paper. Tap the paper lightly. The filings take up positions on lines of magnetic flux (lines of magnetic force). Here, the arrows show the direction of the field, that is, direction in which the north pole of a compass needle would point.

Neutral points can be found (O and P) where there is no magnetic force. At point O, the pull of the magnet's south pole is balanced by the pull of the earth's magnetic field.

There is a magnetic field at the earth's surface even without a magnet being present. It is as if there is a large magnet inside the earth. This is why a magnetised compass needle points towards the north.

There are quite interesting magnetic fields between two magnets placed near one another.

Gravitation

Sir Isaac Newton (1642 – 1727) compared the force acting on the moon to keep it in its path and the force acting on things falling to the ground at the earth's surface. He proved that all objects attract each other with the same gravitational force.

This force, F, between two masses M_1 and M_2 is given by the formula:

$$F = G\frac{M_1 M_2}{d^2}$$

d is the distance apart of the masses and G is a gravitational constant which is the same throughout the universe.

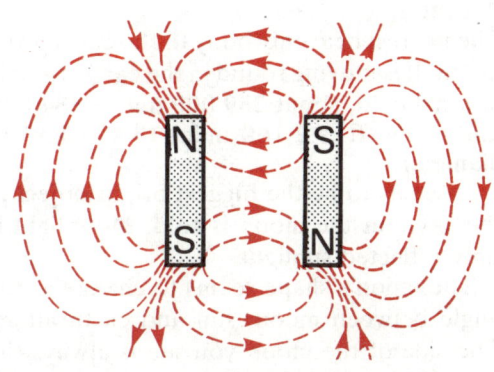

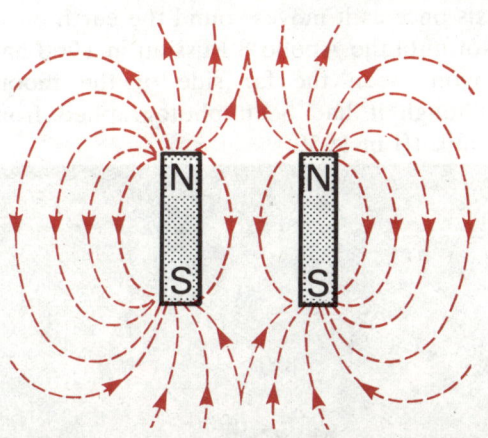

Magnetic fields between two magnets
With unlike poles together (top); with like poles together (bottom)

---

1 If a stone was dropped from the top of a very high building, would it land on the ground (a) at a point vertically below the point from which it was dropped; (b) at a point closer to the building; or (c) further from the building? (Look back to what is mentioned under 'Gravitation' about objects attracting each other.)

2 Draw an outline of a horse-shoe magnet and mark in the pattern of lines of flux you would expect to find around the poles.

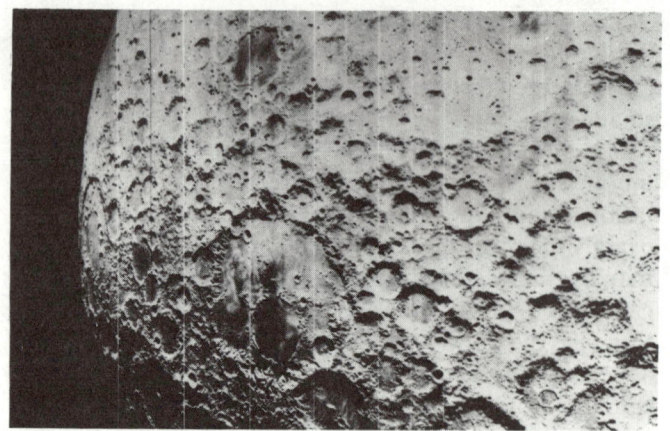

Far side of the moon

Moon

The nearest heavenly body to the earth is the moon. It is moving round in the earth's gravitational field, about 239 000 miles (384 000 km) away. It is 2 160 miles (3 475 km) in diameter.

It seems to be the biggest bright object in the sky, but the moon is cold. Moonlight is only reflected sunlight.

The moon's shape seems to change as the angle between moon, sun and earth alters. The side of the moon you see is always the same. This is because the moon rotates on its axis once as it moves round the earth once. Not until the Apollo 8 Mission in 1968 had anyone seen the far side of the moon, although it had been photographed from Lunik III in 1959.

Planets

In the sky on a clear night you can see many stars. These are so fixed in relation to each other that names have been given to groups that seem to be together. The Great Bear (Ursa major) and Orion are probably the best known of these groups called **constellations.**

But some bright bodies in the sky seem to get out of position as nights pass. This is because they are much nearer to the earth than the stars. They are planets. Like the earth, they move in orbits around the sun, being controlled by the sun's gravitational force.

These planets make up the solar system. They can be thought of in two groups. The major planets, Jupiter, Saturn, Uranus and Neptune are large, but have low densities. They have atmospheres. The earth-like planets, Mercury, Venus, Earth, Mars and Pluto, are denser but much smaller. They have little or no atmosphere (see Table).

Sun

The sun is an enormous mass of incandescent (or glowing) gas. It has a mass 333 434 times the mass of the earth.

The temperature of the sun's surface is about 6000 K, but the temperature inside is probably about 14 million degrees!

The solar system

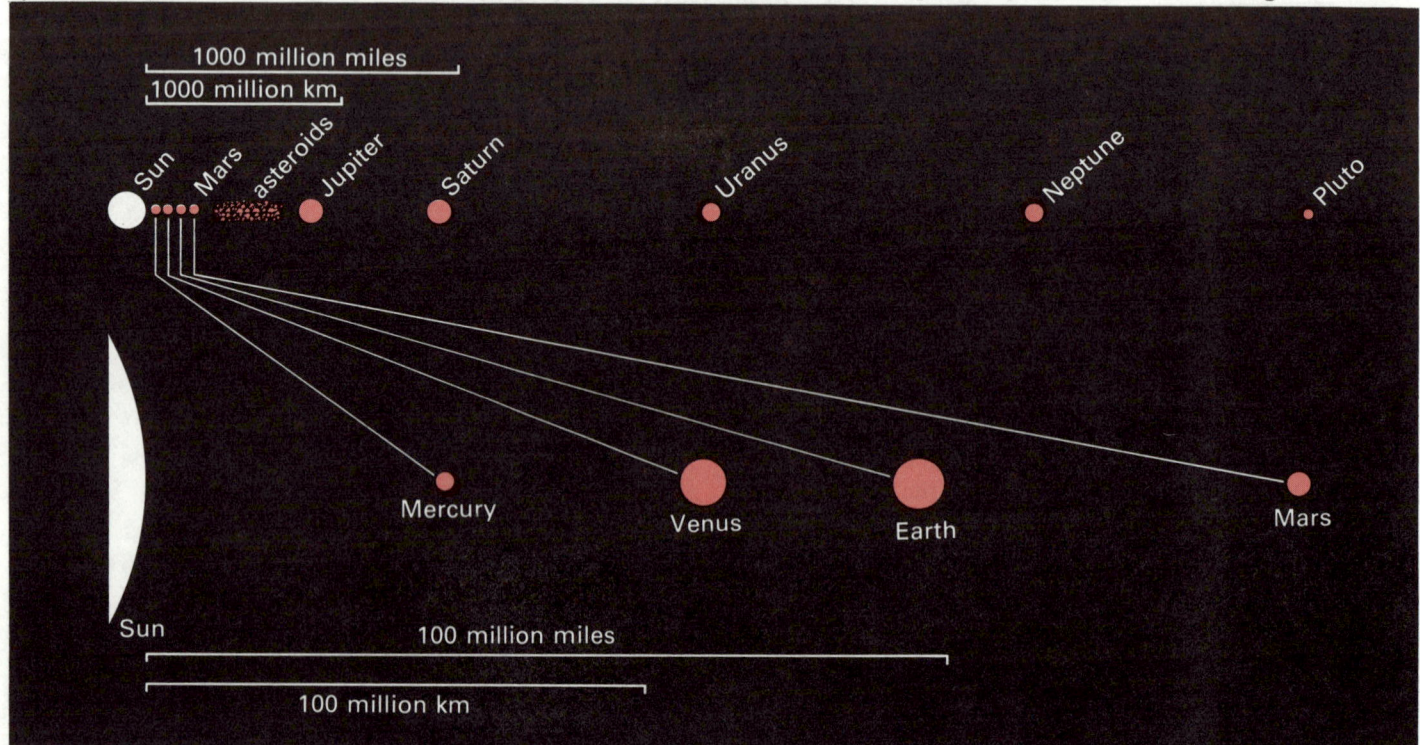

	Diameter	Mass	Volume	Distance from sun (millions of km)	Time to go round the sun	Atmosphere	Number of moons
	(Compared to that of the earth)						
Sun	109.3	333 434	1 300 000	–	–	–	–
Mercury	0.39	0.06	0.06	58	88 days	–	0
Venus	0.96	0.82	0.92	108	225 days	carbon dioxide	0
Earth	1	1	1	150	365.26 days	air	1
Moon	0.27	0.012	0.02	–	–	–	–
Mars	0.53	0.11	0.15	227	687 days	carbon dioxide	2
Ceres	0.06	–	–	430	–		
Jupiter	11.0	318.3	1318	777	11.86 years	methane	12
Saturn	9.0	95.3	736	1425	29.46 years	and	9
Uranus	4.0	14.7	64	2867	84.02 years	ammonia	5
Neptune	3.9	17.3	39	4500	164.8 years		2
Pluto	0.5	1.0	0.1	5905	247.7 years	–	0

The nuclear change producing the energy that causes this temperature has probably been going on for about 5 000 million years. It will most likely go on for another 10 000 million years. Sunspots show on the sun from time to time. They are seen as dark areas. They cause magnetic disturbances on the earth. These reach a maximum effect about every 11 years.

Asteroids

Between Mars and Jupiter there are several thousand relatively small bodies. These are planetoids or asteroids. The largest is Ceres which is only 780 km in diameter.

Comets

These are bodies made up of very low density material. A comet has a trail of gas or dust behind it. The 'tail' always points away from the sun.

Meteors

You might have seen 'shooting stars' at some time. They are pieces of rock or metal entering the earth's atmosphere at high speed. They glow with the frictional heat and are burnt up, usually when they are more than 50 km away above the earth.

Meteorites

On rare occasions, larger pieces of meteors reach the earth, without being burnt up. The few that have been observed leave large craters.

Tides

There are usually two high tides in 24 hours 50 minutes. They are caused mainly by the moon attracting the mass of water in the oceans.

Spring tides at full moon and new moon, are greater than neap tides at the first and third quarters of the moon. Spring tides are due to the moon and the sun both pulling together. For neap tides the moon's attraction is at right angles to the sun's.

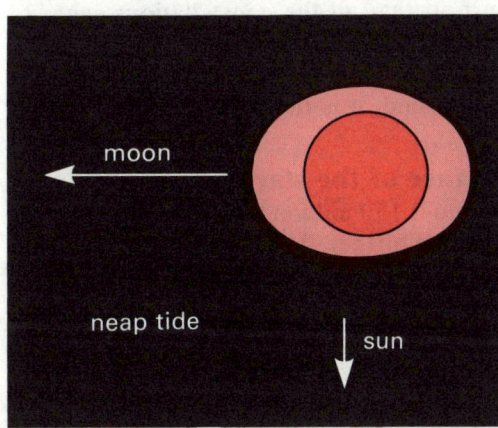
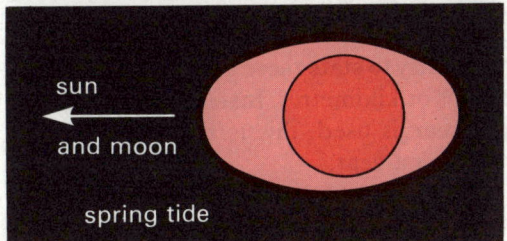

Tides
The gravitational pull of the moon causes the tides

1 Which planet is nearest the sun? Which planet is nearest the earth?
2 Which is the largest planet? Which is the smallest?
3 What is the difference between meteors and meteorites?
4 Use the table to work out the density of Neptune.

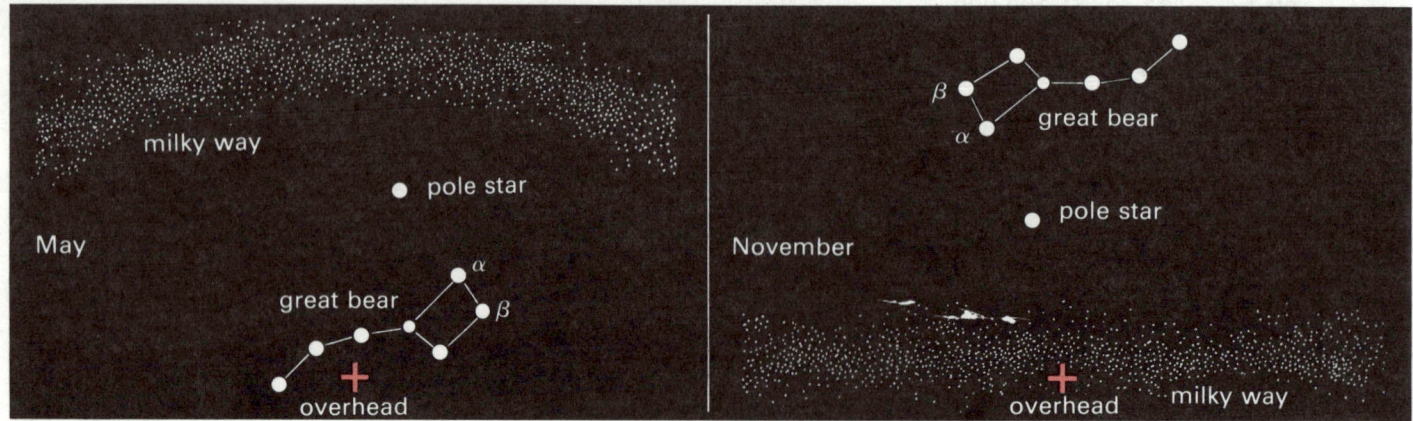

The night sky
The stars appear to rotate around the pole star. The drawing on the left is the situation in May and on the right in November

Parallax
As the earth moves from one side of its orbit to the other the relative position of the stars change

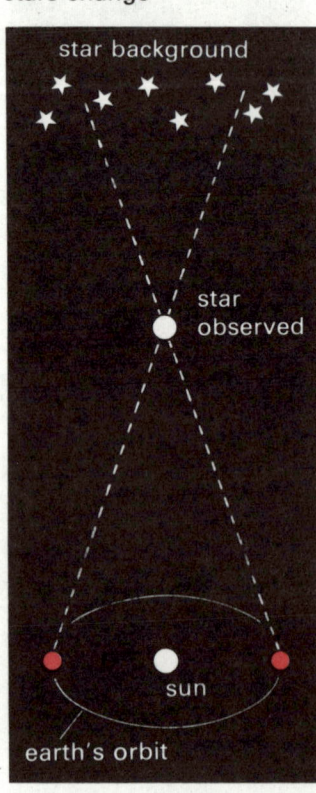

When you look up at the sky on a clear starry night it is easy to see why people living many years ago thought the sky was a great dome with twinkling lights hanging from it.

The Pole star
If you keep looking night after night or if you take a photograph with a long exposure time, you find that all the constellations seem to move around one particular star. This is the North Pole Star (Polaris). The stars do not move round; it is the earth that is rotating.

Distance of the stars
The sun is 150 million km away, but the stars are much further. The nearest star, Proxima Centauri, is so far away that the light from it takes 4.2 years to reach the earth. When you remember that light from the sun reaches us in 8 minutes, you can realize how far away the stars are.

Light year
It is clumsy to state these enormous distances in miles or kilometres. Instead, a unit called a **light year** is used, this is the distance light travels in a year.

The distance of the star 61 Cygni is given as 10.8 light years; This means that this star is 104 260 000 000 000 km away.

Measuring distance
The distance of the nearest stars can be found by **parallax.** You must have noticed the effect of parallax when riding along in a car or train. Suppose you see a cow in a field against a background of trees. As you go forward, the cow seems to move against the trees in the opposite direction. The telegraph poles being

nearer to you move even faster in the opposite direction.

This method of parallax can be used to estimate the distance of a star viewed against a background of other more distant stars. This is possible because in six months the earth has moved to the opposite point on its orbit, 300 000 km away and the shift in the position of the observed star can be measured.

Orders of magnitude
Distances can also be estimated by noting how bright a star is. Some stars are 50 000 times as bright as the sun, but they do not seem as bright because they are so far away.

The temperatures of the stars can be roughly judged from the colour of the light they give out. These range from about 30 000 K for bluish-white to about 3000 K for red light.

Red giants and white dwarfs
Some stars are very bright, although their surface colour is red. These are red giants. Some of these 'supergiants' may be a hundred times as large as the sun. But they are made up of rarefied gas and have a very low density.

White dwarfs are the opposite, with a bluish-white surface colour but not as bright as you would expect. Their density is so high that a teaspoonful of their material would weigh several thousands of kilograms on earth.

The Milky Way
On a clear night a luminous band with irregular edges can be seen across the sky. This is called the Milky Way. This band is caused by the light from stars in our own galaxy. This

galaxy is in the shape of a disc and contains many millions of stars. When you look across the disc you see the light of millions of stars. But when you look at an angle to the disc you see far fewer stars.

There are many galaxies apart from our own galaxy, the Milky Way. There are even clusters of galaxies thousands of millions of light years away.

our galaxy

The dot to which the arrow points is our solar system.

Our solar system is on the edge of a galaxy

Novae and supernovae

From time to time very bright stars are seen in distant galaxies. The brightness seems to be due to a star exploding. The most famous was described by the Chinese in 1054 AD. The remains can be seen today over 900 years later as the Crab Nebula. It is so far away that the original explosion must have happened some 5000 years before 1054.

In recent years more objects have been detected far out in space. This has been possible by using radio telescopes. These pick up radio waves just as ordinary telescopes pick up light waves. **Pulsars** give out radio signals in short pulses. **Quasars** or quasi-stellar radio sources, are even further away. The immense energy they give out is not understood.

1 A unit of distance used by astronomers is called a parsec. It is equal to 3.26 light years. How many million of km is this?
2 Suppose you see a balloon in the sky in line with a church spire. When you move your head to the left, the balloon seems to move away to the right of the spire. Which is nearer to you, balloon or spire?
3 Which of the following is furthest away from the earth? Sun, Moon, Pole Star, a quasar, Jupiter.

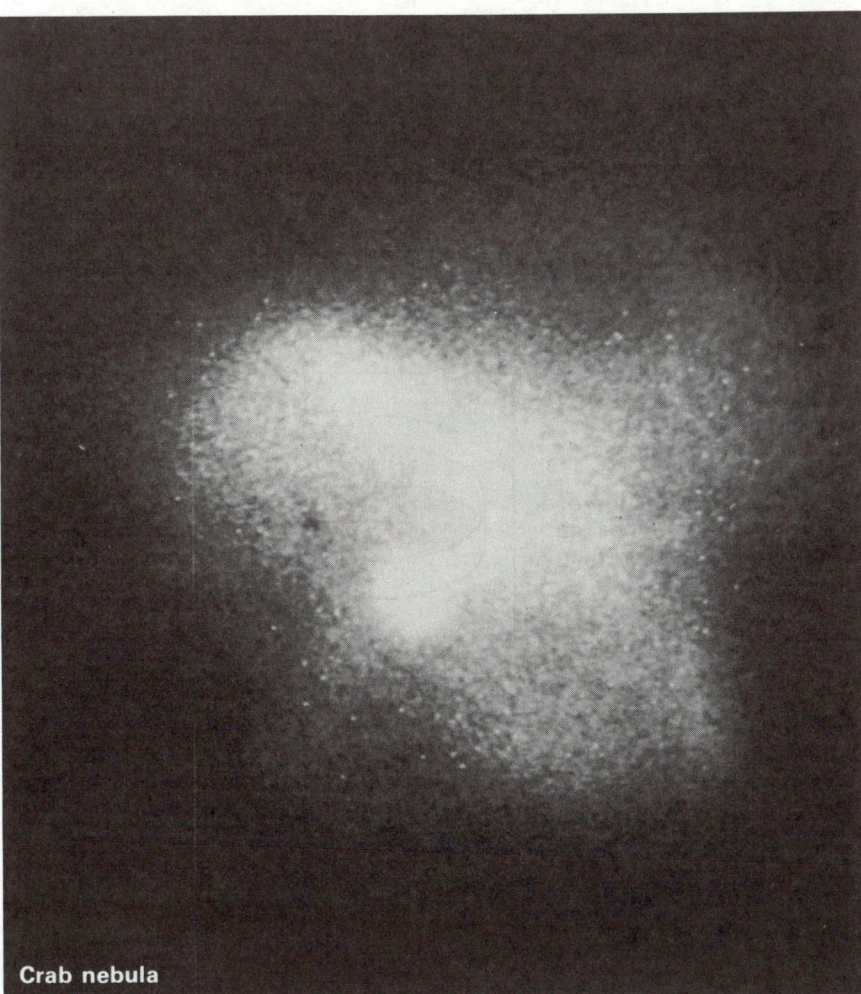

Crab nebula

Spiral nebula

A wire carrying a current produces a magnetic field
The magnetic field can be detected with a compass: the direction it points depends on the direction of the current and whether the wire is above or below the compass

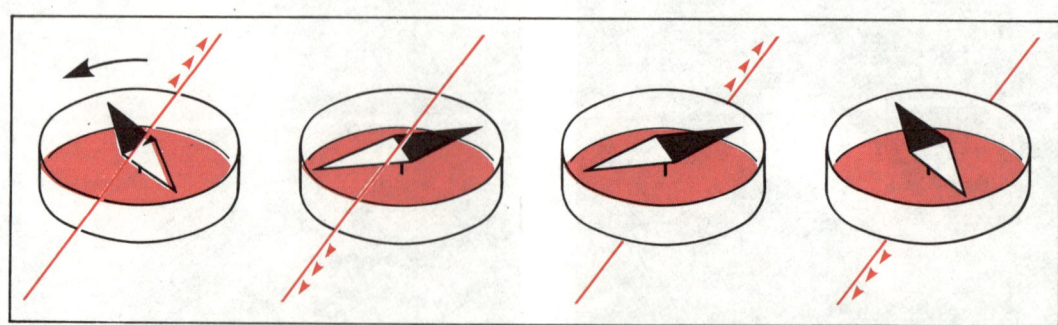

Iron filings show the type of magnetic field produced by the current flowing in the wire

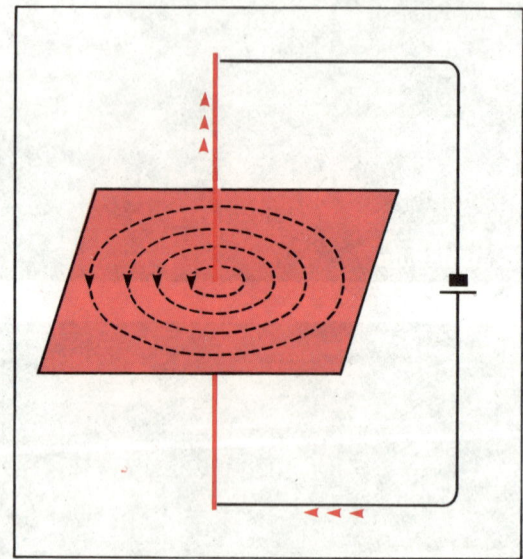

As mentioned in Topic 87, there is a magnetic field in the region of a wire carrying a current.

The way in which the north pole of a compass needle is affected depends on which way the current is flowing and whether the wire is above or below the compass.

The field can be shown by the iron filings method. A fairly large current flows in the wire shown. The lines of magnetic flux are in circles around the wire.

If you point the thumb of your right hand in the direction of the current (that is against the flow of electrons), your clenched fingers show the direction of the magnetic field.

Making a magnet

A piece of steel can be made into a magnet by putting it in a coil of wire. A current is passed through the coil (not through the magnet itself).

The field set up in a coil can be increased:

a by increasing the number of turns of wire in the coil;

b by increasing the current;

c by having a soft iron core in the coil;

d by bringing the poles closer together. This can be done by having two arms of soft iron with coils wound in series. The magnetic field between the tapered pole pieces is very strong indeed (see diagram on far left).

The electromagnet used for lifting heavy steel bars of scrap iron is a squat, circular magnet. Several turns of thick copper wire carry a current of 12 – 16 A from a 220 V direct current source.

Electromagnets

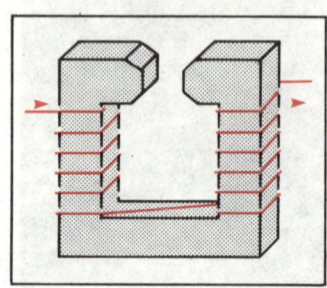

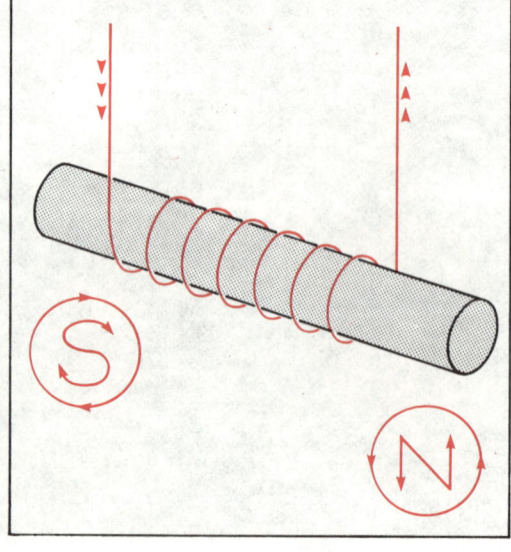

Electric bell

The circuit here is so arranged that current only flows when the light spring L touches the adjustable contact screw S.

But when current passes, the soft iron armature A is attracted towards the coils. So the hammer H strikes the gong G.

Now the contact is broken at L so the armature falls back. The whole process is repeated quite rapidly. This gives the familiar ringing sound.

The electric horn of a motor car and a morse buzzer work in a similar way. The armature is connected to a springy metal diaphragm instead of the bell hammer.

Earphones

In one type of earphone, the thin steel diaphragm is quite near the pole pieces of an electromagnet. The current that flows in the coils of this electromagnet varies many times per second according to the sounds fed into the microphone or telephone mouthpiece at the speaker's end. The changing current causes the earphone diaphragm to vibrate in the same way, so producing the same sounds.

Moving iron ammeter

When current flows through the coil a magnetic field is set up. This field attracts the iron. The amount by which it is attracted depends on the size of the current. So the pointer moves over the scale to show the current flowing.

Tape recorder

A tape recorder makes use of magnetic fields set up by a current. The current fed into the recording 'head' varies according to the sounds fed in at the microphone.

A varying magnetic field is produced at the narrow space between the poles. This causes a varying magnetising effect on a layer of magnetic oxide on one face of the plastic tape.

When the tape is played back the sounds are reproduced by producing current that fluctuates in the same way as the original.

1 Suggest three possible reasons to explain why an electric bell might not ring when the bell-push is pressed.
2 Make a sketch of the magnetic field you would expect to find in and around a solenoid (coil).

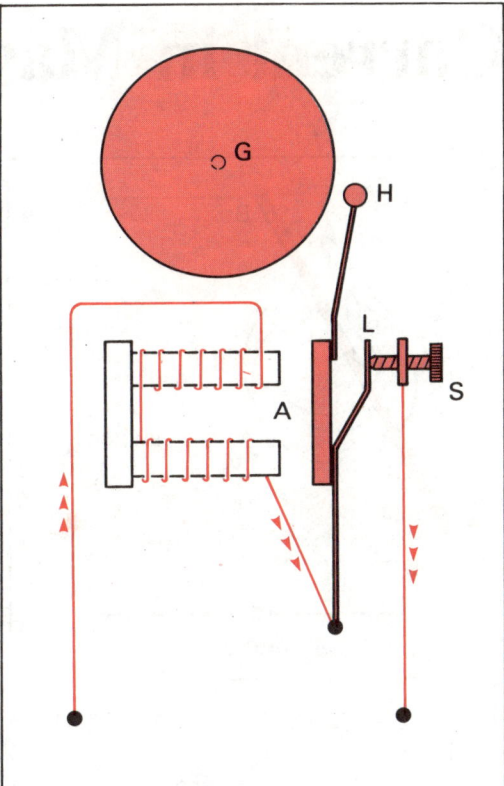

Electric bell

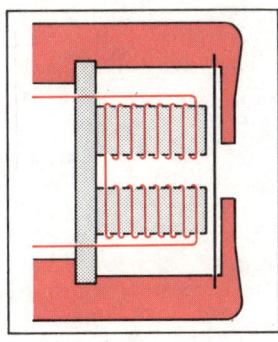

Earphone

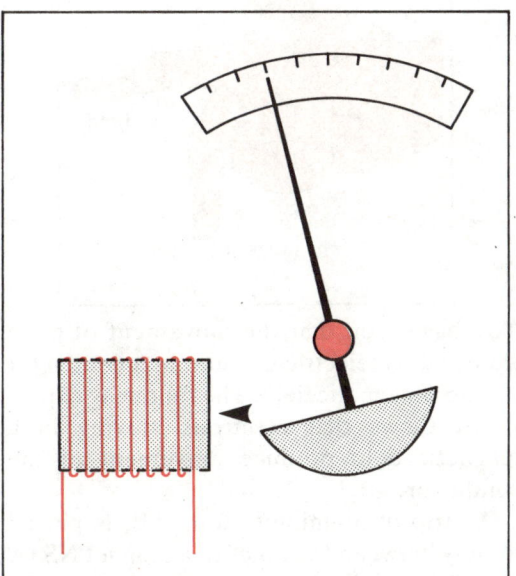

Moving iron ammeter

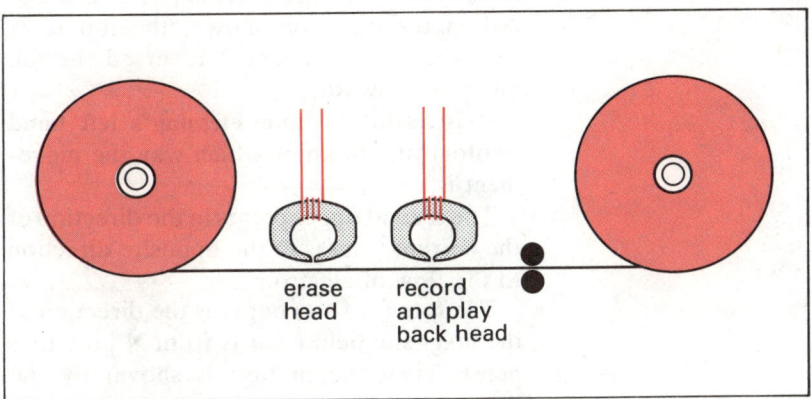

Tape recorder

erase head

record and play back head

Fleming's left hand rule
Check for yourself which direction the aluminium strip should move

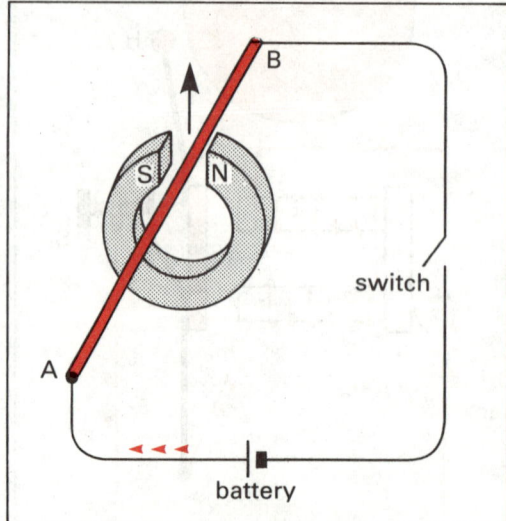

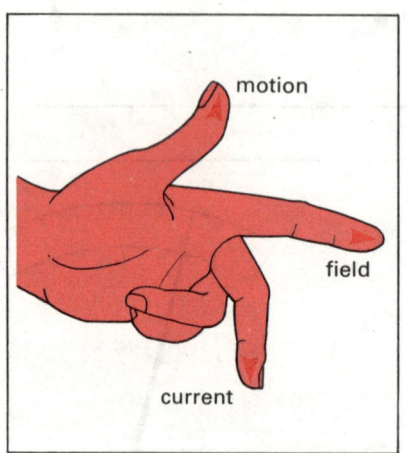

You have seen that the movement of electrons, i.e. an electric current, in a conductor sets up a magnetic field. The following experiment shows that electrons moving in a magnetic field produce movement of the conductor.

A strip of aluminium foil, AB, is placed loosely between the poles of a magnet NS. A and B are joined to a battery. When the switch is closed and a current passes through AB in the direction shown, the foil jerks upwards. If the current is reversed the foil moves downwards.

It is useful to note **Fleming's left hand** (motor) rule to know which way the movement is.

The second finger points in the direction of the current (that is in the opposite direction to the flow of electrons).

The **f**ore (or first) finger is the direction of the magnetic **f**ield (that is from N pole to S pole). Then the **m**otion is shown by the thumb.

Barlow's wheel

Current is passed from the centre of a spiked wheel to a mercury cup at the tip of the spokes. The wheel is in a magnetic field so it moves. But this movement brings the next spoke into the mercury. So there is continuous movement.

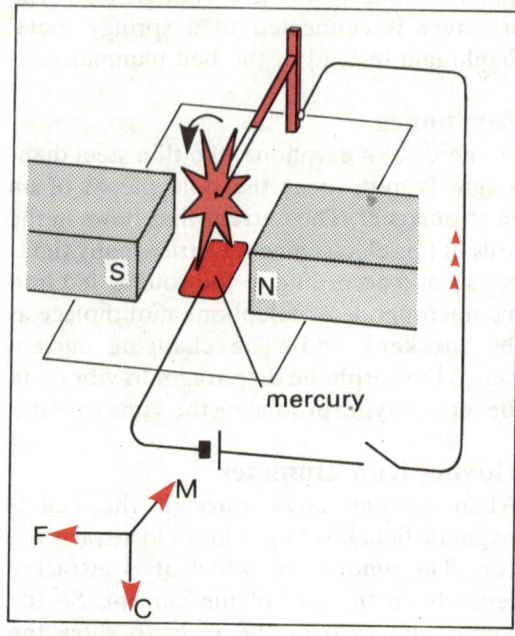

Barlow's wheel

Moving-coil galvanometer

This galvanometer has a coil wound on an iron cylinder suspended between the poles of a magnet.

When current passes through the coil the cylinder rotates by an angle that depends on the size of the current. The current can be shown by a pointer moving over a scale. This could be a milliammeter or even a micro-ammeter used for measuring very small currents.

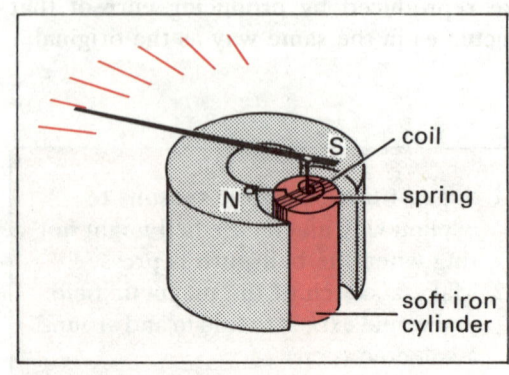

Moving-coil galvanometer

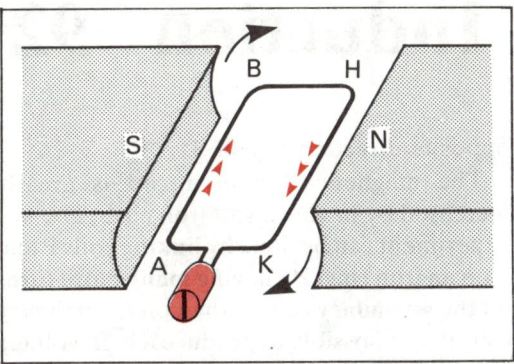

Electric motor

Here, a loop of wire ABHK is shown between the poles of a magnet. A current flows in the direction ABHK.

Using the left hand rule you find that AB is pushed upwards and HK is pushed downwards. So the coil turns. But when the loop reaches the vertical it stops.

However, if the current can be reversed at this point, then AB will be pulled down and HK up, so the loop will go on turning. This can be brought about by using a **commutator**. The current is led in and out through carbon brushes to metal segments. Each segment or half cylinder is connected to the end of the loop. Since the opposite segments make contact when the loop reaches the vertical the current will be reversed in the loop.

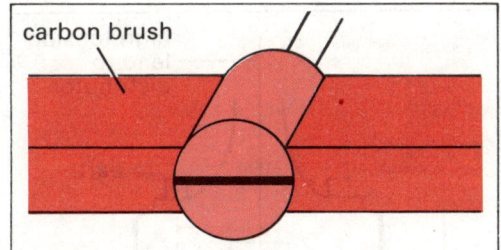

Commutator
This reverses the direction of the electric current as the motor turns

A simple motor

You can make a simple motor from a cork, a few pins and a magnet.

1 Cut grooves in the cork.
2 Wind several turns of insulated copper wire round the cork so as to fill the grooves. Connect the bared ends of this wire to pins at A and B. Note the position of these pins.
3 Larger pins at each end form an axle. The axle rests on crossed pins put in a base board to hold the cork rotor.
4 Fix two thin flexible strips of copper foil (or wire) to the base board at C and D. The strips should just touch the pins A and B as the cork twists round.

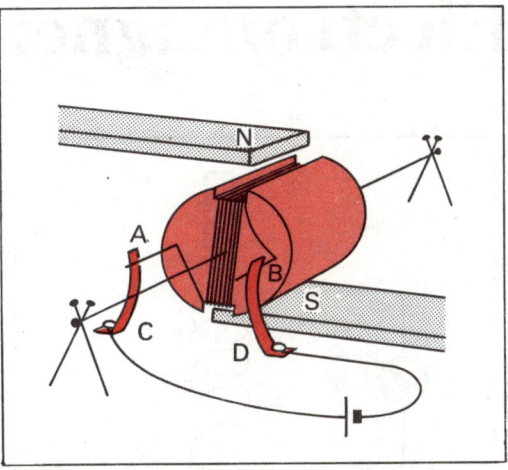

Put the north pole of one magnet above the cork and arrange the south pole of another magnet under the cork (a horseshoe magnet can be used). Connect C and D to a battery. If the cork armature is twisted to start it, it should spin round quite fast.

Electric motors

In practice, electric motors are not so simple. They have many loops. Coils are wound on soft iron. All the coils are on a rotor free to move in a magnetic field. This magnetic field is usually produced by a series of electromagnets.

The commutator is much divided because each coil on the rotor has its own pair of segments.

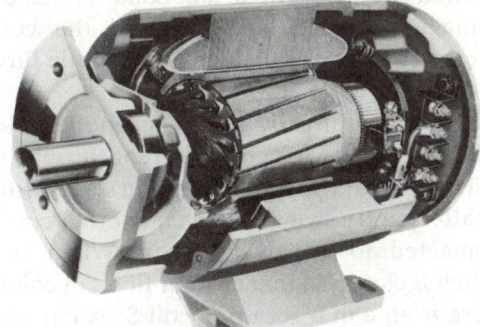

Loudspeaker

Look at the section through a common type of loudspeaker. The current varies according to the sound input. It passes into the coil. But the coil is in the magnetic field of the magnet. So every change in current is followed by movement of the cone and the sound is heard.

1 In a moving coil galvanometer, what advantage is there in having the coil wound on an iron cylinder?
2 In what ways could you improve the 'do-it-yourself' motor so as to make it go faster?

Make this simple electric motor yourself

Electric motor
This motor has many sets of coils

Loudspeaker

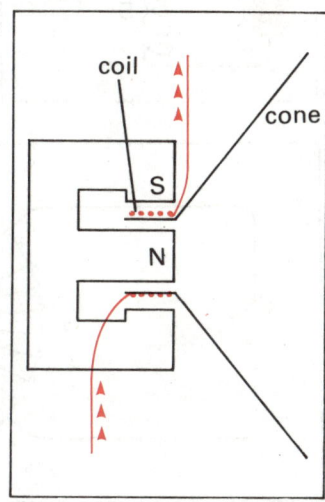

Moving a magnet near a coil produces a current

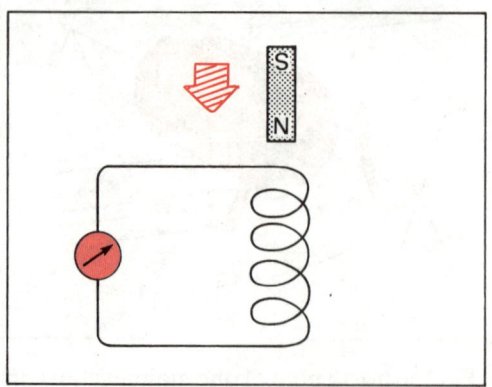

Michael Faraday (1791 – 1867) found that an electric current can be produced by moving a conductor in a magnetic field. His experiments can be repeated.

A magnet is brought up to a coil of several turns of wire. The coil is connected to a sensitive galvanometer. Moving the magnet through the coil causes a potential difference and the meter shows that a current flows. As long as the magnet is held still inside or outside the coil no current flows. When the magnet is taken out of the coil a current flows in the opposite direction. Even moving the magnet near the coil induces an **electromotive force** (e.m.f. for short) in the coil. This effect is called **electromagnetic induction.** An electromagnetic force is induced in the coil whenever there is a change of magnetic flux through the coil.

You know already that a magnetic field can be set up by passing a current through a coil. Note that the coil, P, is put in series with a battery and a switch. P is near another coil S connected to a galvanometer. When the switch is closed so that current flows in coil P there is an e.m.f. set up in coil S. A current flows. If the switch is left closed this current dies down. When the current flow in coil P is stopped by opening the switch, the galvanometer shows an e.m.f. set up in the

Currents can also be induced in a coil by the magnetic field from another coil

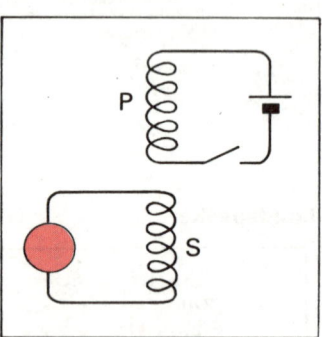

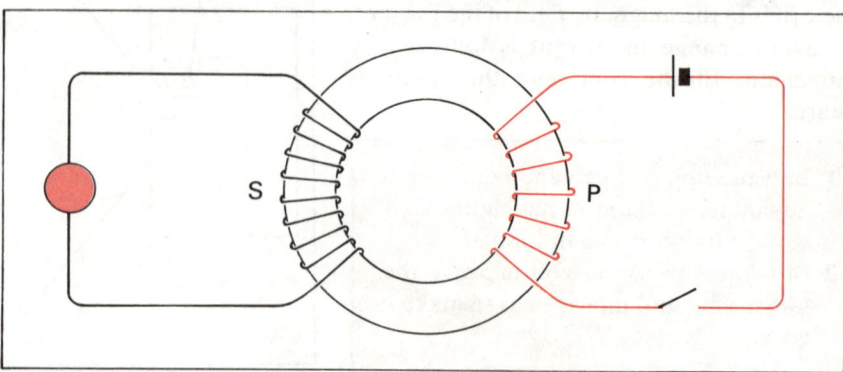

opposite direction in coil S.

The magnetic flux in a coil is greatly increased by having a soft iron core. The last experiment can be done by linking coils P and S on an iron ring. By having many more turns on the **secondary coil,** S, than on the **primary coil,** P, it is possible to produce a high voltage across the ends of the secondary coil.

Induction coil

The primary coil consists of a few turns of thick wire. The secondary coil is made up of several thousand turns of thin wire. Current is made to flow and stop again and again in the primary. This is brought about by a make-and-break device like that on the electric bell.

The electromotive force set up in the secondary coil can be enough to give a spark several centimetres long.

The induction coil was formerly used to produce high voltages for the discharge of electricity through gases and for operating X-ray tubes. You can see a kind of induction coil today in the ignition system of a motor car.

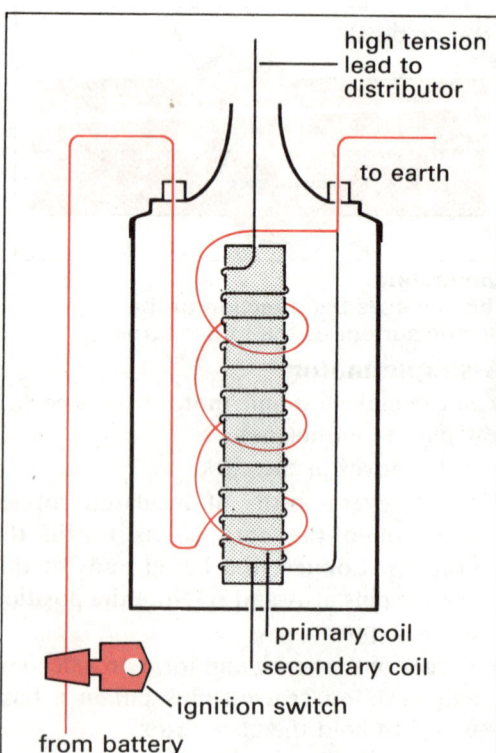

Motor car induction coil
This provides the high voltage needed to make the spark to ignite the petrol

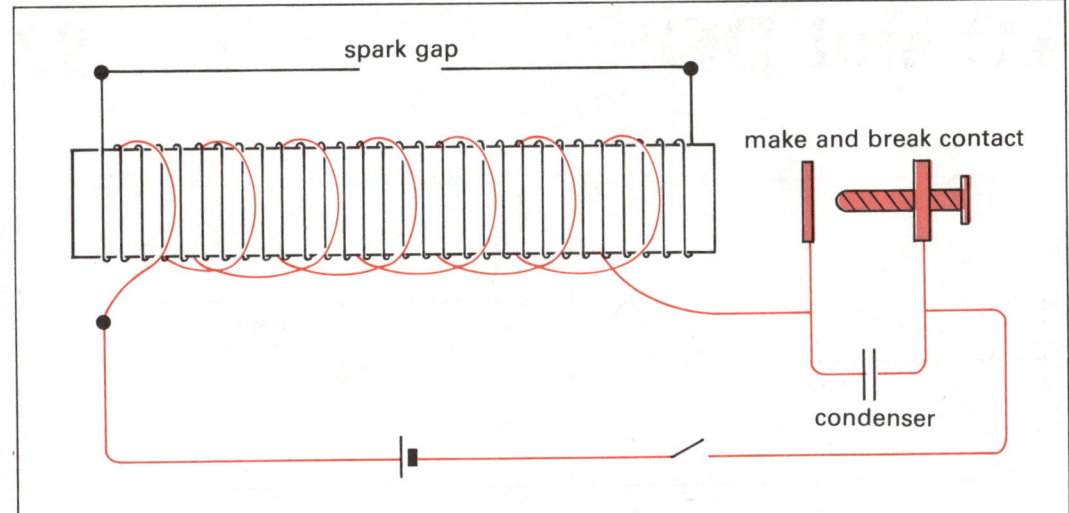

Dynamo

The simplest kind of dynamo is the type often used on a cycle for lighting.

In principle a magnet is rotated near a coil. If the coil is connected to a sensitive meter, a current flows first in one direction then the other. It is an **alternating current.**

With a cycle dynamo the current is used to light a lamp.

Alternatively a loop of wire is rotated in a magnetic field. An electromagnetic force is set up in the loop. If the circuit is completed a current flows.

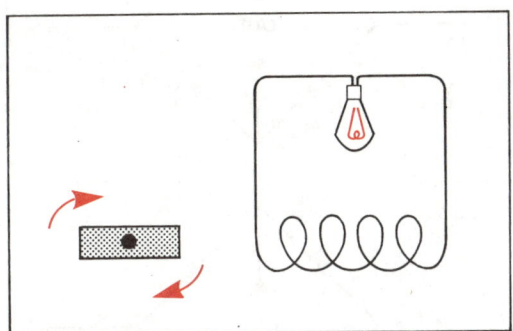

Fleming's right-hand rule (dynamo rule)

The fore finger is put in the direction of the field (north to south pole). The thumb shows the direction of movement. Then the second finger points in the direction of the current induced.

The direction of current is shown by the arrows in the dynamo. As the loop turns, the side AB will reach the top and then move downwards. The current now moves in the opposite direction.

In order to take current from the turning loop, a pair of slip-rings are fitted to the end of the loop. An alternating electromotive force is produced.

In practice, generators have many turns on the coil. The magnetic field is made much stronger by having a soft iron core in the rotating coils (armature). In some generators the coils are kept stationary and the magnet rotates.

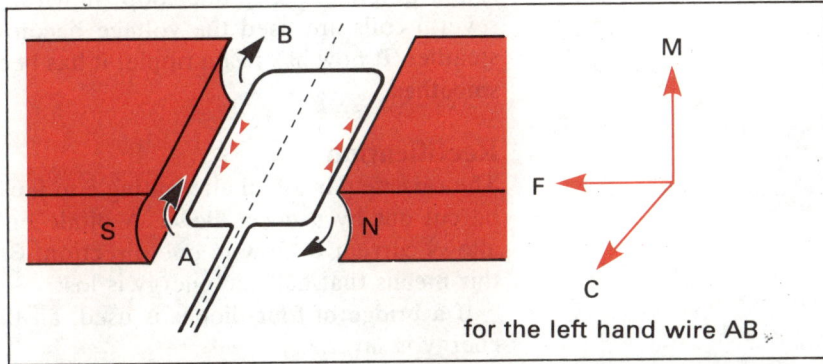

for the left hand wire AB

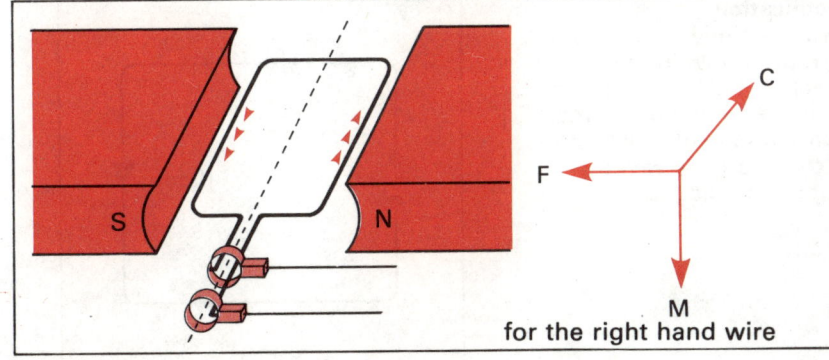

for the right hand wire

Dynamo
Fleming's right-hand rule gives the direction of the current

1 When your television set is being used, a stream of electrons comes from the back towards the screen. What would you expect to happen to the picture if you brought a magnet near the set?
2 What would you expect to see if you brought a compass needle close to a wire carrying an alternating current?

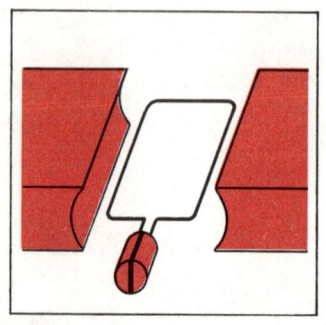

DC generator
The commutator ensures that the e.m.f. produced is in the same direction. One loop produces a varying supply (left). Most generators have many loops that produce a smoother supply (right)

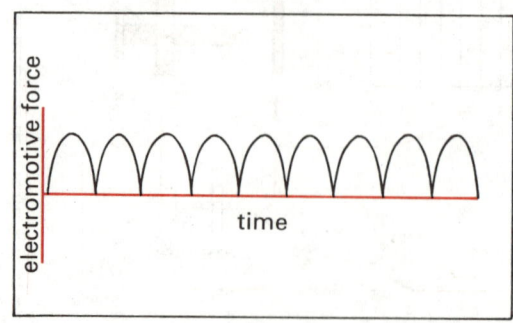

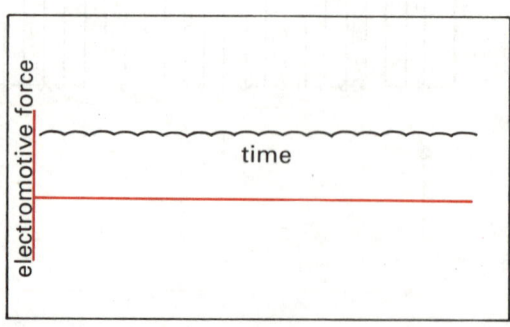

DC generator

A direct current can be produced from the dynamo described in Topic 92. It is necessary to fit a commutator so that each half of the rotation will produce an electromotive force in the same direction.

You need to remember that the wavy line is the graph for only one loop of wire. If several coils are used the voltage becomes steadier. It now has just a ripple; it has been **smoothed.**

Rectification

The reverse part of an alternating e.m.f. can be cut out by using a **diode.** A diode only allows current to flow in one direction. But this means that half the energy is lost.

If a bridge of four diodes is used, all the energy is saved.

Measuring AC voltage

Meters that can be used to measure direct current cannot usually measure alternating current. If the alternating current is rectified, an ordinary voltmeter can be used.

The voltmeter measures not the peak voltage (in this case about 340 V), but the voltage (about 240 V) of DC that would have the same energy.

Rectification
A single diode only allows current to flow in one direction, but half the energy is lost (left-hand diagrams); with a bridge of four diodes all the energy is saved (right-hand diagram)

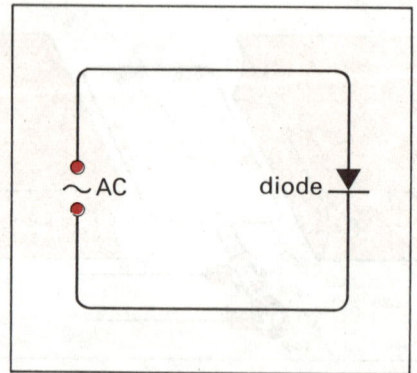

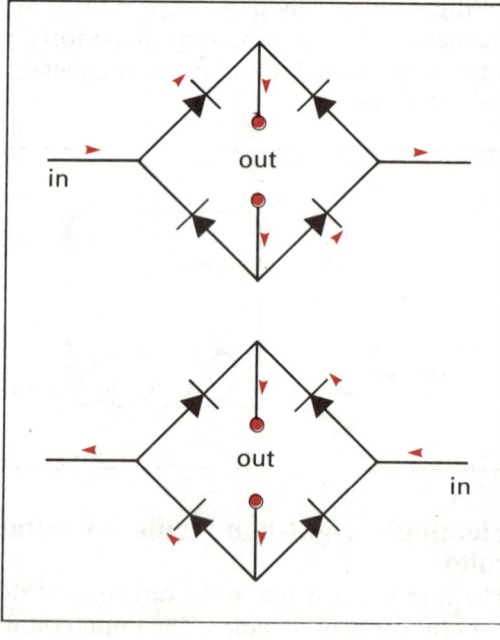

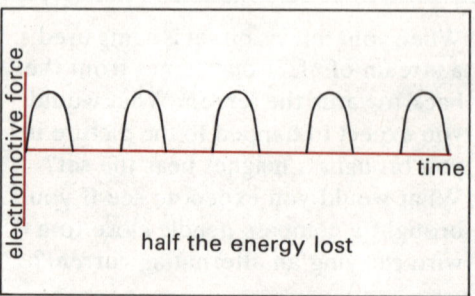

half the energy lost

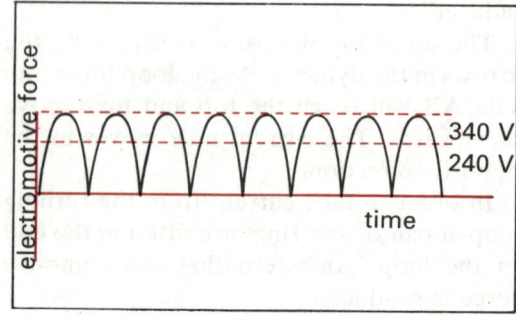

Transformer

One great advantage of AC is that voltages can be changed easily. This is because of the electromagnetic induction effect.

A transformer has two coils of wires in it. One coil is called the primary coil and the other the secondary coil. If an alternating current at a particular voltage is flowing in the primary coil then the voltage set up in the secondary coil depends on the number of turns in the secondary compared with the number in the primary.

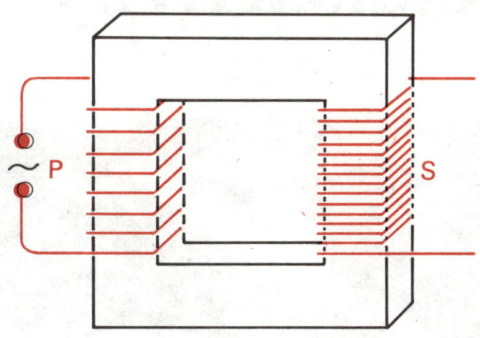

$$\frac{\text{voltage in secondary}}{\text{voltage in primary}}$$

$$= \frac{\text{number of turns in secondary}}{\text{number of turns in primary}}$$

For example, 6 volts in a primary coil of 10 turns would produce a voltage of 600 volts in a secondary coil of 1000 turns.

Electricity is transmitted from place to place by the National Grid system. Less energy is wasted if high voltage (low current) rather than low voltage supplies are used. So the National Grid carries electricity at voltages as high as 400 000 V or 400 kV. This high voltage supply is reduced to the domestic voltage of about 250 V by a succession of transformers. This would not be possible with direct current.

Effects of an electric current

A direct current shows three important effects: heating, magnetic and chemical (see Topic 53). Are these effects shown by an alternating current?

Heating. You know that alternating current produces heat. It is used in homes to produce heat in electric fires, ovens, irons, blankets and so on. The heating effect of a current depends on the **square** of the current, so that even if one half of the AC flows in a different direction, it still produces heat energy.

Magnetic. Because of the rapid reversal of current, an AC cannot be used to magnetize an iron bar. But it is not true to say that it has no magnetic effect. Alternating current will make a loudspeaker hum. This is because of the magnetic effect of AC.

Chemical. AC passed through copper sulphate solution does not deposit copper on the cathode. But when passed through acidified water, equal volumes of a mixture of hydrogen and oxygen collect at the electrodes.

This is because there is a rapid change from cathode to anode. This can be seen in the following experiment.

A sheet of blotting paper is dipped in a mixture of potassium iodide with a little starch. This is placed on a sheet of copper. One wire from a low voltage AC supply (from the step-down transformer) is joined to the copper sheet.

The other wire is now moved on the paper quickly. A broken blue line is produced. The blue is formed where iodine is formed. This only happens when the wire touching the paper is momentarily the anode.

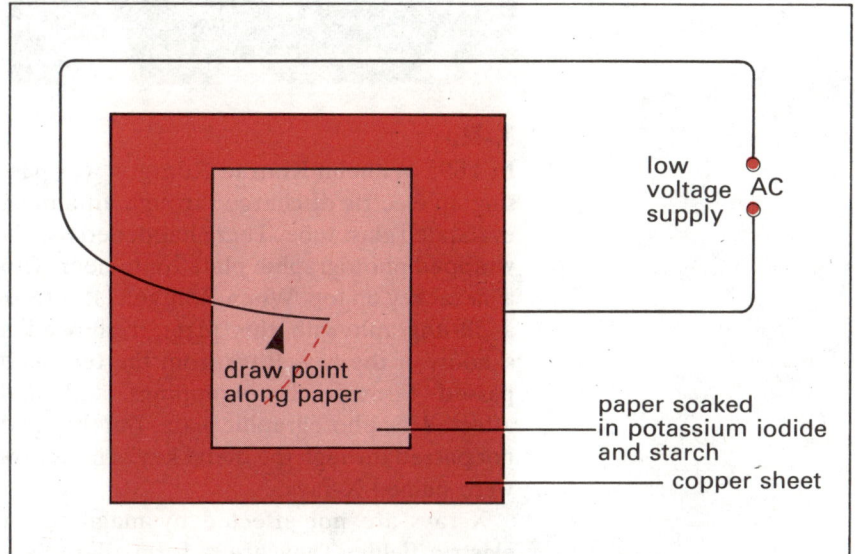

draw point along paper

low voltage AC supply

paper soaked in potassium iodide and starch

copper sheet

Alternating current
This experiment demonstrates the rapid change of direction of AC

1 By what means is energy lost in the transmission of electricity by the National Grid?
2 Draw a diagram of a step-down transformer showing the kind of coils that would be used on the primary and the secondary.
3 Which of the following could be used on AC and on DC? A moving iron ammeter; an electric kettle; a moving coil galvanometer; a transformer.

Radioactivity

X rays
This person has swallowed a
knife and fork

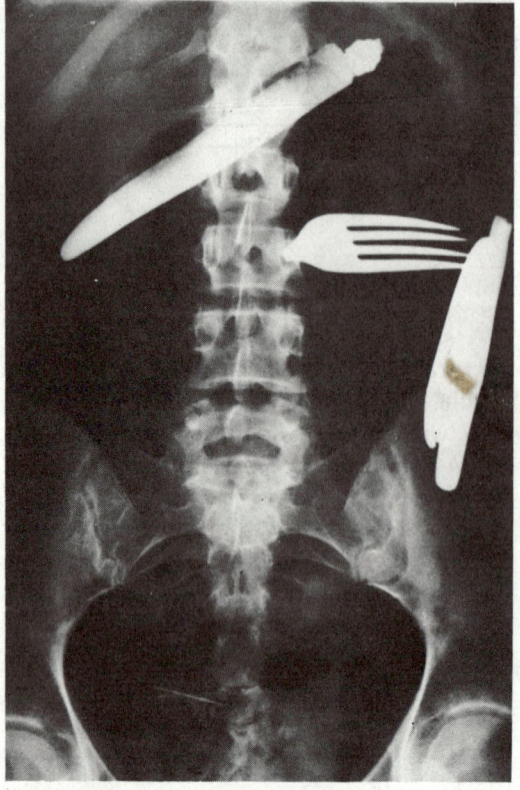

94

X-Rays

In 1895 Wilhelm Konrad Röntgen was passing an electric discharge through an almost
evacuated glass tube. There happened to be a
wrapped photographic plate quite near with
a metal key on top. When Röntgen later took
a photograph with this plate it showed a
shadow of the key. Rays from the tube had
passed through the wrappings and had
affected the photographic plate. But they had
not passed through the metal key. These rays
were named X-rays.

X-rays are not affected by magnetic or
electric fields. They are a form of radiant
energy.

Geiger-Müller tube

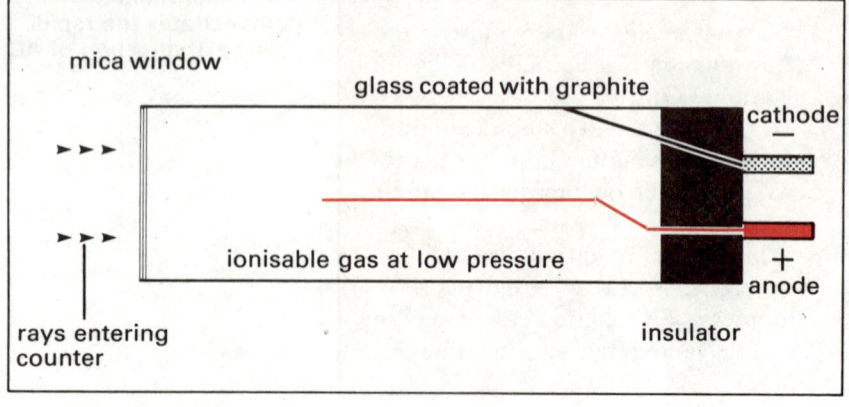

Radioactivity

In 1896, a Frenchman, Antoine Henri Becquerel, was studying phosphorescent substances. Some substances left in the sun and
then put in the dark, gave out light; they
phosphoresced.

Becquerel was examining a uranium ore.
He was surprised to find that a photographic
plate in a dark cupboard placed near the ore
was affected. But the uranium ore had not
been exposed to light. It had spontaneously
given out energy; it was radioactive.

Three different kinds of energy are given
out by radioactive substances.

α-particles. This kind of radiation is
deflected in a magnetic field in such a way
that suggests it is a stream of positively
charged particles. These particles have a
mass of 4 atomic units. They are helium nuclei. The symbol for the helium nucleus is
$^4_2He^{2+}$

β-particles are negatively charged and
have very little mass. They are electrons.
Sometimes their speed is almost as great as
the speed of light.

The γ-rays are not particles. They are electromagnetic waves, like light, radio and
X-rays. But they have a very short
wavelength. γ-rays are like X-rays in being
able to pass through matter.

α- and β-particles and γ-rays all have the
property of ionising gases.

The diagram shows a tube which contains a
gas under low pressure (bromine for
instance). A wire in the centre is positively
charged as the anode. The tube itself is negatively charged as the cathode.

With a high voltage across these terminals
no current flows until charged ions are present.

When radioactive particles or rays pass
into the tube through the thin mica window,
the bromine atoms are split into Br^+ ions and
electrons. So an electric pulse flows across
the terminals.

This is a Geiger-Müller tube and it can be
used to detect radiation. If it is connected in a
circuit having a loudspeaker, clicks can be
heard in the speaker as the electric pulses are
produced. The pulses can also be counted
and recorded on a counter.

α-particles produce many ions but they
have great mass so they do not penetrate far
into substances. Their range is only about 2
cm in air. A piece of tissue paper is enough to
stop them.

β-particles are more penetrating. They have moderate ionising power. But they are stopped by a thick sheet of aluminium.

γ-rays have poor ionising power, but several centimetres of lead are needed to absorb them.

Radioactive substances are breaking down all the time they are giving off radiation. The rate at which they are disintegrating depends on how many radioactive atoms are left. So the rate gets slower and slower. It is convenient to measure this rate by noting the time it takes for half of the original radioactive substance to break down. This is called the **half life.**

Iodine-131 has a half life of 8 days. This means that a mass of this isotope will be halved in 8 days. After the next 8 days only one quarter will remain and after 32 days, there will still be one-sixteenth left.

Plutonium-239 has a half life of 24 400 years. So long-lived isotopes like this cannot be thrown away. Their disposal needs careful planning.

The isotopes mentioned are examples of artificial isotopes made by bombarding some elements with neutrons in a nuclear reactor. These isotopes then give off radiation. They have many uses, e.g. in treating cancer, using γ-rays from cobalt-60. This isotope is also used for sterilization of medical instruments.

1 It is found that the radiation from a radioactive substance is not stopped by a sheet of paper, but is stopped by a sheet of aluminium. What type of radiation is it?
2 The count of the radiation from a radioactive piece of material was 1200 but found to have dropped to 300 when observed 24 hours later. What is the half-life of this material?
3 The following results were found for the decay of protoactinium-234. Draw a graph and use it to find the half life of this element:
You will find it accurate enough if you work to the nearest 100 for the intensity figures.

Time (min)	Intensity of radiation
0	10085
1	5763
2	3426
3	1873
4	1070
5	602
6	406

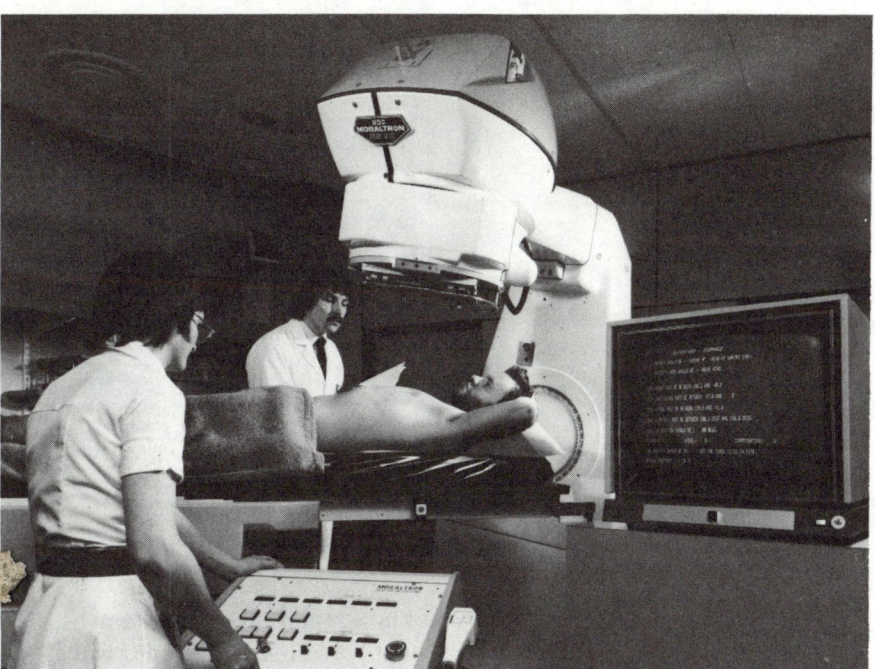

Radiotherapy
Radiation is used in the treatment of cancer

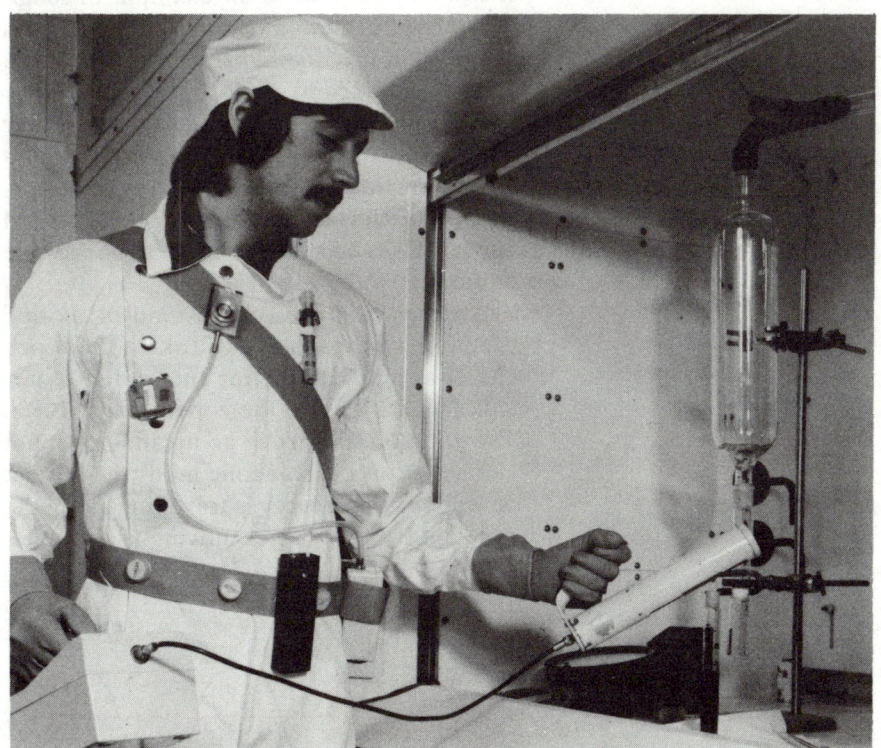

Radiation counter
This technician is checking for radiation leaks with a counter. Note the other devices on his overall that also detect radiation.

Waves

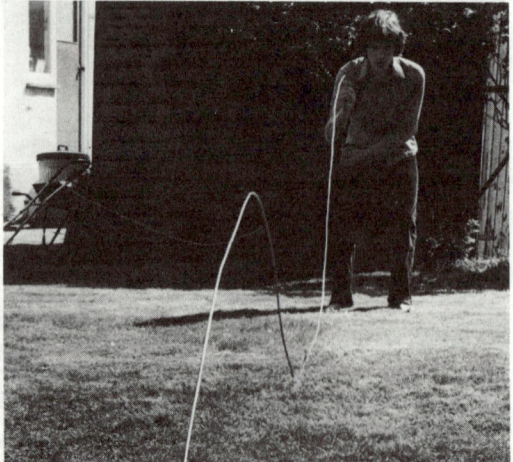

95

As you know, there are sound waves, light waves, heat waves, X-rays, radio waves and others.

Sound waves can travel in solids, liquids or gases. The vibrations of particles that carry sound through the air or through liquid or solid take place in the same direction as the sound travels. This is an example of **longitudinal** wave motion. Sound energy passes along, although the separate particles, e.g. air molecules, come back to the same place.

Waves of the sea are different. The water vibrates up and down. This is at right angles to the direction in which the energy travels. A cork on the surface of the water bobs up and down. This is an example of **transverse** wave motion.

Another example of this kind of motion can be shown with a rope. Take hold of one end of a rope with the far end tied at a suitable height. Waggle the rope up and down. The various bits of rope go up and down not along the rope. If someone holds the far end he will feel that energy is travelling along the rope but the rope is not moving as a whole.

If the distances that any given particle is pushed aside are plotted against the time, a graph is obtained. For longitudinal waves movement to the right is plotted as upwards and movement to the left downwards.

Suppose you connect up a microphone to an instrument called a cathode ray oscilloscope. This instrument is like a miniature television set. A spot can be made to move across (representing the time axis). Now if you sing a note into the microphone, the diaphragm of the microphone vibrates. This in turn causes a varying potential difference inside the cathode ray oscilloscope, forcing the spot up and down. But as the spot itself is moving horizontally, the up and down movement of the vibration produces a wavy trace.

The distance between any point on a wave to the next point in the same position, e.g. AB in the diagram, is the wavelength of the wave. The number of wavelengths that happen in one second is the frequency.

A little thought will convince you that:

$$\text{wavelength} \times \text{frequency} = \text{speed of wave}$$
$$(\lambda) \qquad\qquad (f) \qquad\qquad (c)$$

This is true of all kinds of waves. A note of frequency 256 Hz (middle C) has a wavelength of:

$$\frac{\text{velocity of sound}}{\text{frequency}} = \frac{330 \text{ m/s}}{256 \text{ s}} = 1.3 \text{ m}$$

A radio signal of wavelength 1500 m (Radio 4) has a frequency:

$$\frac{\text{velocity of radio waves}}{\text{wavelength}} =$$

$$\frac{300\,000\,000 \text{ m/s}}{1500 \text{ m}} = \begin{array}{l} 200\,000 \text{ Hz} \\ (\text{or } 200\text{kHz}) \end{array}$$

Waves

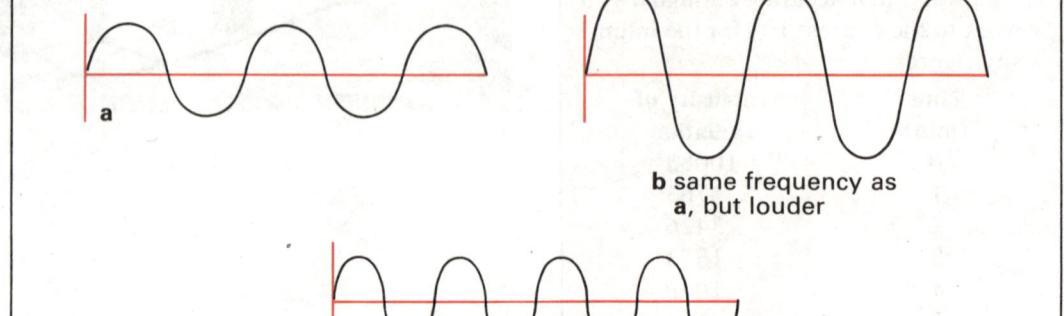

a

A ← wave length → B
b same frequency as a, but louder

c higher frequency than a

In the case of sounds the higher the pitch of the note, the greater the **frequency.** The note 256 Hz is an octave above the note 128 Hz.

The louder the note is, the deeper and higher the waves. That is, the height of the crest of the wave (the **amplitude**), measures the loudness.

Musical notes are regular vibrations that can vary in loudness (amplitude) and pitch (frequency). A noise is different from a note because its vibrations are not regular.

How is it then that a note played on a guitar sounds quite different from the same note played equally loudly on a recorder? The answer is that waves of equal frequency and amplitude may still differ in quality or timbre. The diagram shows two notes of the same frequency and amplitude but different quality.

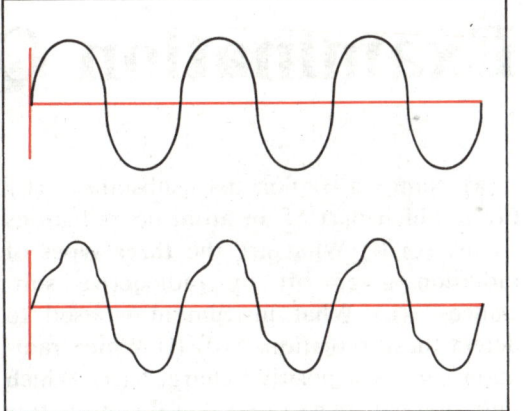

Notes
These two musical notes are of the same frequency and amplitude but of different quality

Electromagnetic radiation

Light, heat, radio and X-rays are all different from sound waves. They do not need a medium to transmit them. They are electromagnetic radiations.

As this kind of radiation passes through space each point on the path undergoes regular changes in intensity of magnetic and electric fields. These variations can be shown as waves. Electromagnetic waves have the same features as the displacement-time waves discussed above. The table shows how the various kinds of radiation differ in wavelength or frequency. All electromagnetic waves travel with the speed of light: 3×10^8 m/s.

γ-rays and X-rays have wavelengths less than one-millionth of a mm. Visible light occupies only a very narrow wave band of the spectrum. The ultra-violet and infra-red regions on either side of the visible light have a much wider range of wavelength. Microwaves and radio waves have longer wavelengths.

1 Which kind of waves have the greatest frequency: red light, X-rays, infra-red, radio waves?
2 Why can an explosion on a distant star not be detected even by the most sensitive sound detector on earth?
3 Work out the frequency of a radio signal of wavelength 300 m. (Velocity of radiation = 300 000 000 m/s.)

Electromagnetic spectrum

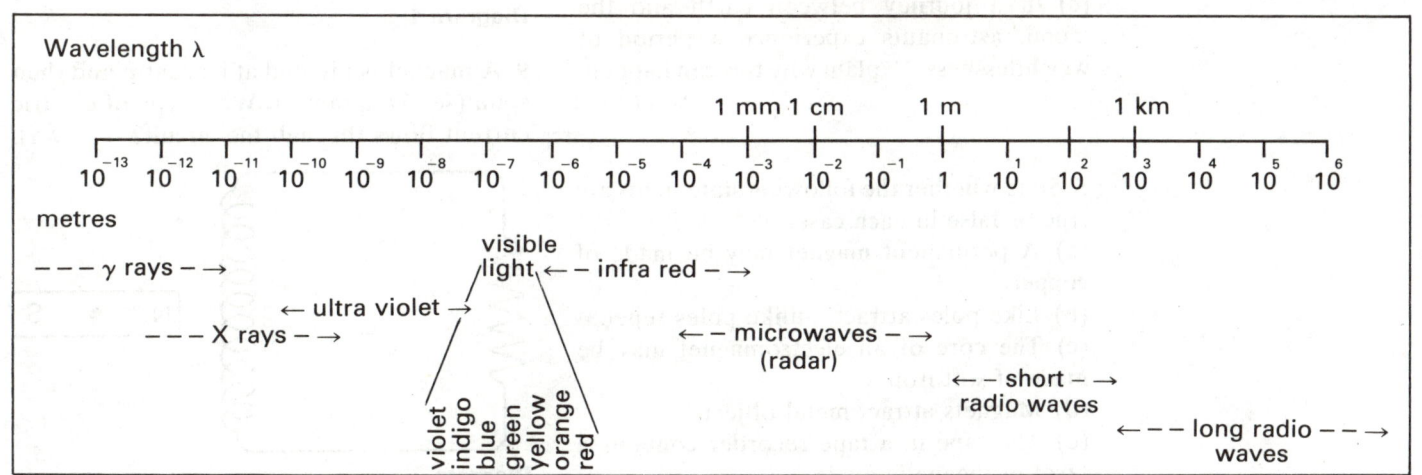

Examination Questions VIII

1 (a) Name a radioactive substance. (b) From which part of an atom do radiations come? (c) (i) What are the **three** types of radiation given off by radioactive substances? (ii) What instrument is used to detect these radiations? (d) (i) Which radiation carries a positive charge? (ii) Which radiation can penetrate several centimetres of lead? (e) What is the attraction of atomic energy as a form of energy for the future? (f) Give **two** possible dangers from increasing the number of nuclear power stations. SW

2 What are the missing words:
(a) The approximate distance of the Earth from the Sun is   *   .
(b) The planet nearest the Sun is   *
(c) The largest planet is   *
(d) The orbit of the Asteroids is between the orbits of the planet   *   and the planet   *   .

Draw labelled diagrams of the Earth, Sun and Moon showing how the following are caused on the Earth. (i) Eclipse of the Moon; (ii) Eclipse of the Sun. WYL

3 Draw a diagram of a circuit to show how the following items could be used to make an electromagnet to deflect a compass needle. Switch, soft iron rod, salt water, piece of copper foil, piece of zinc foil, compass needle, length of insulated copper wire with both ends exposed. EAN

4 (a) What is an annular eclipse?
(b) What are the following thought to be? (i) Comets; (ii) Meteors; (iii) Meteorites.
(c) In a journey between earth and the moon, astronauts experience a period of weightlessness. Explain why this can happen. WM (part)

5 State whether the following statements are true or false in each case:
(a) A permanent magnet may be made of copper.
(b) Like poles attract, unlike poles repel.
(c) The core of an electromagnet may be made of soft iron.
(d) Magnets attract metal objects.
(e) The tape in a tape recorder contains a layer of magnetic oxide. M

10 Give reasons for **each** of the following.
(a) A spacecraft may burn up when entering the Earth's atmosphere.
(b) A spacecraft can accelerate and reach extremely high speeds in outer space without burning up.
(c) The surface temperatures on the moon vary far more than the surface temperatures on Earth.
(d) Frost formation does not take place on the moon. W

11 Choose the correct answer and write down the sentence. The commutator is an essential part of (a) a direct current generator, (b) a transformer, (c) an alternator, (d) an induction coil. NW

8 See Diagram 1 showing the components of a DC bell. Copy the diagram and (i) draw the internal connections; (ii) label the electromagnet and the circuit breaker; (iii) what is the purpose of the spring at A? WYL (part)

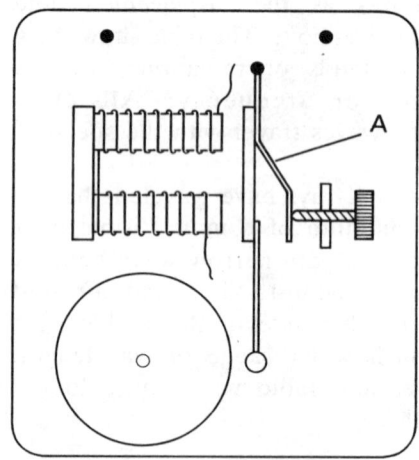

Diagram 1

9 A magnet is pivoted at its centre and then spun (see Diagram 2). What type of electric current flows through the circuit? WYL

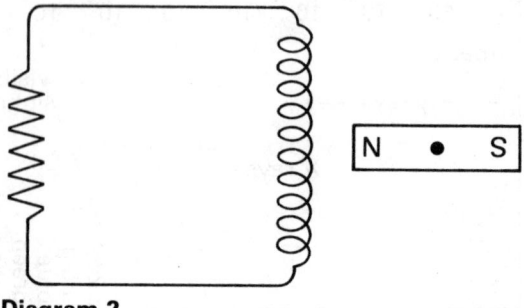

Diagram 2

10 Look at Diagram **3** of the step-down transformer.

(a) If the voltage in the primary coil is 100 V, what will the voltage in the secondary coil be?

(b) What material will A be made of?

(c) Why is this material used for A?

(d) Why will the transformer not work if direct current is passed through the primary coil?   EM

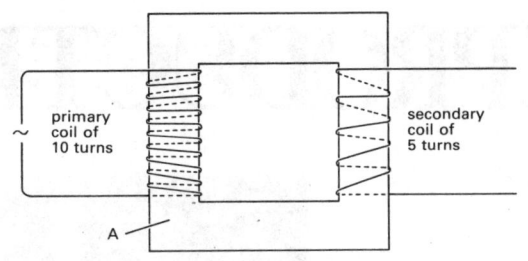

Diagram 3

11 Diagram **4** shows part of an arrangement for producing electricity. (a) State what the parts A, B and C are called. (b) Copy the axis (Diagram **5**). Show how the current generated by such an arrangement would change as the coil is turned.   EM

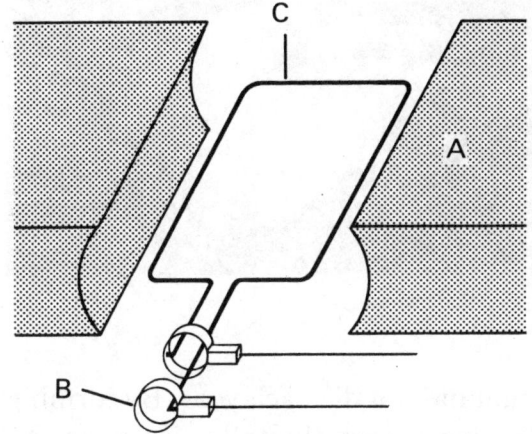

Diagram 4

Diagram 6

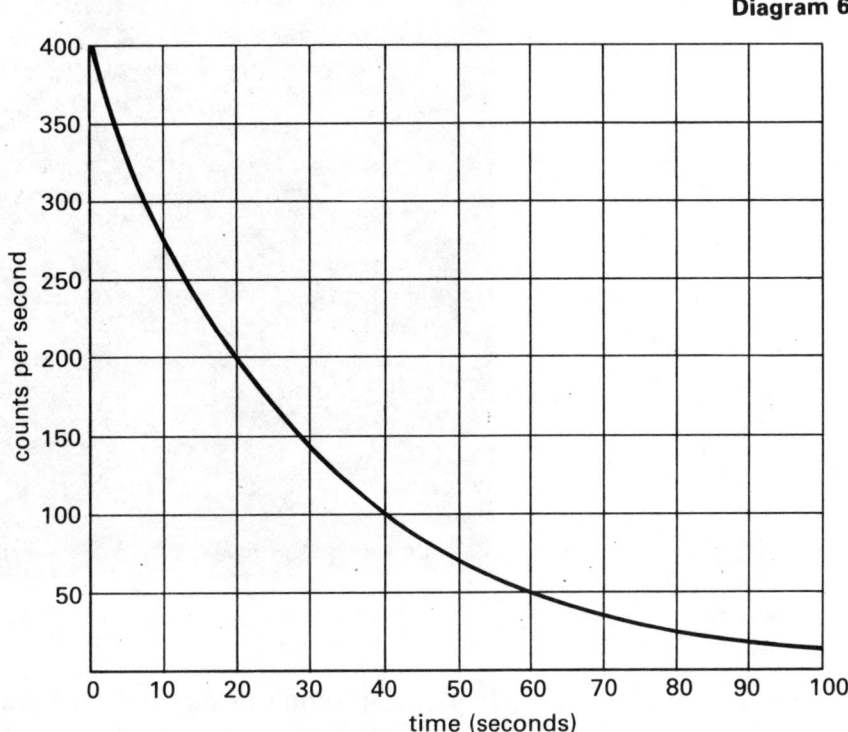

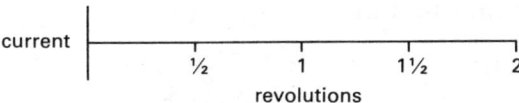

Diagram 5

12 The graph (Diagram **6**) shows the radioactive decay of an element X.

(a) What is meant by the half-life of a radioactive element?

(b) Using the graph work out the half-life of element X.

(c) What are the three types of atomic radiation?

(d) Which form of atomic radiation is the most penetrating?

(e) (i) Which form of atomic radiation is the most ionising? (ii) How does this atomic radiation cause ionisation in a gas?

(f) State one use for atomic radiation. WM

13 Complete the following sentences by using appropriate words from the following list, and write down the complete statements. Week; moon; twelve; spring; attraction; rotation; neap; month; earth; repulsion; twenty-four; fortnight.

Tides are caused by the * of the sun and moon on the waters of the oceans. When the sun and moon are on the same side of the Earth and all three are in line * tides are produced, which occur once in every * . The * rotates once in every twenty-four hours so producing two high tides at any one place in * hours.   Y

213

REPRODUCTION & RHYTHM

In Topic 45 you saw examples of plants that did not die each year. By forming bulbs or corms or by other means, fresh plants grew the following year. All plants and animals must provide for more generations if the species is to survive.

Most flowering plants form seeds by sexual means.

$$\text{Flowering plant} \nearrow \text{pollen} \rightarrow \text{male nucleus} \searrow \\ \searrow \text{ovule} \rightarrow \text{female nucleus} \nearrow \text{seed} \rightarrow \text{plant}$$

Some organisms reproduce without sex, e.g. *Mucor* by forming spores, or *Hydra* by budding.

Some plants and animals have more complicated life-cycles. A cabbage white butterfly lays eggs on the underside of cabbage leaves. The eggs hatch into caterpillars. These feed and change through a pupal stage into butterflies. The striking change from a crawling caterpillar to a flying insect is called **metamorphosis.**

Each new organism is usually like its parents. It is interesting to find out how this likeness is passed on from one generation to another.

Rhythms are to be found also in non-living things. Carbon and nitrogen can be traced through cycles in which they may form the same substances again.

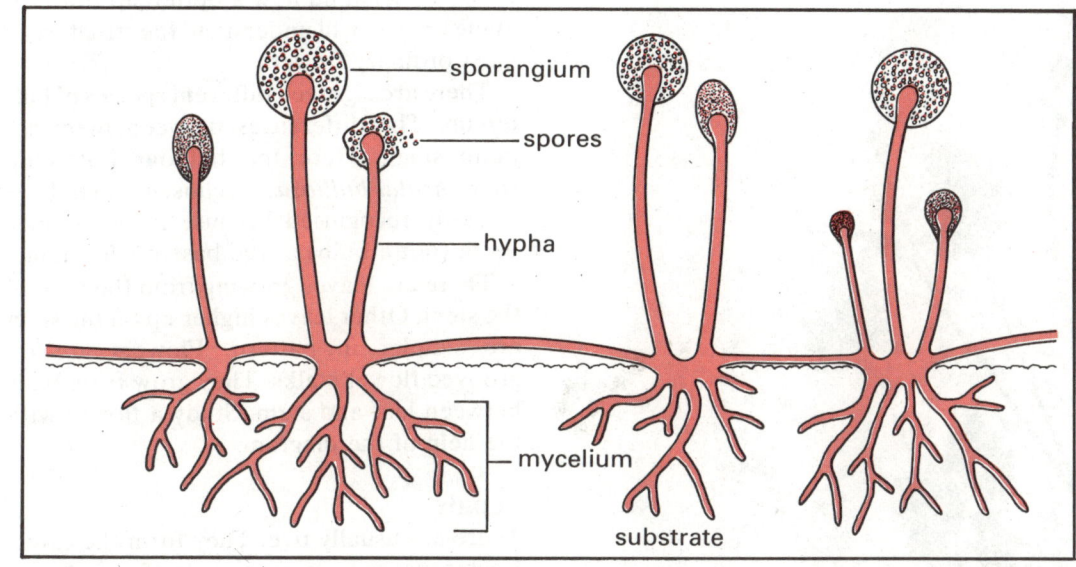

Mucor

This fungus (sometimes called pin-mould) has already been mentioned in Topic 66. It reproduces both sexually and asexually.

Asexual reproduction

Black swellings form at the top of vertical hyphae. Each swelling, or **sporangium,** is cut off from the rest of the hypha by a cross-wall. Inside the sporangium a large number of spores form.

The sporangium breaks and the spores are set free. They are so light that they are carried about on air currents. Large numbers are likely to be present in ordinary air.

When a spore settles on a suitable food it sends out a tube that grows into a mycelium. So a new mould is formed.

Sexual reproduction

The sexual method of reproduction is not so common in *Mucor*. In this process one hyphal thread comes near another of a different type (called + and −). Swellings grow where they touch.

The swellings grow into short side branches. These swollen tips (gametangia) join together to form a **zygospore.** The zygospore sends out a hyphal thread and usually a sporangium forms.

More plants come from the spores produced by the sporangium.

1 Name one plant and one animal that can reproduce by non-sexual methods.
2 Explain why cheese and some soft fruits left exposed for a short time to the air are almost certain to show a growth of mould. What conditions favour this formation of mould?
3 Pick out the correct answer: Black spots can be seen on bread on which pin-mould has grown. These are: hyphae; the mycelium; spores; sporangia; seeds.

Sexual reproduction of Mucor

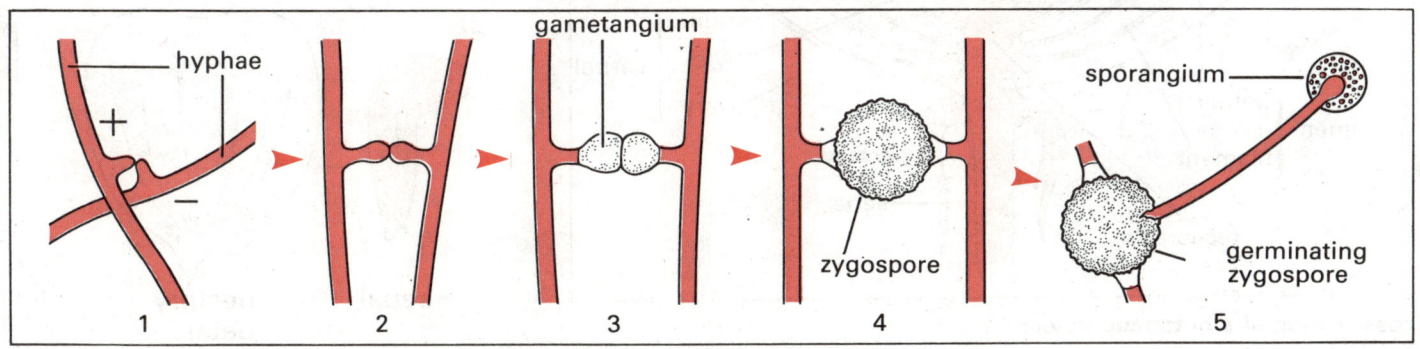

If you understand how a buttercup flower is made up, you will understand the structure of most ordinary flowers.

There are eighteen different species of buttercup. The differences between them are quite small. Here the bulbous buttercup (*Ranunculus bulbosus*) is chosen for study. It is easily recognised because it has a small corm (not a bulb) at the base of the stem.

There are leaves growing from the base of the stem. Other leaves higher up on the stem are simpler in pattern. Flowers are on grooved flower stalks. They grow from buds between leaf and stem. Study a flower with the help of the diagram.

Sepals

There are usually five. They form the calyx. Each sepal is boat-shaped. In the 'bulbous' buttercup, the sepals bend down against the stem.

Look for the sepals on a bud. Note how they protect the young developing flower.

Petals

These bright yellow petals, usually five, make up the **corolla.** They are very glossy inside. Each petal is separate, but overlaps the next one. At the base of a petal where it is joined to the swollen top of the stem, called the **receptacle,** is a small pocket. This is the nectary. It contains a sugary liquid called nectar. Honey guide lines on the petal point the way towards the nectary.

Insects are attracted by the bright shining petal and guided to the nectary for nectar.

Stamens

These form the male part of the flower. Each stamen is club-shaped. The clubbed end, the **anther**, produces pollen. There are numerous stamens.

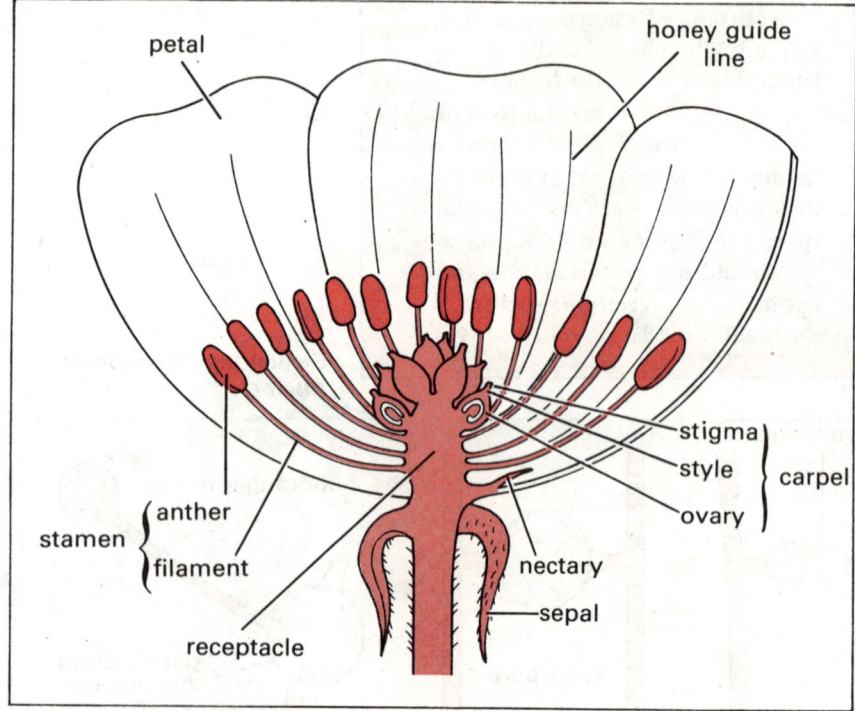

Cross section of a buttercup flower

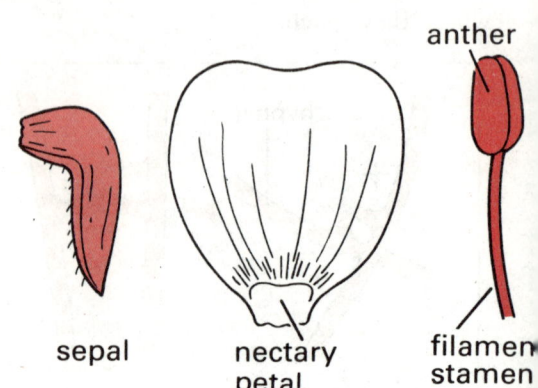

Carpels

Inside the spiral of stamens is a collection of little pips called carpels. Each carpel contains a single **ovule,** inside an ovary. At the top is a stumpy **stigma** on a short neck called the **style.**

Pollination

Bees and some other insects are attracted to the buttercup flower mainly because of its bright colour. The insect moves into the flower in search of nectar. It brushes against the anthers. So pollen grains may be caught in the hairy body of the insect.

The insect flies away and visits other buttercups. The pollen on its body may be brushed against the stigmas of the carpels.

If these stigmas are ripe they are sticky and the pollen grains are left on them. This moving of pollen from stamen to stigma is called **pollination.** If the stigma receives pollen from a different flower, it is called cross-pollination. Sometimes pollen is transferred from the anther of a flower to the stigma of the same flower. This is called self-pollination.

Some plants depend on the wind to carry pollen to the stigma. These plants do not have brightly coloured flowers. They do not need to attract insects. The anthers of grasses are often most delicately poised at the top of long slender filaments. So they are well suited for having pollen shaken out of them in a breeze.

Fertilization

Pollination is only the first step towards reproduction.

The pollen grains that land on a ripe stigma develop little tubes that grow down the style.

You can see the pollen tube under the microscope if you look at a few pollen grains on a drop of sugar solution on a slide. When the pollen tube reaches the ovule, the male nucleus from the pollen fuses with the female nucleus in the ovule. This joining together, or fusion, of male and female nuclei is **fertilization**.

Fruit

This fertilized ovule becomes a seed. The collection of seeds stays on the flower as a fruit after all the other flower parts have died away. Fertilization stimulates the seeds to grow. The ovary itself grows much larger.

A seed can be distinguished from a fruit. It has only one scar; where it was attached to the ovary. A fruit shows two marks, one where it was attached to the receptacle and another where the style was.

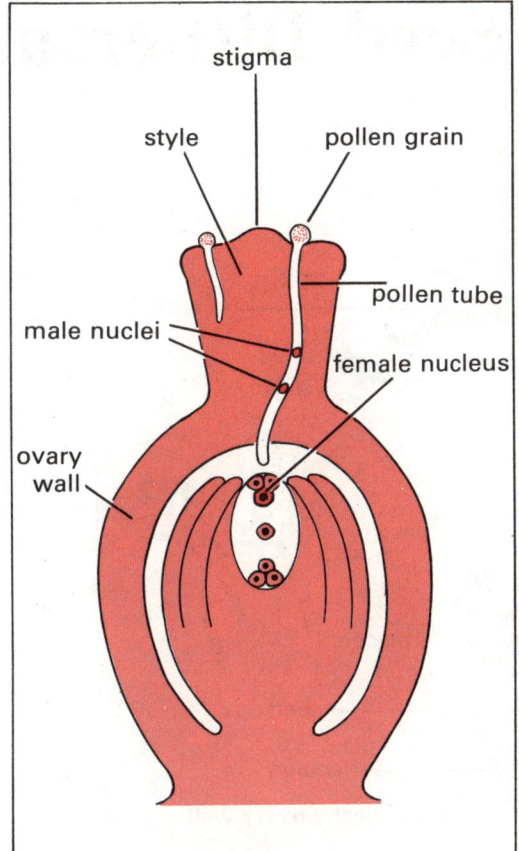

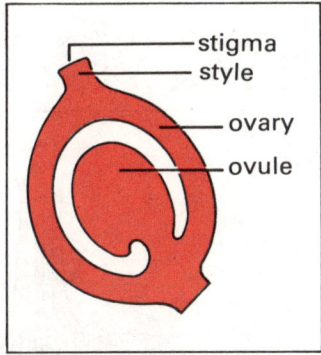

Cross section of ovary
Long tubes grow from the pollen grains down to the ovule, where fertilization takes place

Buttercup fruit

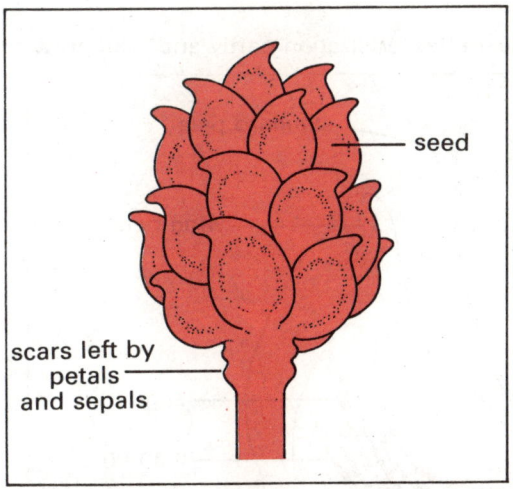

1 What is the difference between a spore and a seed?
2 Name the parts of a flower that (i) produce pollen, (ii) protect the bud, (iii) produce nectar, (iv) show insects the way to nectar, (v) form seeds after fertilization.
3 Name two plants that are pollinated by insects and two plants that are pollinated by wind.
4 Distinguish between pollination and fertilization.

Seed Dispersal

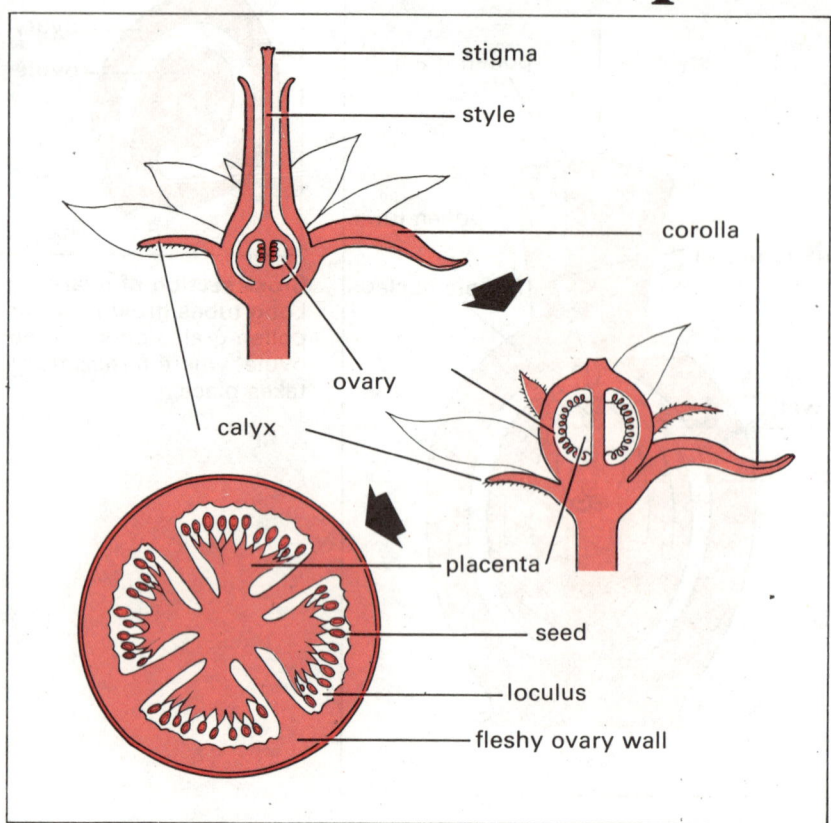

98

Tomato fruit
The drawings show the fruit just after fertilization partly and fully grown

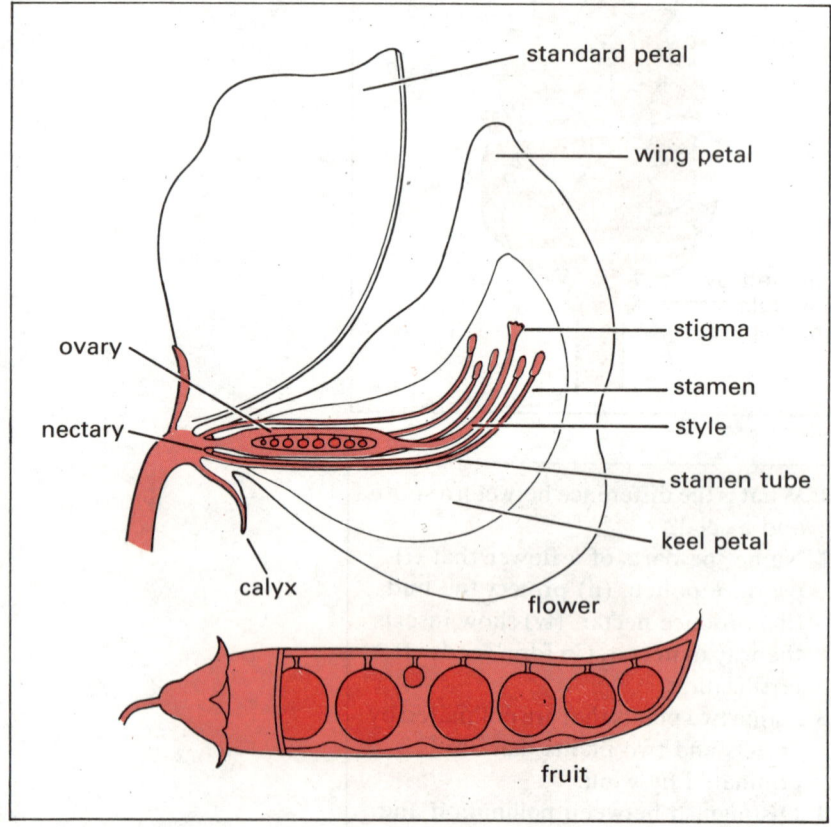

Pea
Cross section of the flower and the fruit

Fruit structure

The shape of a fruit depends on the flower.

The tomato forms from the ovary of the flower, as shown in the diagram. (You may not have thought of the tomato as a fruit, but it comes from a flower and contains fertilized seeds.)

In the pea, you see that the familiar pod fruit grows from what was the ovary in the flower.

Many flowers like daisy, dandelion and dahlia have a complicated structure. Each flower is really a colony of florets.

The fruit of the dandelion grows a little parachute, so that it is carried along by the wind.

Need for dispersal

It is vital for a plant to have some way of spreading its seeds widely. If all the seeds germinated close to the parent plant, they would be overcrowded and could not get enough room to grow. Many of them would die.

So there are a number of ways in which the fruits and seeds have become adapted to meet this need.

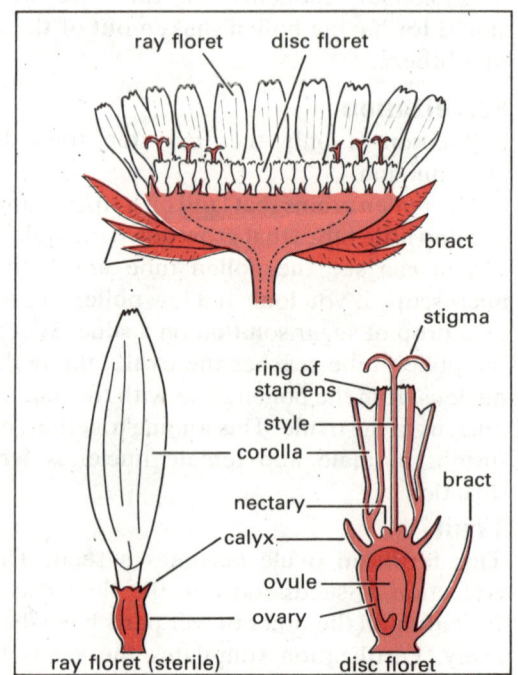

Dandelion
Cross section of the flower, also showing the structure of the ray floret and the disc floret

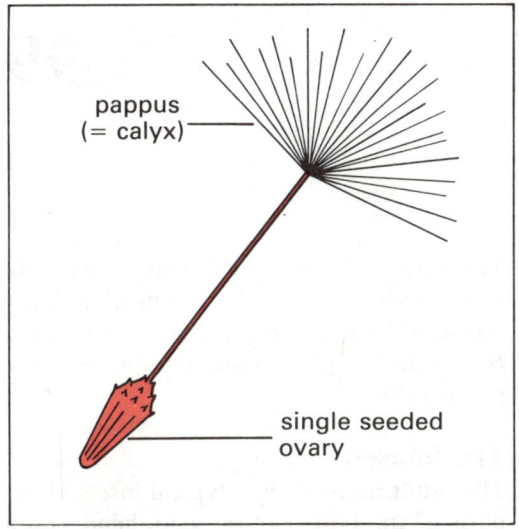

Dandelion fruit

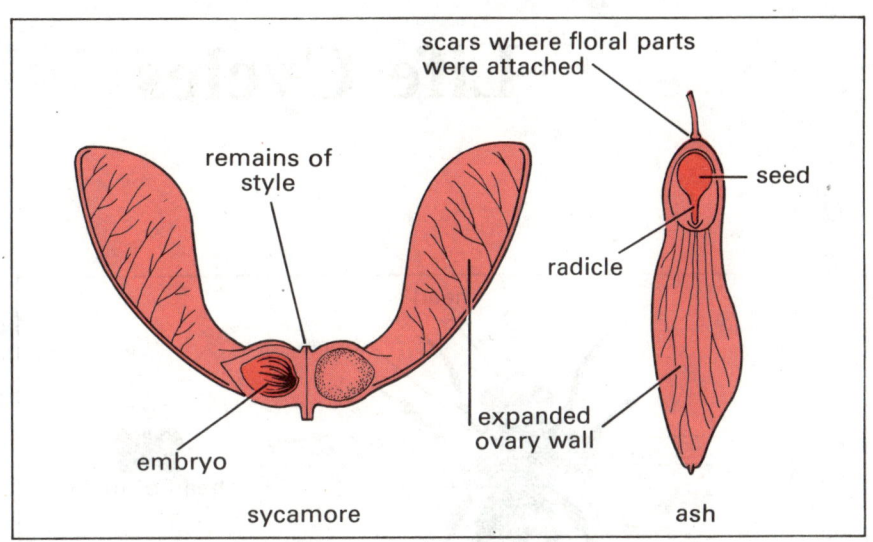

Winged seeds

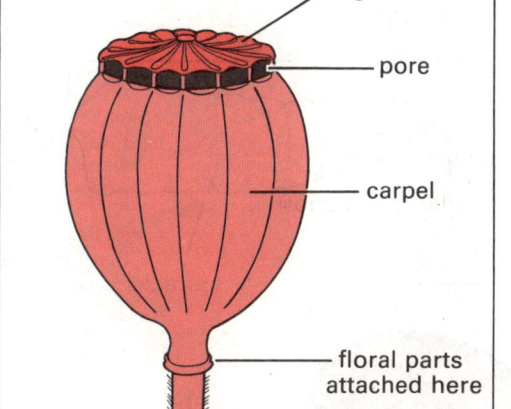

Poppy capsule

Some plants, e.g. the dandelion and the clematis, have feathery styles or tufts. These seeds are blown great distances by wind.

You see a similar device in the seeds of the sycamore, the elm, the ash and the hornbeam. Here the wall of the ovary has grown out into a wing-like attachment and the seeds flutter down quite a distance away from the parent tree.

The fruit of the poppy is a kind of pepper pot. The tiny seeds are shaken out of pores at the top when the wind blows the stem about.

Several fruits are shaped like the pea or bean pod. As the pod dries, strains are set up until suddenly the pod splits. This jerks out the seeds quite a distance. You may have heard the clicking noise of gorse and broom pods splitting during a hot day.

Other plants form fruits that are juicy. Birds and animals, including man, find these fruits good to eat. They eat the juicy part, but the seeds or 'stones' are dropped far away. The apple, strawberry and plum are good examples of these juicy fruits. If the animal swallows the whole fruit, the seeds still survive, because they are not attacked by digestive juices. They pass out with the faeces.

Water can help to disperse some fruits. Water lily seeds are very light and float away. Coconuts, the fruit of the Coco palm, can be carried away to distant islands by floating on the sea.

Apple

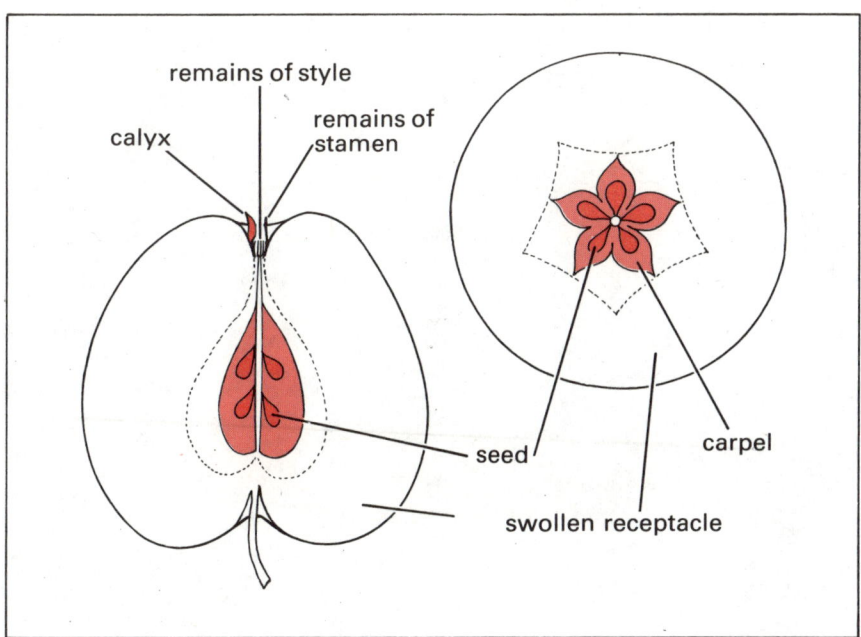

1 Name three plants whose seeds are dispersed by wind.
2 Name three plants whose seeds are scattered by animals eating the fruits.
3 Why is it important to a plant that its seeds should be scattered?

Life Cycles

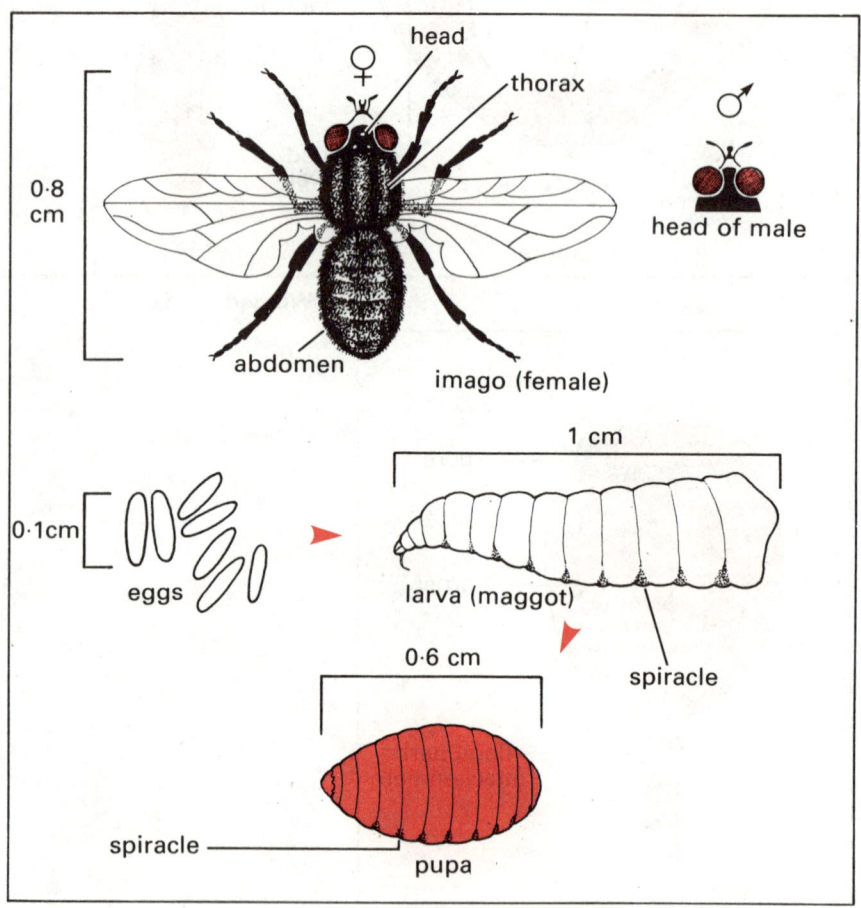

99

Life cycle of the housefly

The fertilized ovule (seed) of a pea grows into another plant. The ovum of a *Hydra* fertilized by a sperm cell grows into another *Hydra*. But life cycles are not always as simple as this.

The housefly

The adult housefly is a typical insect. Three parts of the body can be seen, head, thorax and abdomen. Three pairs of legs and one pair of wings are attached to the thorax. Where you might expect to see a second pair of wings there is a pair of stumps, called halteres. These help in balancing.

Houseflies are common enough, but there are many different species. Many people mistake smaller flies *Fannia canicularis,* as young members of the commonest species, *Musca domestica.* But adult flies do not grow in size because the body is bounded with a firm skeleton of chitin.

The fly has large compound eyes. These are rather larger in the male than in the female.

After mating masses of eggs are laid.

Eggs These are usually laid on manure or rubbish heaps. They are only about 1 mm long. They hatch into tiny larvae or maggots.

Larvae These grow to about 1 cm long. They feed greedily on waste matter.

Pupae After some days, the larval body darkens and contracts into a barrel-shaped puparium. The puparium contains the pupa inside it.

Imago (adult insect) The adult fly hatches out, depending on temperature, in anything from three days to three weeks.

On average, during summer, the whole life cycle takes two to three weeks.

The housefly feeds on anything from jam and sugar to manure. It is because of this that it can spread disease so easily.

Germs that cause typhoid, cholera and dysentery may be carried by the fly.

The frog

In the spring large numbers of frogs, after hibernating during the winter, travel to ponds. Here mating takes place. It is said that the males arrive first and wait for the females. The male frog, at this time can be distinguished from the female because he has a large pad on the 'thumb'. Females are much larger because of the bulk of eggs contained in their ovaries.

The male mounts on the back of the female and grips her firmly with the forelimb. The female is stimulated to shed the eggs into the water. The male then sheds sperm into the water. So **external** fertilization takes place. Each egg contains a large amount of yolk. There is a jelly-like covering round the egg which absorbs water and swells. The eggs hatch into larvae called tadpoles, about two weeks after the eggs are laid. The tadpoles have external gills. They attach themselves to weeds by the 'cement gland'. Development is as follows:

1st week after hatching The mouth develops and also eyes. External gills are replaced by internal gills.

2nd week Operculum closes over gills but leaves the 'spout' on the left side.

4th week Tadpole now has the typical shape. Its tail is quite recognizable.

6th week Hind legs begin to grow.

7th week Fore legs appear.

8th week Lungs develop. The tadpole comes to the surface occasionally to breathe. Tail is slowly absorbed. Tadpole stops feeding.

12th week Young frog hops about on land.

So the life cycle is:

frog → egg → tadpole → frog

1 Suggest ways in which you could protect yourself against dangers of disease that houseflies can cause.
2 The life-history of the housefly takes less time in warm weather. Explain why this is.
3 Pick out the larval forms from the following lists: tadpole; butterfly; puparium; caterpillar; chrysalis; earthworm; frog; maggot.

Life cycle of the frog

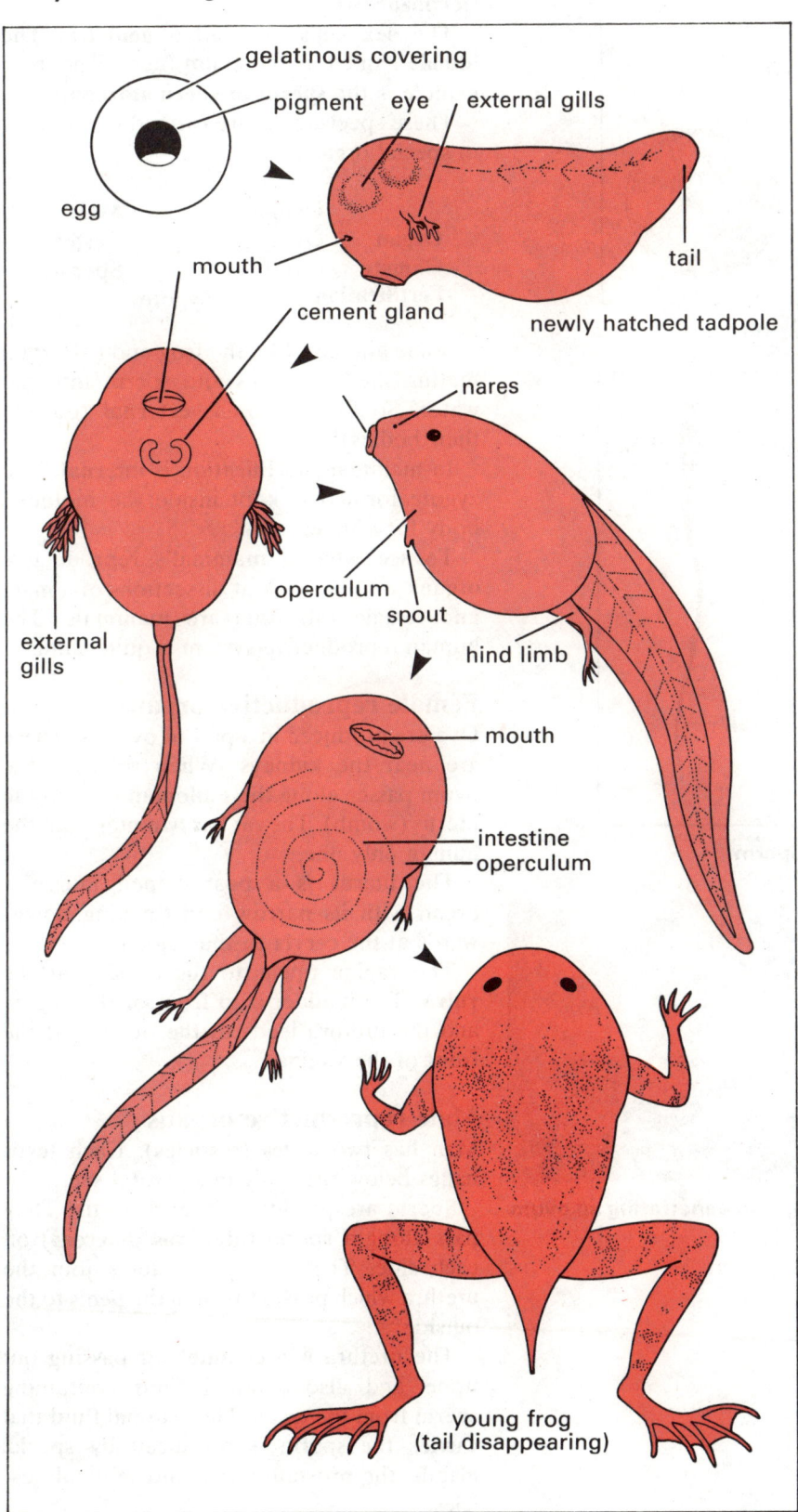

gelatinous covering
pigment
eye
external gills
egg
mouth
cement gland
tail
newly hatched tadpole
nares
operculum
spout
external gills
hind limb
mouth
intestine
operculum
young frog
(tail disappearing)

221

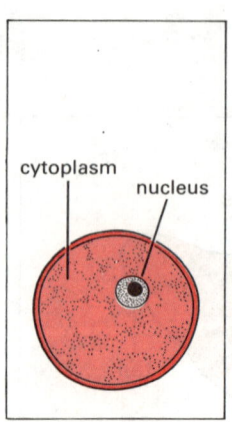

Ovum

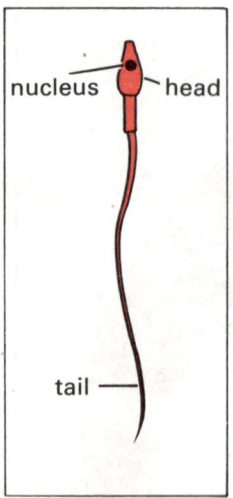

Sperm

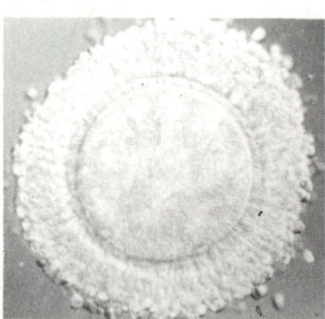

Sperm penetrating an ovum

For a new individual organism to be formed sexually, two sex cells must join together (**fertilization**).

The sex cells are called **gametes.** The female gamete is the **ovum** (egg). The male gamete is the **sperm** or **spermatozoön.**

These special cells are formed in the body in special organs:

	Female	Male
Organ	Ovary	Testes
Gamete	Ovum	Sperm
Fertilization	Zygote	

Some animals, like the frog and fish when mating shed the eggs and sperm into the water. So fertilization is **external** (outside their bodies).

In mammals, fertilization is **internal.** The zygote formed is kept inside the mother's body for a time.

To see what a mammal's reproductive organs are like, look at dissections of a male and female rat. Rats are mammals. The human reproductive system is quite similar.

Female reproductive organs

Ova are produced in a pair of ovaries. These are near the kidneys. When released the ovum passes along the Fallopian tube to the uterus (womb). The rat has two uteri, but the human only one.

The uterus is a pear-shaped muscular organ with its narrow end opening downwards at the **cervix** to the **vagina.**

The vagina opens to the outside at the **vulva.** The bladder is in front of the vagina and the urethra leads to the outside at the front of the vagina.

Male reproductive organs

Man has two **testes** (testicles). Each testis hangs below the body in a scrotal sac.

Sperm are produced in each testis. They pass along a sperm tube (**vas deferens**) on each side. The two sperm ducts join the urethra which passes through the **penis** to the outside.

The urethra is a channel for passing out urine and also seminal fluid containing sperm from the testis. The seminal fluid that carries the sperm is produced by special glands, the prostate gland and seminal vesicles.

Fertilization

The male passes sperm into the female by inserting his penis into the vagina. The penis becomes stiff enough by blood flowing into it.

The sperm are very small. They may reach the Fallopian tube within a few hours. Here, there may be an ovum passing down from the ovary.

Although there are several million sperm in the few cm^3 of seminal fluid, normally only one sperm fuses with the egg. The fertilized egg passes down the Fallopian tube to the uterus.

Oestrous cycle

Human females do not release eggs from the ovary all the time. One ovum is released about every 28 days. This is called **ovulation.** As an ovum grows it becomes enclosed in a capsule known as a **follicle.** The ovum is set free when the follicle bursts.

The wall of the uterus goes through a cycle of changes during the 28 days menstrual cycle. Around the time of ovulation, the wall of the uterus becomes thicker. This provides a suitable place for the fertilized egg to attach itself, in order to grow into an embryo. But if no ovum is fertilized, this lining of the uterus is cast off and passes out through the vagina.

Puberty

The menstrual process does not happen until girls are about 11 to 13 years of age. At the same time other changes happen. The breasts grow larger and pubic hair grows round the vulva.

Boys also become mature around 12 years old. The penis grows larger. Pubic hair appears and also hair on the chest and the face. The voice breaks about this time.

1 What gametes are concerned in sexual reproduction? Where are they produced?
2 Name any four mammals.
3 How are mammals different from other animals?

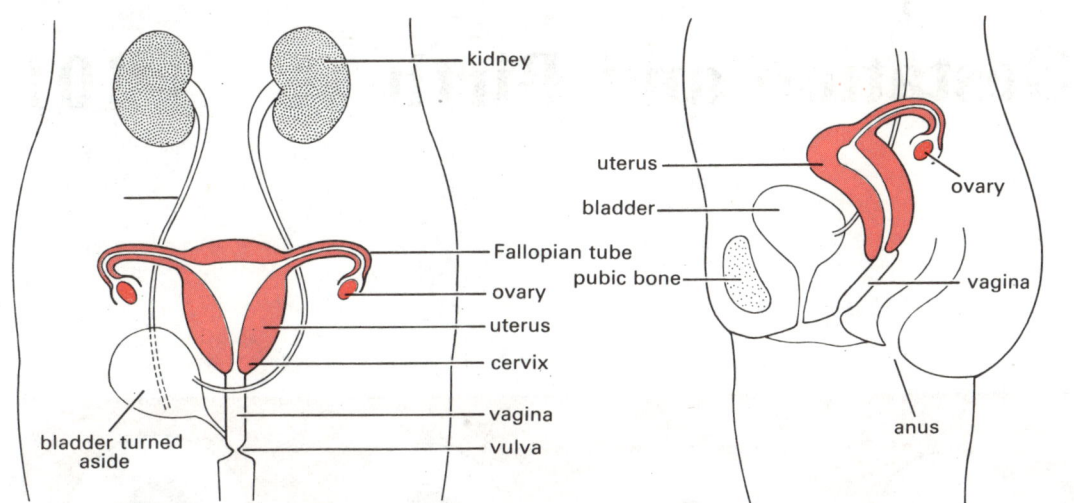

Human female reproductive system

kidney

uterus
bladder
pubic bone

ovary

vagina

Fallopian tube
ovary
uterus
cervix
vagina
vulva

bladder turned aside

anus

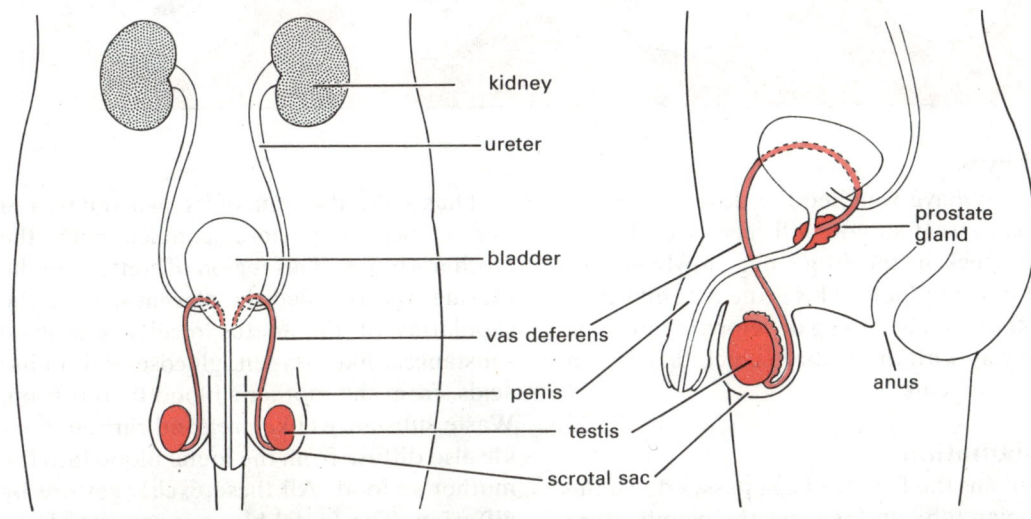

Human male reproductive system

kidney

ureter

bladder

vas deferens

penis

testis

scrotal sac

prostate gland

anus

 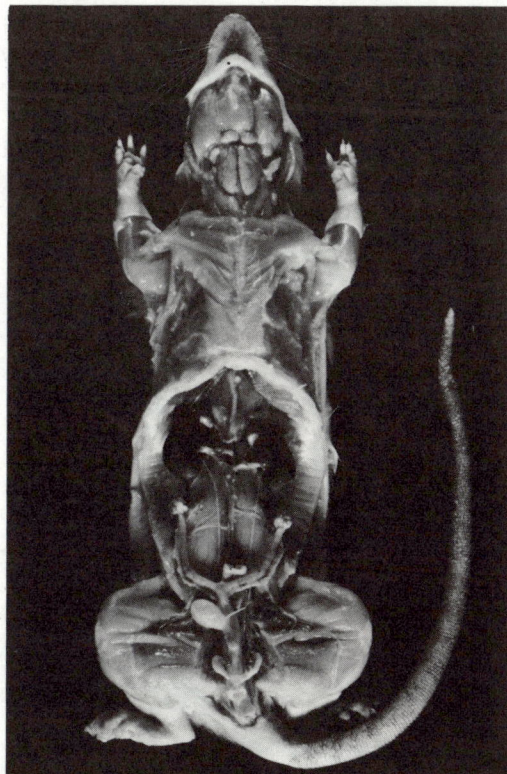

Dissection of a male rat showing the reproductive system (left)

Dissection of a female rat showing the reproductive system (right)

Frog spawn development

Embryos

You may have watched the development of frog spawn. If so you will have seen how a black speck in the frog's egg slowly shapes itself into a tadpole. This is the frog embryo.

A fertilized hen's egg develops in much the same way with the embryo chick floating on the yellow yolk.

Implantation

In humans the fertilized egg passes down the Fallopian tube and reaches the womb. Here it becomes implanted on the swollen tissues.

The zygote divides again and again forming many cells. Different groups of cells grow into different tissues and a fish-like embryo forms.

The embryo grows larger and after about two months, head, arms and legs can be seen. It is now called a **foetus.**

It has a blood system of its own and part of this comes into close contact with the mother's blood. This region of contact on the uterine wall is called the **placenta.** Here the capillaries of the foetus receive dissolved substances, like oxygen, glucose and amino acids, from the mother's blood by diffusion. Waste substances like urea and carbon dioxide also diffuse from the foetal blood into the mother's blood. All these exchanges are by **diffusion.** The foetal blood is separate from the mother's blood.

Amnion

The foetus is now in a liquid, the amniotic fluid, contained in a sac called the amnion, that has grown round it. It is attached to the placenta by the **umbilical cord.** This consists only of arteries and veins. The amnion protects the foetus from possible damage while delicate tissues and organs are growing.

The time it takes for the foetus to develop inside the human womb is about nine months. During this time it has grown to a mass of about 3 kg. The womb has become very much larger, but its muscular walls are able to stretch.

Birth

When the baby is due to be born, it is normally lying in the uterus with its head downwards.

Muscular contractions of the uterus force the baby down through the vaginal canal. The child is delivered head first. The amnion bursts during this time and the amniotic fluid

Fertilized hen's egg

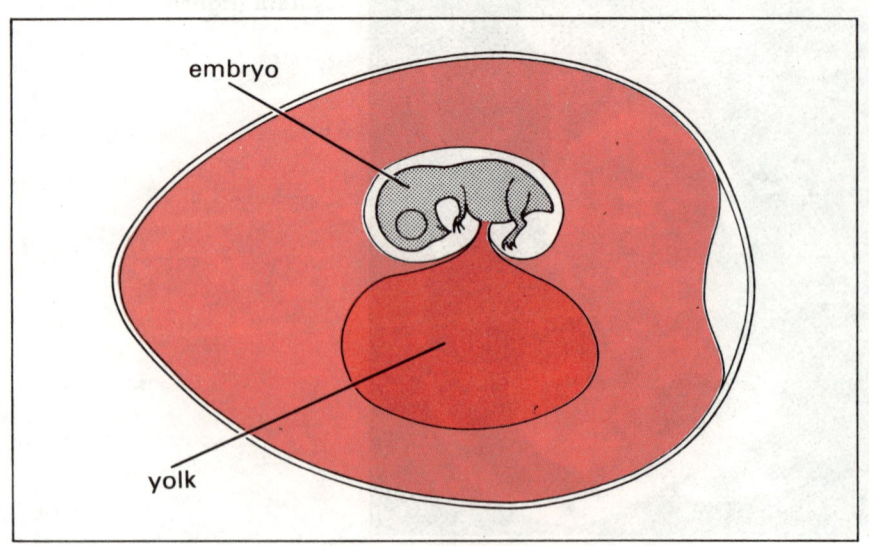

embryo

yolk

flows out. The umbilical cord is cut to separate the baby from the placenta. The cut end shrivels up and remains as the navel.

The placenta then comes away from the uterus and is expelled. This is the 'afterbirth'. The uterus soon contracts to its normal size.

Parental care

For the first few days the baby actually loses weight, but it feeds at its mother's breast and soon grows bigger.

The mother's breasts have been developing during pregnancy. They are a series of glands, mammary glands, that yield milk. All mammals suckle their young. Most mammals look after their young until they are able to look after themselves. In the case of humans this parental care may go on for many years.

Twins

Sometimes a fertilized egg will divide into two separate cells each of which develops into a baby. These identical twins are very much alike and always of the same sex.

But in some cases two ova are fertilized and develop at the same time. These twins may not be like each other nor need they be of the same sex.

1 How is a zygote formed?
2 What advantages can you think of for internal fertilization over external fertilization?
3 The terms uterus, urethra and ureter are similar and it is easy to confuse them. Can you say what each term means?

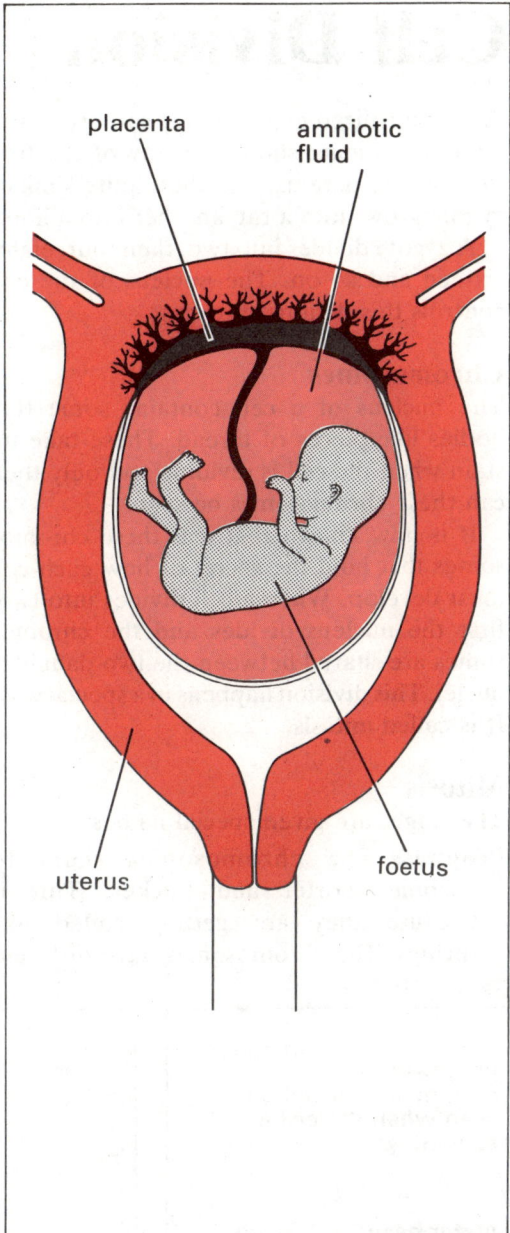

Human foetus developing in the womb

placenta — amniotic fluid

uterus — foetus

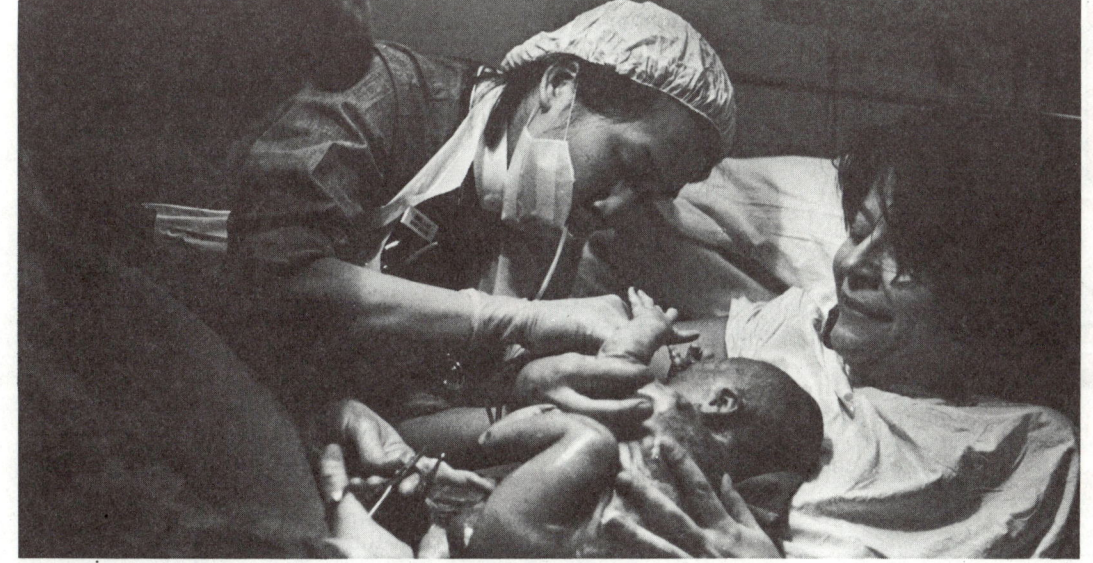

The moment of birth
The umbilical cord is just about to be cut

Cell Division

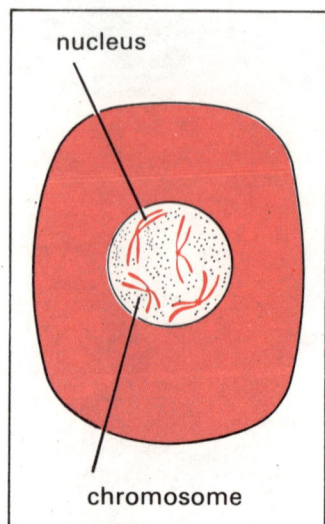

102

Chromosomes
The nucleus of the cell contains chromosomes

Mitotic division (left)

Meiotic division (right)

A tiny fertilized egg (zygote) can grow into an adult human, showing many of the features of the parents. Another, quite similar, zygote grows into a rat, another into a lion.

A zygote divides into two, then four, eight, sixteen and so on. The nucleus of the cell controls this dividing process.

Chromosomes

The nucleus of a cell contains some tiny bodies like pieces of thread. These take up stain when the cell is dividing and only then can these **chromosomes** be seen.

It is now known that it is these chromosomes that hold the secret of how each cell must develop. When a cell divides into two, first the nucleus divides and the chromosomes are shared between the two daughter nuclei. This division happens in a special way. It is called **mitosis.**

Mitosis

The stages are given special names.

Prophase The chromosomes seem to become shorter and thicker. This is because they are getting coiled like springs. The chromosomes have different shapes and sizes. But each one has a similar one in the nucleus so that there are pairs of chromosomes present.

Metaphase The nuclear membrane disappears. Small bodies called centrioles move to opposite ends of the cell. The chromosomes move to the 'equator'. Each chromosome is now seen to be a pair of **chromatids,** joined together at a **centromere.**

Anaphase The chromatids separate and move to the centriole 'poles'.

Telophase A nucleus is re-formed by a membrane growing round each group of chromosomes.

Each nucleus contains the same number of chromosomes as the original nucleus. So two daughter cells are formed and each has the same number of chromosomes as the parent cell. This is called the **diploid number**.

Meiosis

This is a reduction division and must not be confused with mitosis. When sperm and ova are being produced in the sex organs, the chromosomes divide in this way.

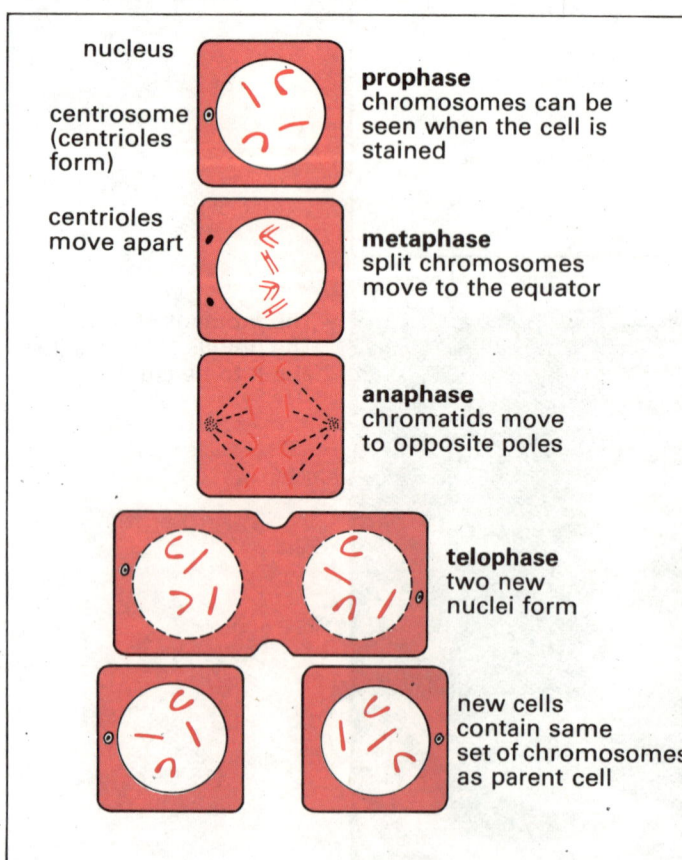

nucleus

centrosome (centrioles form)

centrioles move apart

prophase
chromosomes can be seen when the cell is stained

metaphase
split chromosomes move to the equator

anaphase
chromatids move to opposite poles

telophase
two new nuclei form

new cells contain same set of chromosomes as parent cell

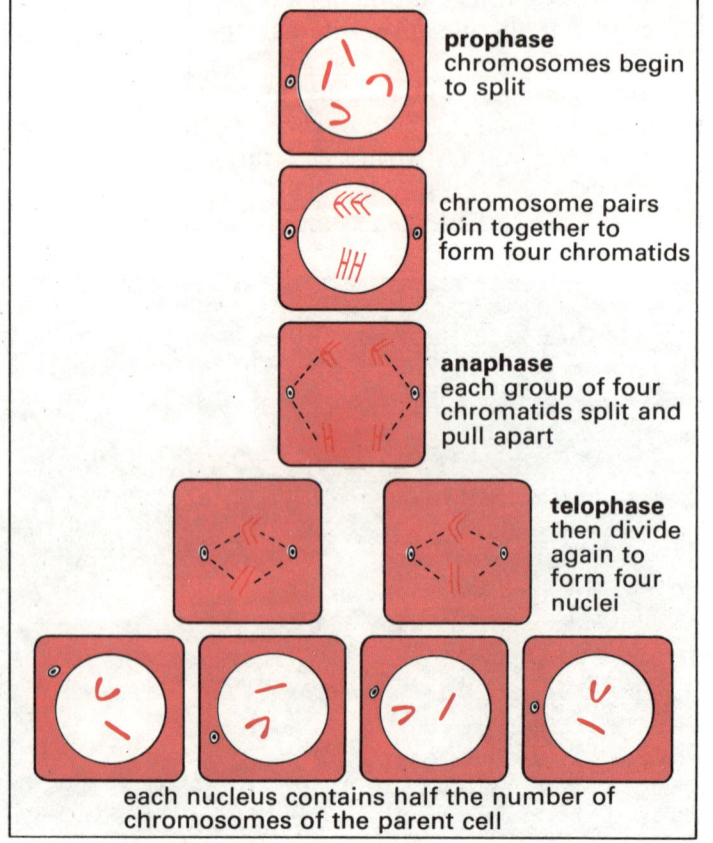

prophase
chromosomes begin to split

chromosome pairs join together to form four chromatids

anaphase
each group of four chromatids split and pull apart

telophase
then divide again to form four nuclei

each nucleus contains half the number of chromosomes of the parent cell

Prophase Similar chromosomes come together in pairs. They split to form four chromatids in each pair.

Anaphase They pull apart and move to the poles. But because the chromatids become twisted round each other, parts may have been exchanged. Only half the number, the **haploid** number, go to each pole.

Telophase These then divide to form four new groups. So each gamete contains only half the number of chromosomes. This is called the **haploid number**.

Fertilization

When sperm and ovum fuse together, each brings the haploid number of chromosomes to make the diploid number of chromosomes in the zygote.

Sex chromosomes

It is not quite true to say that all chromosomes are in pairs, or are homologous. The female human has 23 pairs, but in the male the 23rd chromosomes are not a pair. In females each chromosome of the 23rd pair is designated by the letter X. In males it is either X or Y.

So in an ovum the 23rd (sex) chromosome is X but in a sperm, the sex chromosome can be either X or Y.

A baby will be a girl if it receives the X chromosome, but a boy if it has a Y chromosome from the father.

Haploid generation

In some life cycles, the haploid number remains in the cells throughout a generation.

For example a fern produces spores on the underside of its fronds (leaves) These spores grow into small prothalli not the usual fern plants.

The prothallus has sex organs which produce sperms and eggs. Fertilization results in a new young fern. We call this fern a sporophyte because it produces spores. The sporophyte is the diploid generation. The gamete-bearing **prothallus,** the gametophyte, is the haploid generation.

Chromosome numbers

The number of chromosomes in a species is always the same. But the number differs from one species to another.

Man	23 pairs (diploid)
Mouse	20
Drosophila (fruit fly)	4
Wheat	8
Tomato	12

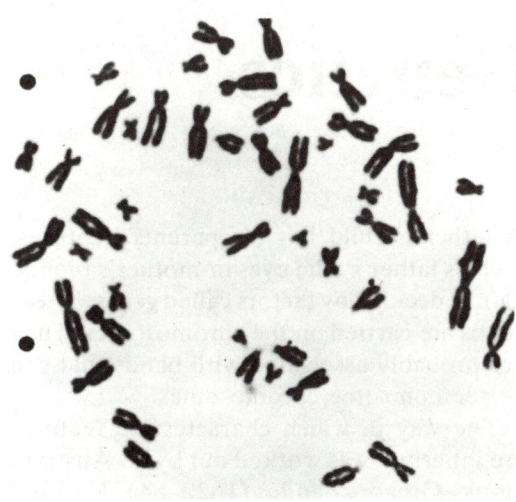

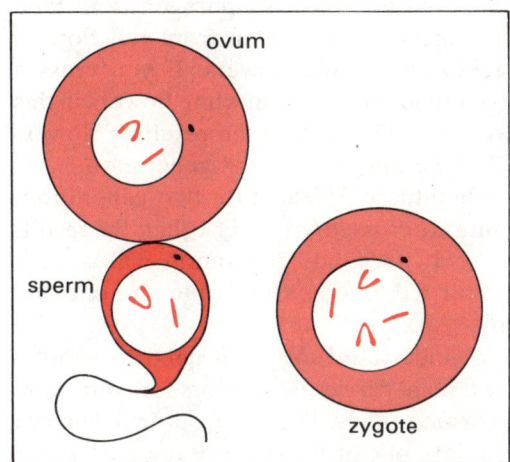

Fertilisation
The sperm and ovum fuse

ovum

sperm

zygote

1 What difference would you expect to find between a batch of tulips grown from bulbs and another batch grown from seed?

2 Select from the list the correct terms for the following: meiosis, fertilization, mitosis, sporophyte.
(a) Cells dividing to give daughter cells containing the same numbers of chromosomes as the parent cell.
(b) Cells dividing to form gametes.
(c) A plant that produces spores.
(d) The process of gametes joining together.

3 What differences are there between mitotic and meiotic division?

Genetics

103

Whether a child has his parents' features, such as father's blue eyes or mother's blonde hair, is decided by factors called **genes.** These genes are carried on the chromosomes. They are probably associated with bands that can be seen on some chromosomes.

The way in which characteristic features are inherited was worked out by an Austrian monk, Gregor Mendel (1822–84). He used plants in his experiments.

Suppose you grow snapdragons and then take care that the pollen from red flowers reaches only white flowers. If you cross a large number of red with white flowers in this way, you will find that the resulting flowers will all be pink in the first generation.

When these F_1 (short for first generation) plants are crossed with each other, the results in the F_2 (second) generation will be one quarter red, one quarter white and one half pink.

To understand this, you need to assume that a plant carries two factors for colour, on its chromosomes. The male and female germ cells have one or the other of the pair. If the factor for red colour is R and for white colour is W, then for the snapdragon:

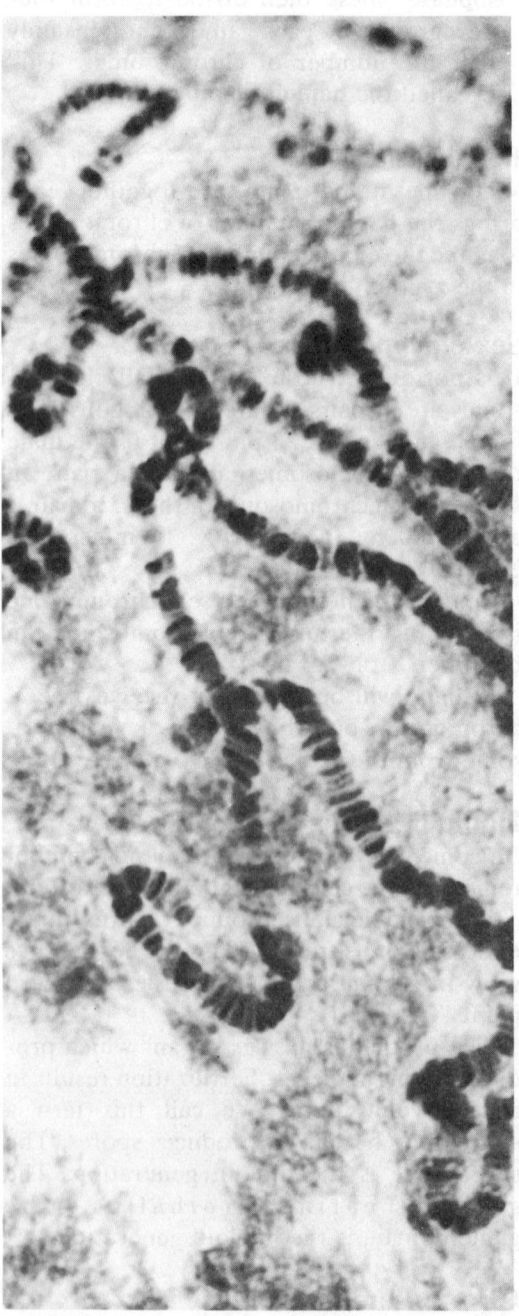

Chromosomes from the salivary gland of a fruit fly

Parents		Red RR	White WW
	gametes	R	W
F_1 generation	**genotype**	all RW	
	phenotype	all pink	

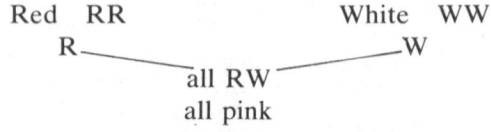

The term genotype means the description of the genes but the phenotype refers to the type of plant judging from its appearance. In the next generation:

F_1 parent	**phenotype**	Pink		Pink
	genotype	RW		RW
	gametes	R or W		R or W
F_2 generation	**genotype**	RR	RW	WW
	phenotype	pure red (25%)	pink (50%)	pure white (25%)

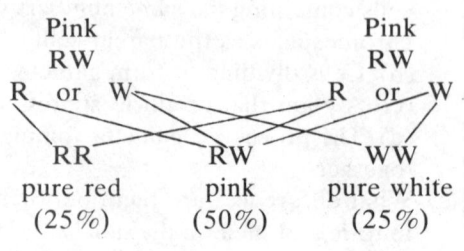

So although the white factor was masked in the F_1 generation, it made its appearance in the F_2 generation.

If you crossed red and white sweet peas in the same manner, the results would be different. In the first generation the flowers would not be pink but you would get all red flowers. However, in the F_2 generation you would find red to white in the ratio of 3:1. The red factor is **dominant;** when it is present the white colour does not show at all. The white factor is **recessive.**

You can see this if R stands for red and r for white.

Parents	genotypes	RR	rr
	phenotypes	red	white
	gametes	all R	all r
F_1 generation	genotypes	all Rr	
		(hybrid)	
	phenotypes	red	

Crossing these hybrid (containing both dominant and recessive factors) flowers:

F_1 parents	phenotype	Red		Red	
	genotype	Rr		Rr	
	gametes	R or r		R or r	
F_2 generation	genotypes	RR	Rr	rr	
	phenotypes	pure red	hybrid red	pure white	
		(25%)	(50%)	(25%)	

So three quarters of the F_2 generation are red, but only one third of these are pure breeding (we say they are **homozygous** because they have two similar genes).

Most features of animals and plants depend on more than one set of genes. Features in humans that depend on only one pair of genes are rare. Eye colour seems to be decided by one pair of genes with brown eye colour being dominant to blue.

Some people are able to roll the tongue. This ability is related to genes. Also being able to recognize a taste of phenylthiourea depends on a gene. Although some people cannot taste anything, others find the taste very bitter and strong.

The type of blood you inherit is also dependent on genes handed on from your parents.

Interesting cases arise when the genes concerned are on the sex chromosomes. An example is the condition **haemophilia.** Clotting of blood is slow and anyone suffering from haemophilia is prone to excessive bleeding which might be serious. The gene is a recessive one. It is sex-linked because it is carried on the sex chromosome.

People with the disease are always male. Females can be carriers. There is a 1:1 chance that a carrier with the haemophiliac X* gene will pass it on to the children.

	mother	father
	XX*	XY
gametes	X or X*	X or Y

Children could be: XX normal daughter
XY normal son
XX* carrier daughter
X*Y haemophiliac son

1 About 70% of people find that phenylthiourea tastes bitter, but the remaining 30% cannot taste it. Does this show that being able to taste it is a dominant or recessive character? Give your reasons.

2 What results would you expect from the crossing of hybrid Rr red sweet peas with pure-breeding peas RR?

3 In tomato plants, tallness is dominant to dwarf character. What would be the result of crossing tall plants with dwarf plants?

Family tree showing the inheritance of haemophilia

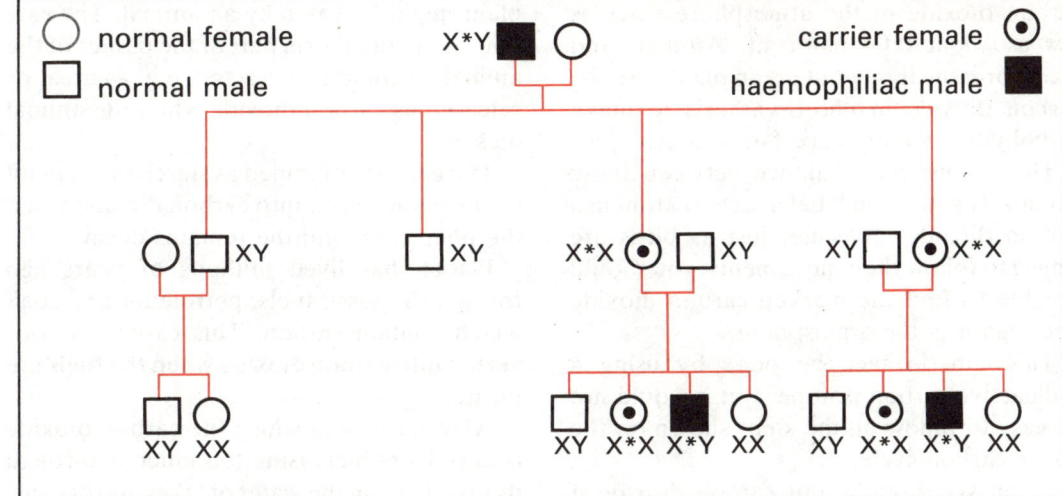

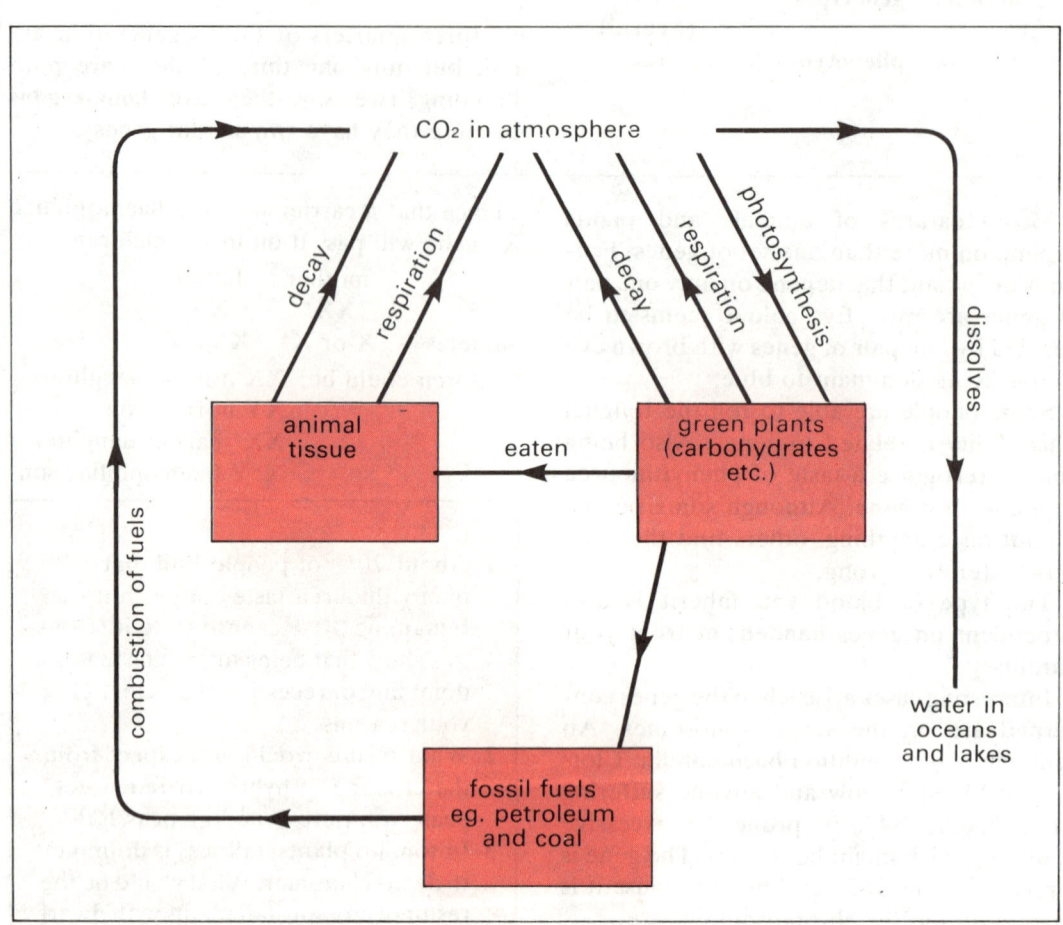

Carbon cycle

Carbon cycle

You may have wondered why the amount of carbon dioxide in the atmosphere stays as low as about 0.04 per cent. Animals and plants breathe it out, but green plants use the carbon dioxide in photosynthesis to make carbohydrates and hence fats and proteins.

There must be a balance between these actions. If you could label a carbon atom in a carbon dioxide molecule, just as birds are ringed to follow their movements, you should be able to find the marked carbon dioxide once again in the atmosphere.

This can, in fact, be done by using a radioactive carbon isotope. But it would not be easy to follow all the steps shown on the above carbon cycle.

When you breathe out carbon dioxide it may not stay long in the atmosphere. It might be used by a plant in photosynthesis. The plant might be eaten by an animal. The carbon would then form part of the tissues of the animal. It could be excreted, e.g. as urea, or released as carbon dioxide when the animal dies.

The carbon combined as starch in the plant could change back into carbon dioxide when the plant dies, and the remains decay.

Plants that lived millions of years ago formed the fossil fuels, petroleum and coal, which contain carbon. This carbon is converted into carbon dioxide when the fuels are burnt.

Another way in which the carbon dioxide is kept from increasing too much is through its dissolving in the water of lakes and oceans.

Nitrogen cycle

The cycle of nitrogen compounds in nature is most important. Nitrogen is an essential element in building proteins. So it is needed for growth.

There is about 79 per cent nitrogen in the atmosphere. But the most vital part of the nitrogen cycle takes place in the soil.

Plants take up the nitrogen they need in the form of nitrates. From these, proteins are formed. When an animal eats a plant, this plant protein becomes animal protein.

The nitrogen may leave an animal during excretion. It leaves the plant or animal when it dies and the remains decay. Ammonia and ammonium compounds are formed. You may have noticed a smell of ammonia in stables. The ammonia is formed by the action of bacteria on excreted nitrogen compounds.

Ammonia is changed to nitrites and these in turn change to nitrates. Both these oxidising actions are carried out by bacteria.

Perhaps you think of bacteria as enemies. As Topic 111 shows, this is often true, but the bacteria that change waste ammonia back to nitrates are most 'friendly'. Without them growth of plants, and therefore of animals, would not be possible. Bacteria have names: *Nitrosomonas* change ammonia to nitrites; and *Nitrobacter* oxidise nitrites to nitrates. But it is more important to remember the changes they bring about than their names.

Denitrifying bacteria set free nitrogen gas into the atmosphere. So this nitrogen is lost from the soil.

Some leguminous plants, such as peas, beans and others have small lumps, called nodules, on their roots. These root nodules have some useful bacteria inside which can fix nitrogen. That is, they change nitrogen from the atmosphere into nitrogen compounds.

There are some chemical processes that use atmospheric nitrogen, such as the Haber process in which man combines nitrogen and hydrogen to form ammonia. Thunderstorms bring about the combination of nitrogen and oxygen in the air to form oxides. These oxides dissolve in the rain to form nitric acid. This in the soil forms nitrates.

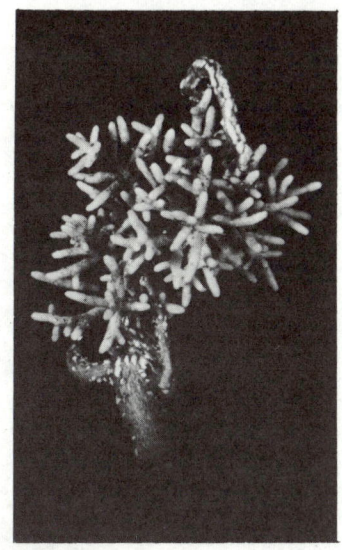

Root nodules
Nitrogen is 'fixed' by bacteria that live in the roots of some leguminous plants

1 How would you expect the carbon cycle to be affected by a great increase in world population? How could you compensate for this effect?
2 State three important changes brought about by useful bacteria.
3 Explain the advantage of (i) putting farmyard manure on the garden; (ii) leaving the roots of bean plants in the soil when the crop is finished.

Nitrogen cycle

denitrifying bacteria → nitrogen in the atmosphere → thunderstorms

nitrogen fixation

proteins in plants — eaten → proteins in animals

root nodules of legumes (pea family plants)

$N_2 + O_2$ form oxides of nitrogen

decay · excretion and decay · Haber process

organic nitrogen compounds

bacteria

nitrates ← nitrites ← ammonia

nitric acid

231

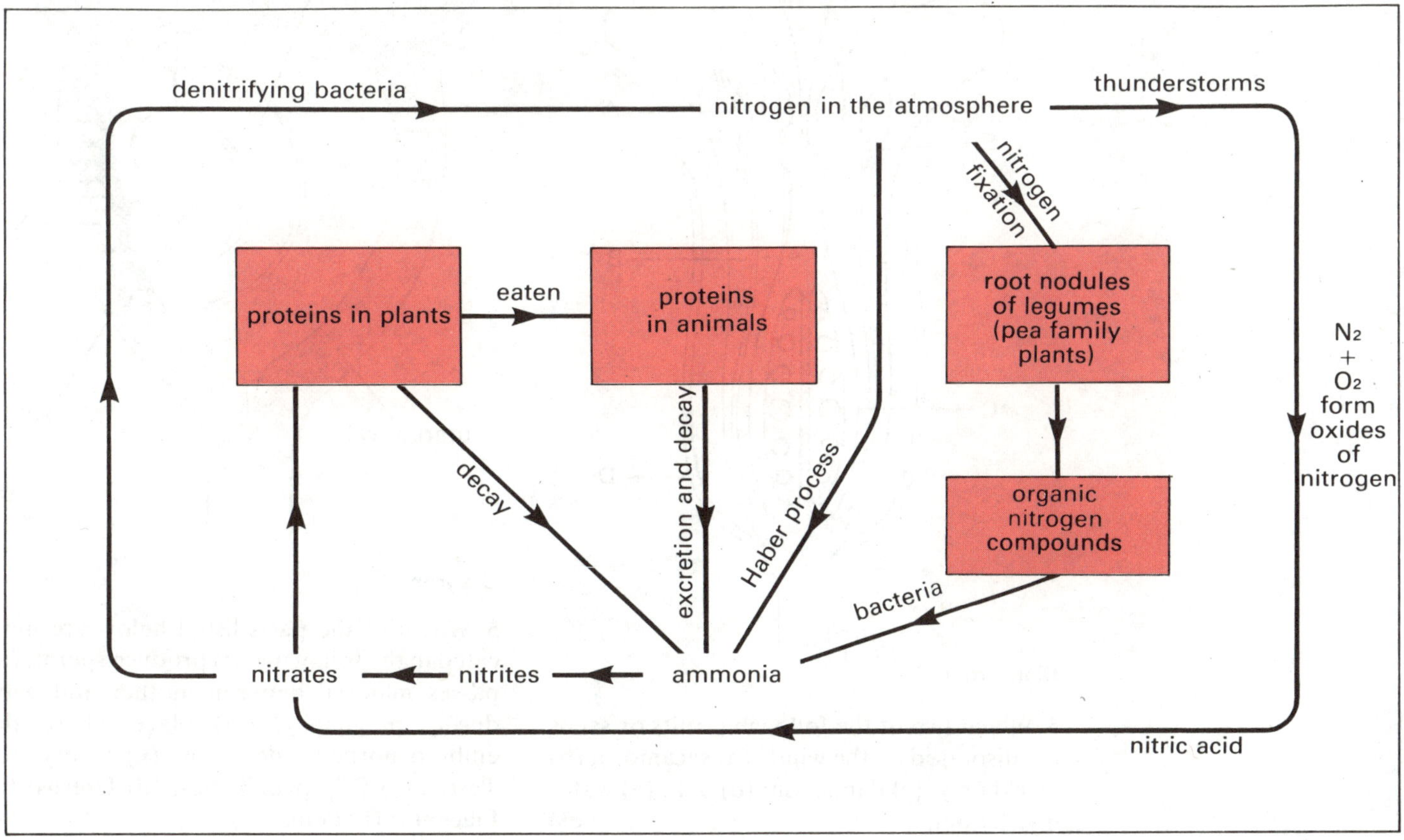

Examination Questions IX

1 The left-hand column below contains a list of plant structures and the right-hand column a number of descriptions:

Plant structures Descriptions

A Sepal   1 is the seed coat
B Stigma   2 protects the flower in bud
C Testa   3 receives the pollen
D Seed   4 contains food in the seed
E Radicle   5 develops from the ovule
F Cotyledon   6 is the young root
          7 attracts insects

Write down the number of the descriptions that best fits the letters A, B, C, D, E, F. Y

2 (i) In Diagram **1** give the names of the parts labelled A, B. C, D, E and F.
(ii) Which two parts form a stamen?
(iii) Which part of the flower produces pollen?
(iv) What is pollination?
(v) Explain how the flower is fertilized.
(vi) Name one flowering plant and state how it disperses its seed.   SE (part)

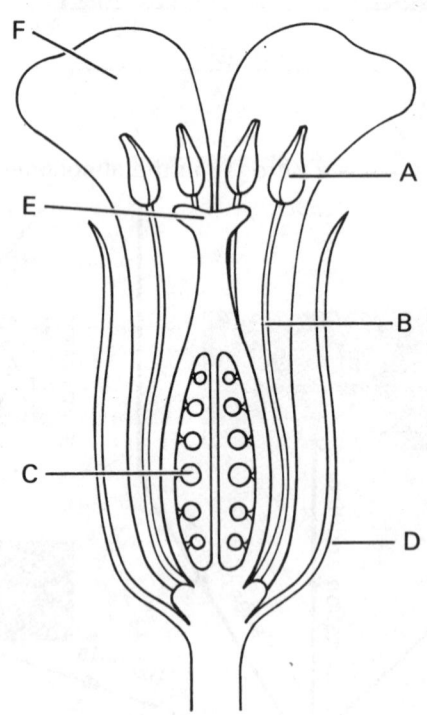

Diagram 1

3 Which two of the following fruits or seeds are dispersed by the wind? (a) sycamore; (b) blackberry; (c) dandelion; (d) oak; (e) water lily; (f) oat.   EM

4 (a) Look at Diagram **2** representing a sperm.
(i) What is the function of part A of the sperm? (ii) What is the function of the part B of the sperm. (iii) Where, in the female's body do the ovum and sperm join?
(b) Look at Diagram **3** showing an embryo developing in the uterus.
(i) Name structure X. (ii) Why does the baby not feel anything when structure X is cut when it is born? (iii) Give three functions of the placenta during pregnancy. (iv) Give two functions of the amniotic fluid which surrounds the embryo.   Y

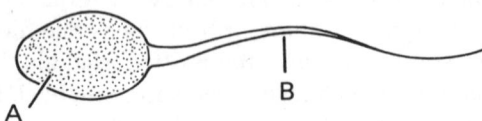

Diagram 2

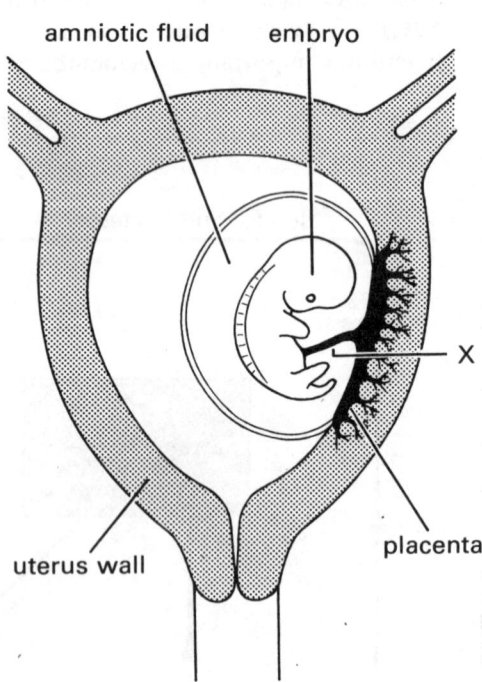

amniotic fluid   embryo

uterus wall   placenta

Diagram 3

5 Which of the parts listed below are indicated in the following: (i) produce sperm; (ii) passes material between mother and baby during pregnancy; (iii) place where the embryo normally develops: (a) Ovary; (b) Testis; (c) Fallopian Tubes; (d) Uterus; (e) Placenta; (f) Penis.   WM

6 From the list below select words that should be in the spaces marked A, B, C, D, E, F.

In the reproductive system the gametes (sex cells) of the male are called ...A... They are formed in organs called the ...B... During reproduction these gametes are deposited in the ...C... of the female and passed into a muscular organ called the ...D... From here they pass into a narrow tube called the ...E... where one may join with the female gamete, which is called the ...F...

testes; uterus; oviduct; ovum; sperms; vagina; egg; womb; fallopian tube; penis.   EM

7 State whether each of the statements is correct or incorrect.
During menstruation:
1 the lining of the uterus is shed;
2 an ovum is fertilized;
3 sperms are released;
4 ovulation is possible.   WM

8 Diagram **4** shows part of the carbon cycle:

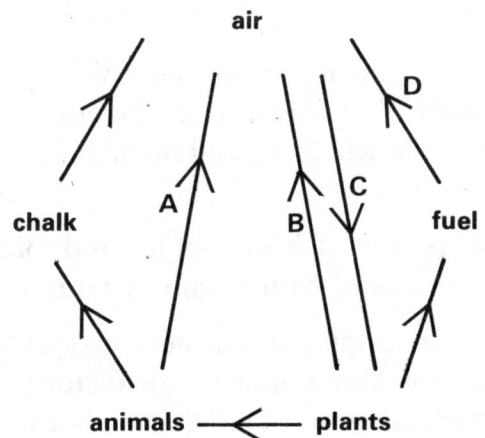

Diagram 4

(i)   Name the processes A, B, C, D.
(ii) Name a fuel formed from plants.
(iii) What type of rock is chalk?
(iv) Explain how chalk is formed.
(v)   Name an industrial process in which chalk is used.   SE (part)

9 (a) Look at the sound wave (Diagram **5**). (i) What is the amplitude? (ii) What is the wavelength?
(b) A, B and C are three sound waves. (i) Which note is the highest? (ii) Which note is the softest?   M (part)

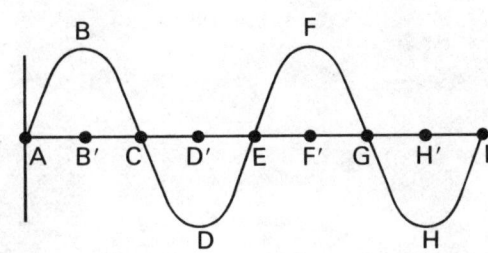

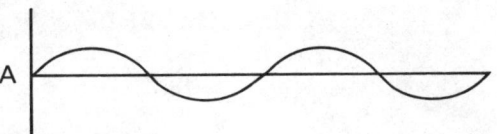

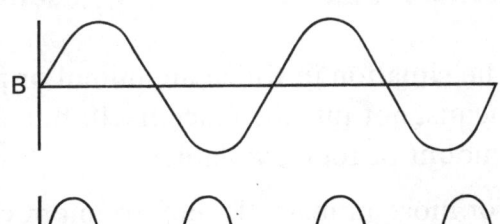

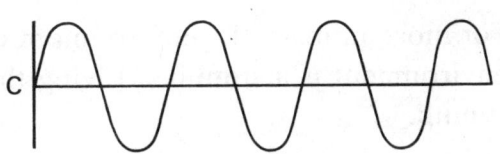

Diagram 5

10 (a) What are the main characteristics of the insect family?
(b) Give the stages in the life-cycle of an insect you have studied, naming the insect.
(c) Name **three** insects which man might consider 'foes' and in each case give **one** reason why they are 'foes'.
(d) Why is it that the use of chemicals to kill insect pests can lead to a reduction in the yield of some crops?
(e) Insects lay an enormous number of eggs. Give **two** examples of methods by which the insect population is controlled in nature. SW

11 There are two kinds of twins – identical and fraternal. Describe how each kind is formed and explain what this means concerning the sex of the twins.   M

MAN & HIS ENVIRONMENT

The situation in which an animal or plant lives is called its environment. This means, not just the place itself, but all the factors like temperature, climate, amount of food available.

For most animals, the environment changes from time to time. A change of environment is a stimulus. Living things are irritable, that is, they react to stimuli.

In the case of higher animals, including man, reacting to stimuli involves a nervous system. This is controlled by the brain. You are aware of the outside world because of nerve endings that respond to touch, taste, and smell, besides light and sound.

The skin houses the nerve endings or receptors, for touch, and for hot and cold. The skin is also important in controlling the loss of heat from our bodies.

The internal organs must be included in your environment. The nervous system is involved in keeping your internal organs working together satisfactorily. After all if you do not digest food properly the stimulus of indigestion is a very real one.

Harm can come to the body in several ways. Some tiny organisms like bacteria, cause greater trouble than their size might suggest. The body needs protection against them. Sometimes man's own efforts bring danger, as for instance in the hazards linked with electrical apparatus or the dangers from radioactive materials. Increasingly nowadays there are threats from pollution caused by discharge of poisonous chemicals into the air or into rivers and seas.

Man, like a motor car, is made up of many parts which must all work together smoothly. You could think of the driver of a car as its brain. He can turn switches and press pedals to start, stop and control the speed of the car's engine. Also, on the dashboard are some indicators that show speed, how much petrol is in the tank and so on. This overall control of the parts of a car depends on electrical circuits.

A man also depends on 'signals' being sent to and from his brain. The 'signals' pass along nerves. These are the very thin fibres that carry impulses. A nerve can mean a single fibre, which is better called a neuron or neurone. It can also mean a bundle of neurons making up a nerve cord, as might be seen in a dissection.

Neuron

When a short piece of nerve cord is teased out a single nerve cell can be separated and examined under the microscope. A neuron is seen to be made up of a cell body with a long fibre attached, ending in a branching process.

Several different shapes of neurons are found. Some have two or more fibres coming from the cell body. Sometimes the axon is very long, sometimes short. Many fibres bunched together form a nerve cord. The

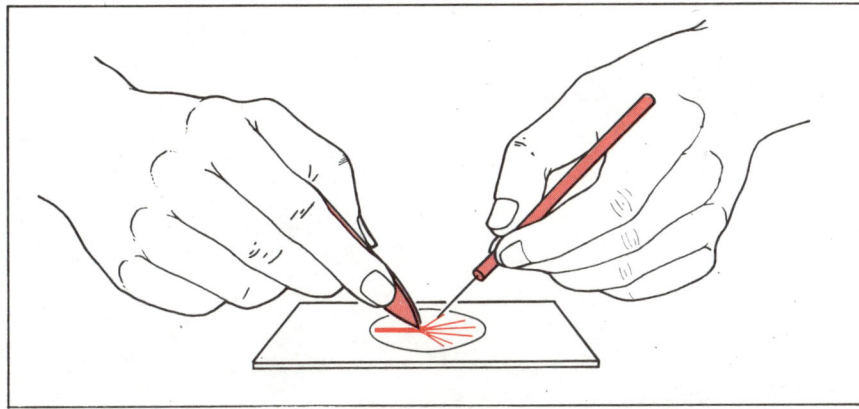

Teasing a piece of nerve cord

swelling where all the cell bodies are grouped together is called a **ganglion** (plural, ganglia).

1 Where would you expect to find nerve endings (receptors) that make you aware of (a) a flash of light, (b) a sudden noise, (c) the scent of a flower, (d) a current of cold air?
2 Describe what each of the following terms refer to: axon; neuron; ganglion; dendrites.
3 Explain briefly the meaning of the terms: environment; irritable; stimulus.

Neuron

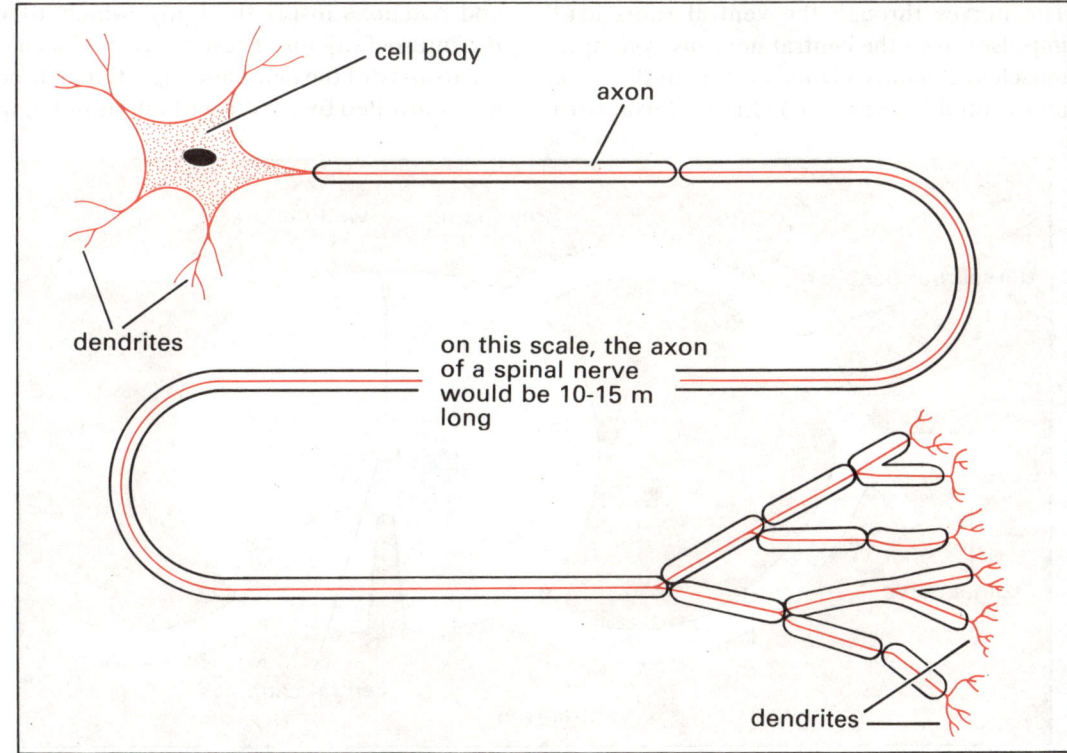

cell body

axon

dendrites

on this scale, the axon of a spinal nerve would be 10–15 m long

dendrites

Central nervous system

The brain and the spinal cord together make up the **central nervous system.** The brain is housed in the **cranium.** The spinal cord is almost as well protected, because it runs through a canal in the vertebral column.

The peripheral nerves

Nerves link the central nervous system with the outside world. In the backbone between one vertebra and the next there are two pairs of nerves at each level. The **dorsal** and **ventral** nerve cords on each side join together to form a common nerve cord passing outwards.

The nerves of the dorsal roots lead nerve impulses from the body into the spinal cord. The nerves through the ventral roots lead impulses from the central nervous system to muscles or glands. (Dorsal refers to the back and ventral to the front.) On the dorsal cord near the spinal cord is the dorsal root ganglion.

Spinal cord

In the spinal cord itself, there are both fibres and cells. Look at the transverse section of the spinal cord. There is a roughly H-shaped middle portion of 'grey matter'. There are many cell bodies in this region. The outer 'white matter' is made up mostly of cell fibres. The fibres are protected by a sheath of fatty material and they reflect light more readily. So the fibres seem lighter.

The autonomic system

This is a system linked with the central nervous system. It is concerned only with stimuli and reactions inside the body, which affect the internal organs. These reactions, such as the beating of the heart and digestion of food, are controlled by nerves without us noticing.

Transverse section of spinal cord

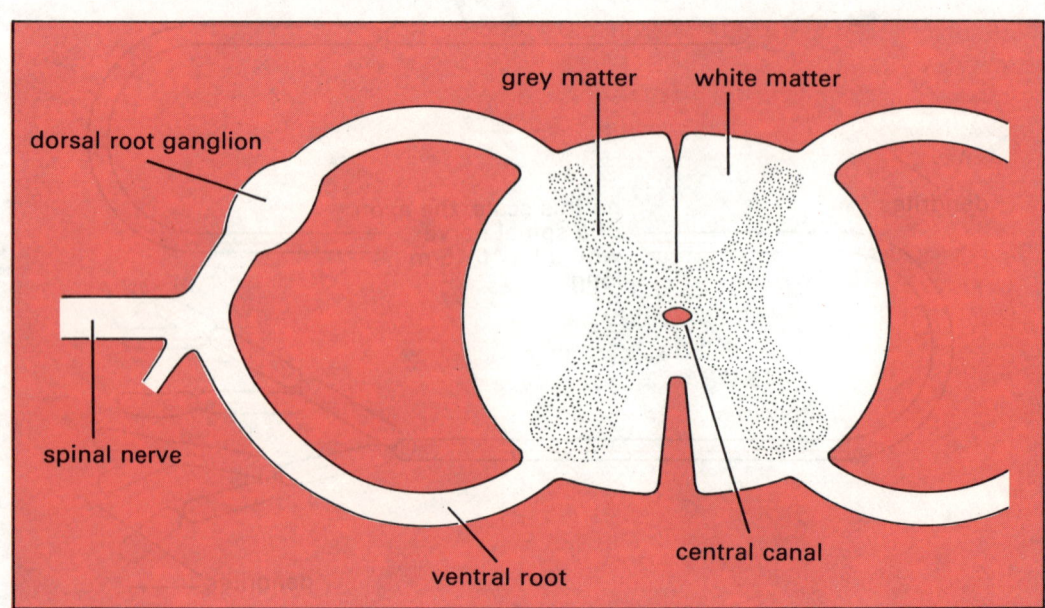

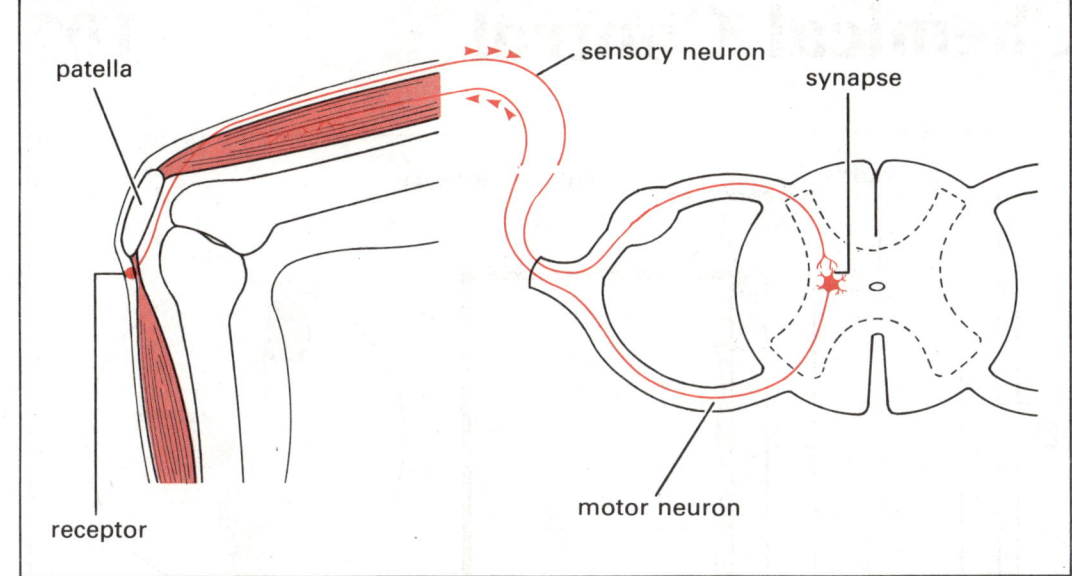

patella

sensory neuron

synapse

motor neuron

receptor

Reflex action

This is the simplest kind of nerve response. If a doctor taps a patient's leg just below the knee, the healthy patient automatically kicks up his leg. This is called the knee jerk reflex.

Reflex actions of this kind are immediate and automatic responses to a stimulus. Two neurons are involved. The stimulus (the tap on the patella tendon of the knee) excites a nerve ending in the thigh muscle. An impulse is set up in a sensory neuron, leading to the spinal cord. The cell body of this neuron is in the dorsal root ganglion. Inside the spinal cord the dendrites of the sensory neuron signal across a small gap, called a **synapse**, and the impulse passes into a motor neuron. This neuron carries the impulse to the muscle in the leg and makes it contract.

Note that the brain does not actually take part in this reflex arc (as it is called), but the man does know what is happening, because of other nerve paths. This reflex is one of the quickest (it takes only about 25 thousandths of a second). Most reflex arcs involve more than one synapse, with intermediate neurons linking the sensory and motor neurons.

The reflex has evolved as a kind of defence mechanism. It is perhaps not easy to see how kicking up our leg when something touches the knee is a defence action. If you accidentally touch something hot, you do not wait while the brain considers the sensation and thinks out a course of action. A reflex action comes into play immediately and your hand is taken away automatically. You can see, too, the value of blinking by reflex action when something is coming near your eyes.

Other reflex actions are the salivary reflex, which produces a flow of saliva when food is seen, and the iris reflex. This is easy to study. Sit in a dim light looking into a mirror. Look at your pupils. Note their size. Now shine a light into your eyes and note what happens to the size of the pupils. The reflex in this case reduces the size of the pupil so that the retina of the eye is not damaged by the extra light that would otherwise enter.

It is known that when an impulse passes along a nerve, there are changes in electric potential. Nerves can be stimulated by an electric shock.

A doctor tests the knee-jerk reflex

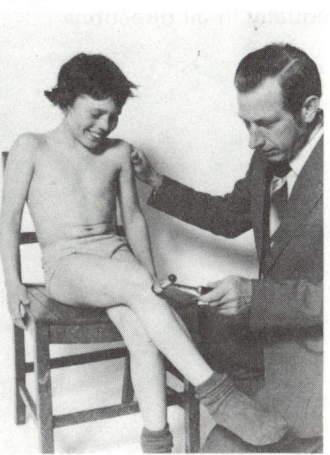

1 Which of the following actions are reflex actions? eating; sneezing; biting; swallowing; speaking; blinking.
2 What is the autonomic nervous system? How is it different from the central nervous system?
3 Where would you find a synapse?

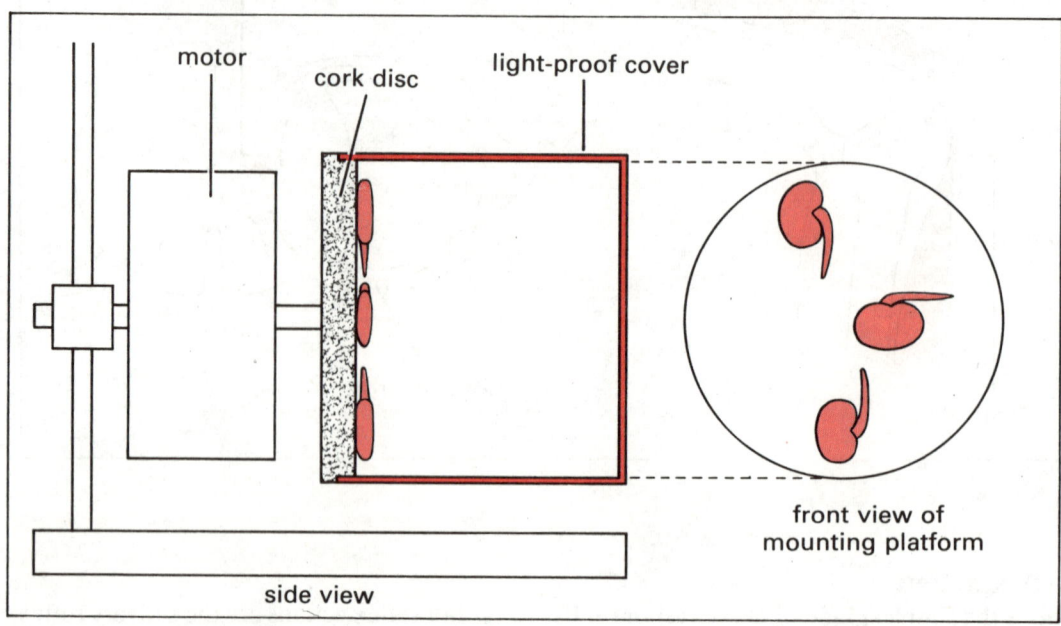

Clinostat
When rotating the effect is as if gravity acted on the seeds equally in all directions

Plants respond to stimuli, but not by reflex action. They respond to the stimuli of light, gravity and water, but these reactions are not rapid. The response is by growing towards or away from the stimulus. Such a growth response is called a **tropism**.

Geotropism

You know that roots grow downwards. It is thought that this is due to gravity. How can this be proved? Gravity can only be removed by going far out in space.

You can rule out its effect by keeping the plants going round so that gravity acts equally in all directions in turn. When this is done, e.g. by having bean seedlings rotating on a clinostat, the roots no longer turn down.

Shoots are negatively geotropic, the growing tip grows upwards. Roots are positively geotropic.

A tropism is a response to a stimulus made by a part only of the plant or animal. If an animal moves as a whole away from a stimulus, e.g. a spider running away from the light, it is called a **taxis.**

Other factors that affect the growth of a plant are light (**phototropism**) and the nearness of water (**hydrotropism**). Roots grow towards water. This is the reason why it is best to water the ground thoroughly near a parched plant. If the watering is only on the surface of the soil, the roots will turn upwards to reach it.

Auxins

The way in which plants react to stimuli can be revealed in experiments.

Some young seedlings of oat (coleoptiles) are put in a box with light coming in only from one side. If the tips are cut off or if they

Effect of auxins

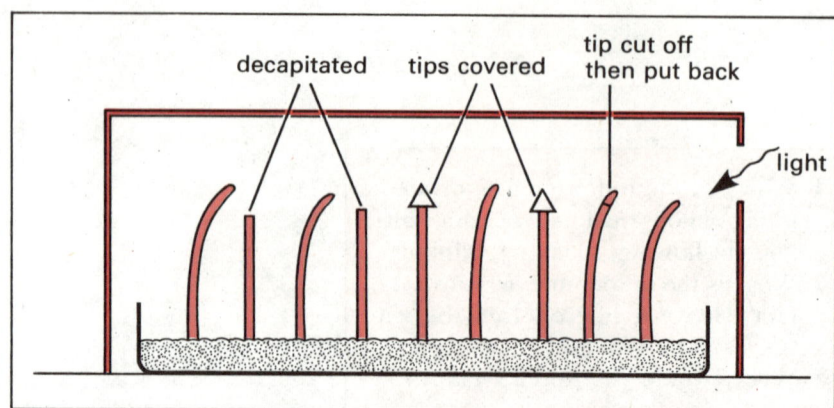

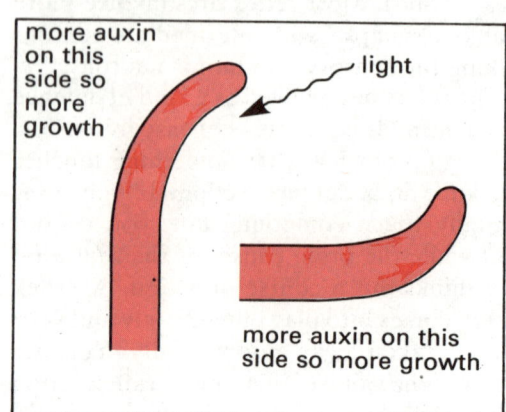

are covered with paper cones the seedlings do not grow towards the light. Seedlings that are not so treated, bend over towards the light, as they grow.

Next the tip of a seedling is cut off and put back again with a slice of gelatine between. The shoot bends over towards the light in the normal way.

The experiments suggest that a chemical substance is involved. This chemical can diffuse across the gelatin. Substances of this kind are called **auxins** or **plant hormones.**

In the experiment above, more auxin is available on the side away from the light. So the shoot grows towards the light.

Plant hormones can be put to good use. Gardeners dip the end of cuttings into a hormone to encourage root formation. Hormones are used to help fruits form or to stop them falling from the tree.

Weedkillers, such as 2,4-D (2,4-dichlorophenoxyacetic acid) are used to kill broad-leaved weeds. They kill the weed by making it grow too quickly. Grasses and cereals with narrow leaves are not affected.

Hormones in animals

Animal hormones were known before auxins were discovered. They are chemical substances made in special glands called **endocrine** organs. The substances pass straight into the blood stream. They can therefore act quickly. They have been called 'chemical messengers'. Each one has a specific action and gives results only on a particular group of cells or organs.

Secretin is made by glands in the wall of the duodenum. When food passes into the intestine, secretin goes into the blood stream and stimulates the pancreas to secrete digestive enzymes into the duodenum.

The hormone **insulin** is made in the Islets of Langerhans, a part of the pancreas. It affects the cells of the liver so that glucose is changed to glycogen. If insulin is not present (as in the disease diabetes), glucose is lost from the body.

The **thyroid** glands produce a substance called **thyroxine.** This hormone controls the basal rate of metabolism. A person with a very active thyroid is vivacious and uses up food without saving some of it as fat. An underactive thyroid makes its owner sluggish both in mind and in body. The thyroid itself is controlled by the pituitary gland, a master gland that has been called the 'conductor of the endocrine orchestra'!

The medulla region of the adrenal gland makes adrenaline. This flows into the blood stream and plays an important part in co-ordinating the body's activities. In cases of emergency caused through cold, fear or hunger, extra adrenaline is formed. As a result blood pressure increases. The heart beat becomes more rapid and the smaller arteries are constricted. More glucose is discharged into the blood stream. All these effects prepare the body for dealing with emergency; that is to prepare the body for 'fight or flight'.

Reproductive organs produce hormones. The **oestrogens** from the ovary control the development of the secondary sexual characteristics (see Topic 100). this includes the growth of the mammary glands and the events during the oestrous cycle. The male sex hormone is **testosterone,** produced by the testis.

1 What tropism shows itself when a potted plant is put near a window?
2 Mention one way in which plant hormones are different from animal hormones.
3 Which of the following is involved in a nerve reflex action? auxin; tropism; synapse; endocrine gland; thyroxine.

Human endocrine glands

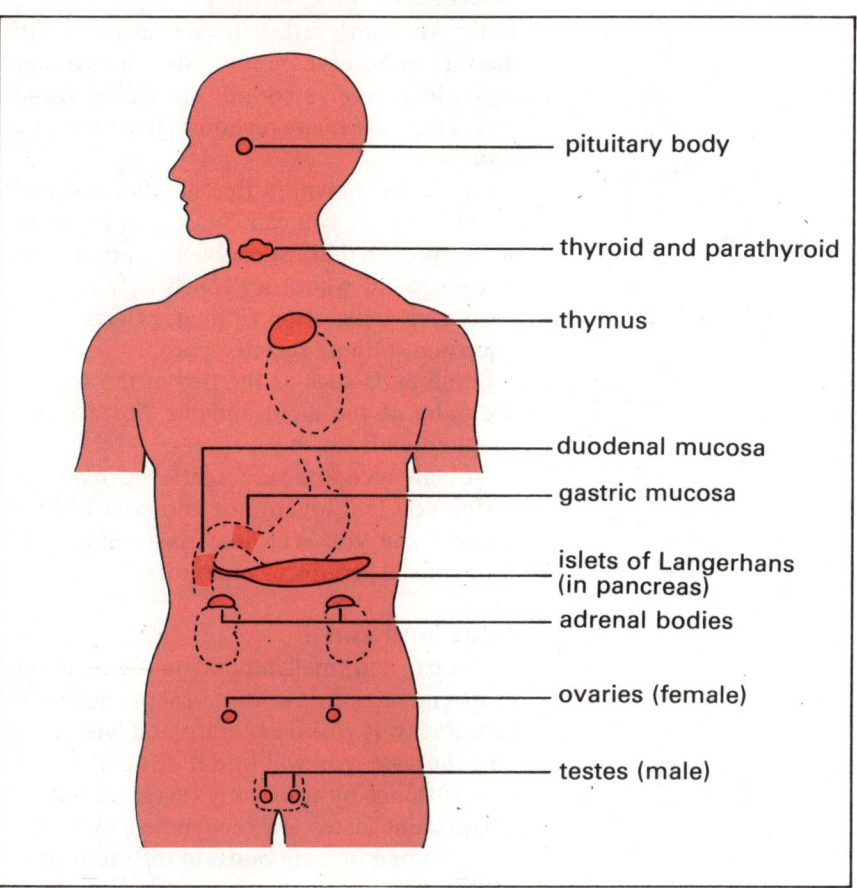

- pituitary body
- thyroid and parathyroid
- thymus
- duodenal mucosa
- gastric mucosa
- islets of Langerhans (in pancreas)
- adrenal bodies
- ovaries (female)
- testes (male)

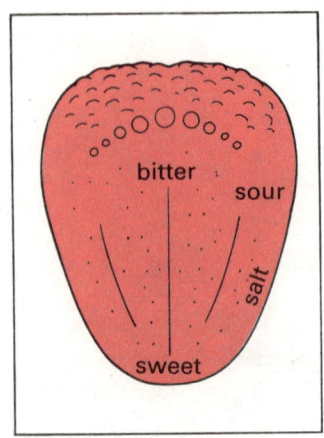

Taste regions of the tongue

Taste bud (right)

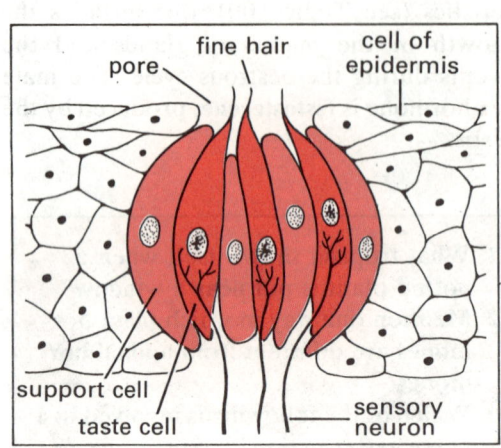

Receptors

In the knee jerk reflex the action starts with the tap on a nerve ending. Such nerve endings which receive stimuli are called **receptors.** There are many receptors just under the skin.

By exploring with a fine needle, touching the skin lightly you can check that there are some spots that are sensitive but other spots can be found where no touch is felt. Some spots give a sensation of heat, others a cold sensation. Others register pain.

Some parts such as the tips of the fingers, the palm of the hand and the lips are well supplied with touch receptors.

Not all receptors are scattered over the body. You feel touch, hot and cold in most regions, but you see, hear, taste and smell with special groups of receptors.

Taste and smell

Both taste and smell depend on the reactions of special receptors to chemical properties of substances. If you are blindfolded with a clip on your nose, you will find it difficult to recognize things by taste only, even onions.

Different tastes are recognized by receptors grouped in taste buds, in different parts of the tongue.

Ear

The nerve endings for detecting sounds are grouped together in the ear. They are quite close to the brain, as if priority is given to the hearing sense.

The only part of the ear that shows is the flap called the **pinna.** This flap is absent in some animals, like birds and frogs, but these animals still have ears. The human ear varies from one person to another. Before the days of finger-print records, the police used photographs of ears to help identify criminals.

The external ear or pinna seems to have been intended to direct sound waves to the ear drum. Some animals, e.g. rabbits, dogs can use their ear flaps in this way. Most of the ear lies out of sight inside the head.

The ear has three parts.

1 The outer ear. A channel leads from the pinna to the ear drum (tympanum). This is slightly conical in shape. On the other side is:

2 The middle ear. This is a space between the outer bone and the bone of the inner ear. Across the space is a chain of three little bones or ossicles. From ear drum inwards they are the hammer (**malleus**) the anvil (**incus**) and the stirrup (**stapes**). These bones are a system of levers that take vibrations of the ear drum, set up by sound waves, to the oval window (fenestra ovalis) of the inner ear. From the middle ear there is a passage, called the Eustachian tube, leading to the mouth. This is normally closed, but opens when you swallow and allows the air pressure in the middle ear to be equalised with the external pressure. This prevents distortion of the ear drum.

3 The inner ear. This is in the bone of the skull. The oval window is the entrance to some little caves hollowed out of this bone. One of these channels is a spiral one called the **cochlea.** This is the real organ of hearing. It is filled with liquid (the endolymph). It connects through two bulb-like parts, the

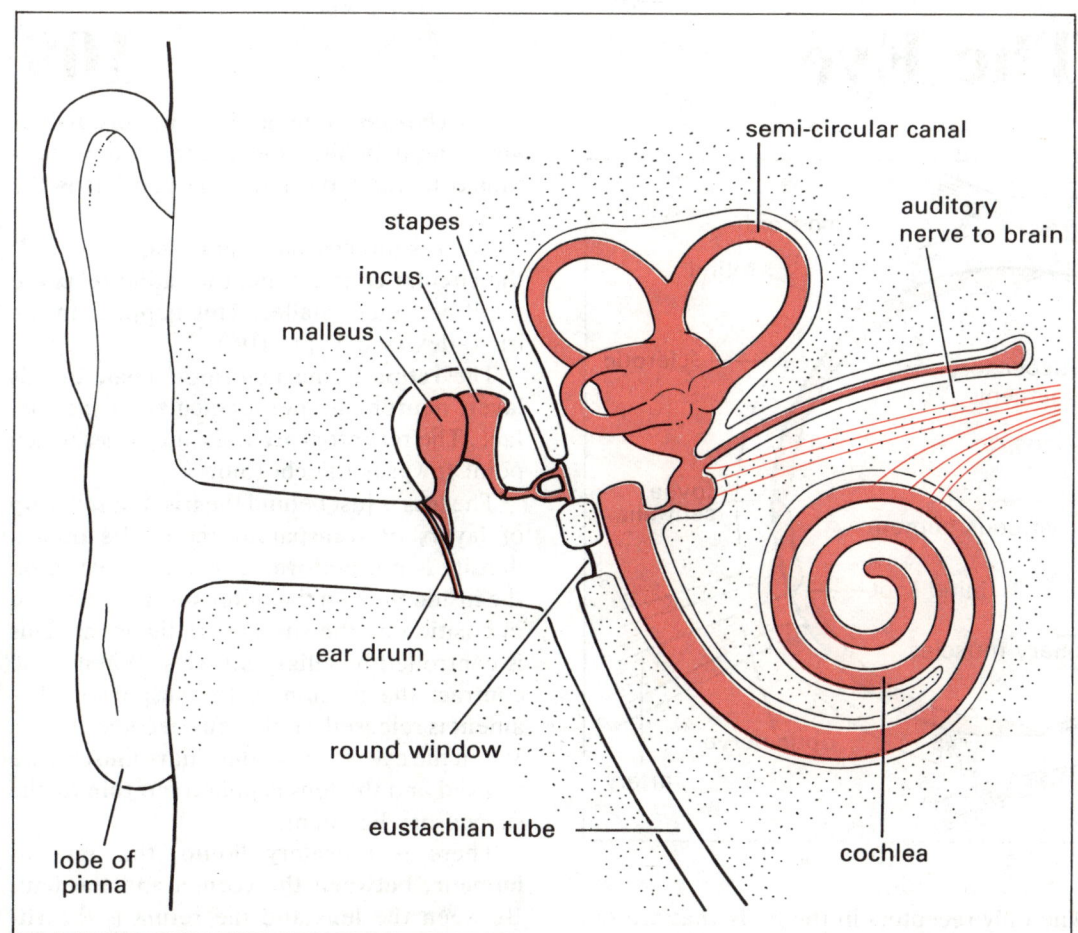

Human ear

saccule and utricle, with three semi-circular canals. These are three loops set at right angles to each other. They form the organ of balance. These also contain liquid.

Hearing

When you hear a sound, the sound waves passing into your ear make the ear drum vibrate. The little bones in the middle ear transmit the vibration to the oval window.

The liquid in the inner ear is set vibrating. Along the length of the cochlea there are tiny nerve endings called hair-cells. These respond to the vibrations and pass impulses along the auditory nerve to the brain. It is because you have two ears that you are able to form a rough idea of the direction from which the sound comes.

Balance

There are three semi-circular canals at right angles to each other, so any movement of the head will affect at least one of these canals. Nerve endings make you aware of any movement.

When you turn round quickly and then suddenly stop, you feel dizzy. This is because the liquid in the canals goes on moving.

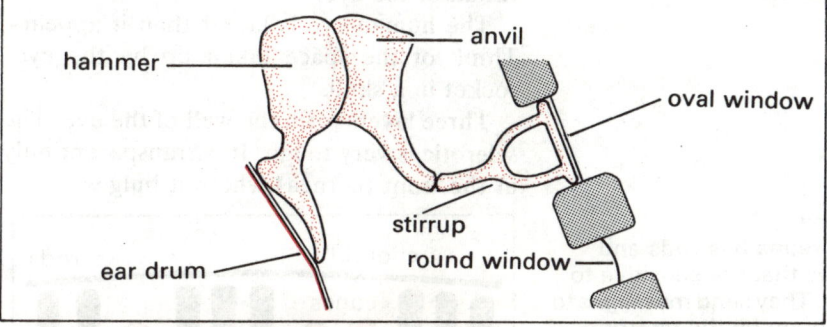

Bones from the middle ear

1 It is more difficult to keep your balance standing on one leg if your eyes are closed. What does this tell you about your sense of balance?
2 Why do you feel discomfort in the ears when your plane is taking off or about to land? Why does sucking a sweet help to ease this feeling?
3 Some deaf people can hear a noise when a vibrating tuning fork is placed on the bone behind the ear. Can you explain this?
4 What is the difference between a noise and a musical note? (Look back to Topic 95.)

The Eye

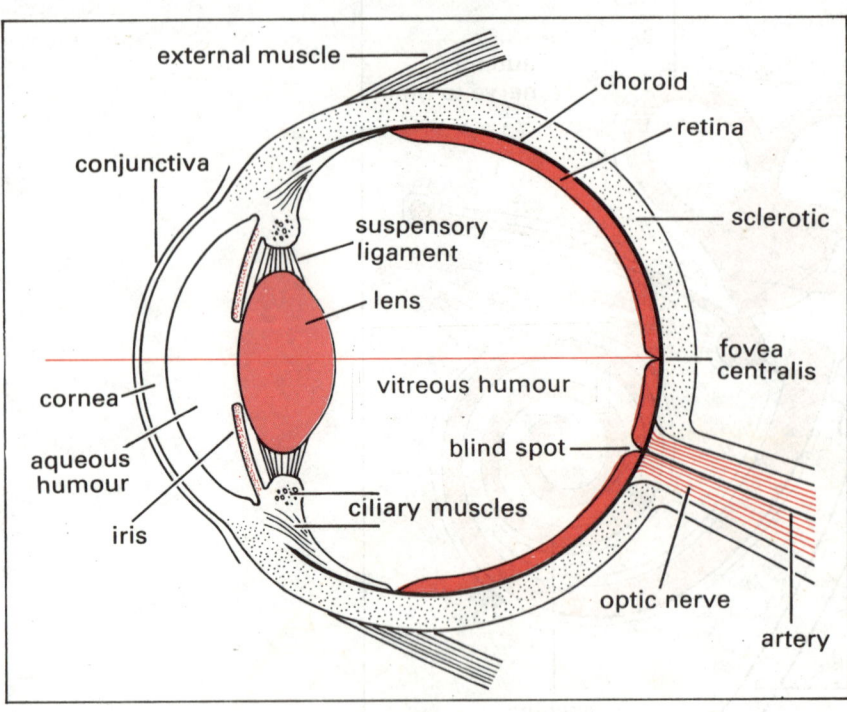

Human eye

Retina
The retina has rods and cones that are sensitive to light. They send messages to the brain by the neurons

The only receptors in the body that are sensitive to light are grouped together in the retina of the eye.

The human eye is larger than it appears. Think of the space taken up by the eye-socket in a skull.

Three layers form the wall of the eye. The **sclerotic** is very tough. It is transparent only at the front (cornea) where it bulges.

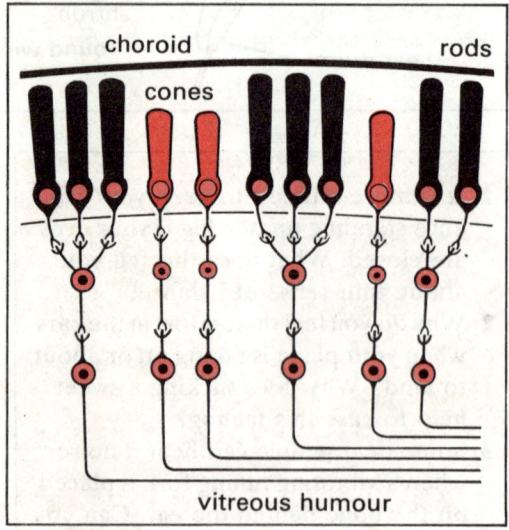

The **choroid** contains blood vessels. It does not extend behind the cornea but is continued to form the **iris.** The gap left is the pupil.

The iris has circular and radial muscles. If the circular contract and the radial relax the pupil becomes smaller. This happens in the iris reflex (see Topic 106).

The **retina** is the innermost coat. In this layer, neurons lie perpendicular to the surface. The receptors are rods and cones which point towards the choroid.

The **lens** is just behind the iris. It is made up of layers of transparent tissue. Its optical density is not uniform. It is more curved on the inside than on the outer surface. It is held in position by the suspensory ligament. This is controlled by ciliary muscles. When these contract the tension in the suspensory ligament is released, so the lens becomes fatter. When the eye is at rest, the ciliary muscles are relaxed and the lens is pulled out thin by the suspensory ligament.

There is a watery liquid, the **aqueous humour,** between the cornea and the lens. Between the lens and the retina is the **vitreous humour,** which is more jelly-like.

The nerve fibres from the rods and cones pass to the brain by the optic nerve. Where this nerve cord leaves the retina there are no rods or cones, so you have a blind spot.

Look at the X— ● line. Cover your left eye and look straight at the X. If you try moving away from and closer to the book, you will find a position where you cannot see the dot on the right.

The image of the dot falls on the blind spot, even though the image of the X falls on the retina.

When you look at an object, rays of light from the object pass into your eye. These rays are refracted first at the cornea then when they reach the aqueous humour and again at the lens. An image of the object is formed upside down on the retina.

Accommodation
If the object is near to your eyes, the lens must become fatter in order to focus the image on the retina. This changing the shape of the lens by the action of the ciliary muscles is called accommodation. Some people can see distant objects clearly but have difficulty in accommodating for near vision. They are long-sighted. The defect is called **hypermetropia.** It might be due to failing muscle

X   ●

action through old age, and then it is called **presbyopia**. Wearing convex lenses as spectacles can help overcome the defect.

A short-sighted person suffering from **myopia**, has difficulty in seeing distant objects clearly. Using concave lenses helps in this case.

Colour

It is not easy to recognize the colour of objects when lighting is poor, although you see them clearly in 'black and white'. This is because the 'cones' which respond to colour are not as sensitive as the 'rods'. Cones are mostly concentrated around the fovea centralis (central spot). Colour is most easily seen when you are looking straight at the object so that the cones at the centre of the retina are stimulated.

Persistence of vision

The effect of light on the retinal nerve endings does not die away immediately. So a series of quick flashes is seen as a continuous line. This is why you see the succession of dots on the television screen as a picture.

Binocular vision

You need two eyes rather than one to judge distance. Each eye forms its own image, but the brain interprets the two images as one.

The brain sometimes has difficulty in making sense of what the eyes report. Look at the pictures below and see if you can believe your eyes!

---

1 Write down three ways in which your eye is like a camera. In what ways is your eye different from a camera?
2 What part of the eye gives it its colour?
3 Draw up a family tree of the colour of your relatives' eyes, and try to decide if eye colour is inherited.

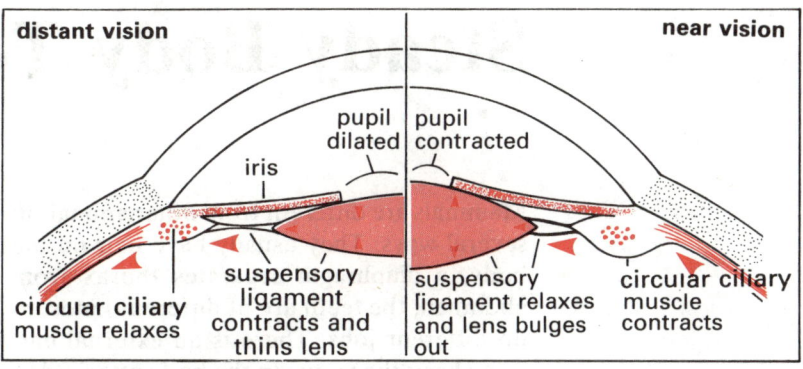

distant vision · near vision

pupil dilated · pupil contracted · iris · suspensory ligament contracts and thins lens · suspensory ligament relaxes and lens bulges out · circular ciliary muscle relaxes · circular ciliary muscle contracts

Accommodation
The shape of the lens changes to allow light from near or distant objects to be focused on the retina

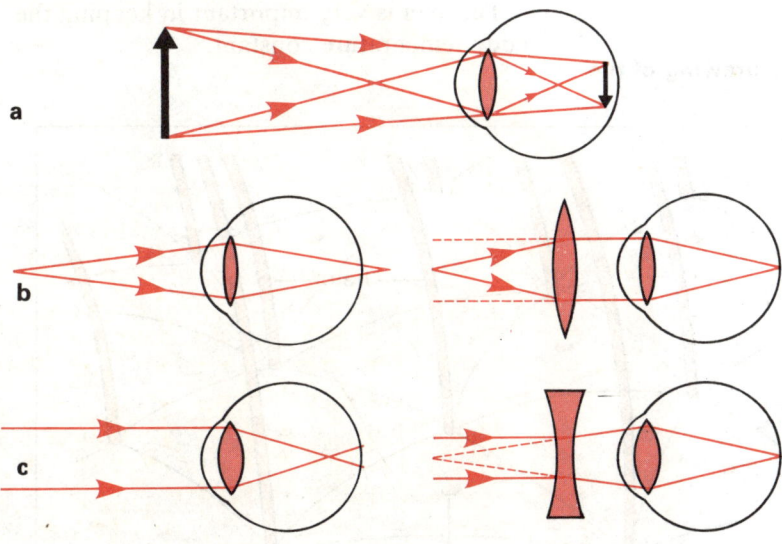

Defects of vision
a Normal eye: images are formed upside down on the retina
b Longsight and correction with a convex lens
c Shortsight and correction with a concave lens

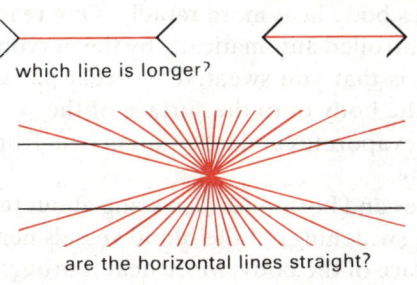

which line is longer?

are the horizontal lines straight?

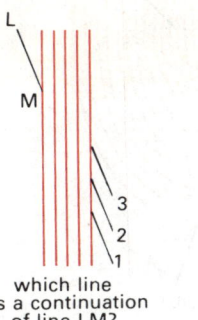

which line is a continuation of line LM?

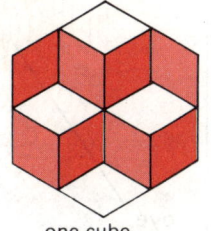

one cube standing on three others or two cubes jutting out over another — try half closing your eyes

do you see a cake-dish or two faces?

Mammals are different from other animals in several ways. They usually have hair on the body, a diaphragm separates thorax from abdomen, the teeth are of different kinds, to do different jobs. There is an external ear. They keep the young in the body attached to the mother's placenta. The young are fed with milk from mammary glands.

A most important difference is that the body is kept at a fairly constant temperature, and, probably because of this, mammals, especially man, have better developed brains than other animals.

The skin is very important in keeping the body temperature constant.

Magnified drawing of the skin

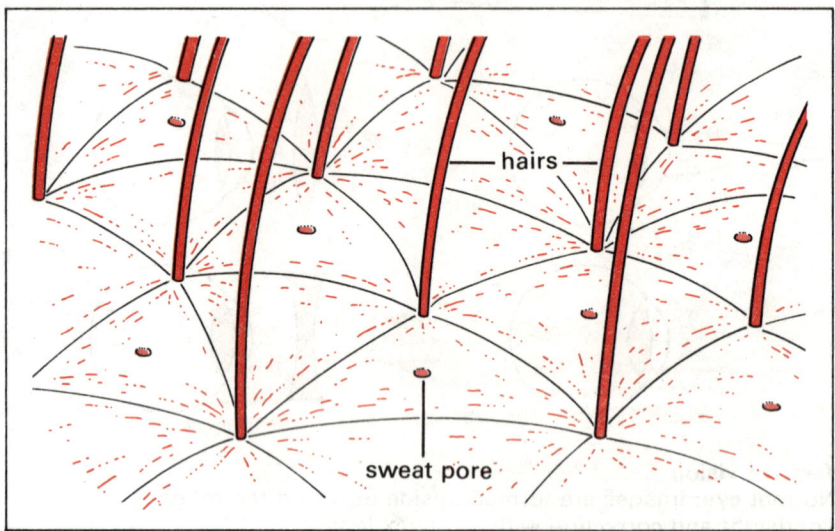

hairs

sweat pore

Fingerprint
There are sweat pores in the furrows

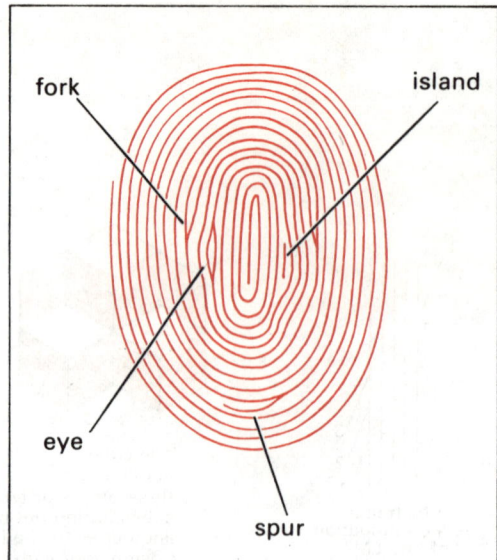

fork

island

eye

spur

When you look at the skin on your arm you can see openings of sweat pores between the hairs. Even on the finger tips there are sweat pores in the furrows of the 'finger-print' ridges.

The sweat glands can be seen under the microscope in a thin section of skin cut at right angles to the surface. You can see a layer called the **epidermis** or outer skin. From this layer dead cells flake off.

Under this is the **dermis.** In this layer you find:

1 **Sweat glands** with ducts leading to the surface pores.
2 **Hairs** that grow from the papillae at the base. Note the small muscles that pull the hair up. This accounts for the 'goose-flesh' appearance of the skin when you are cold.
3 **Sebaceous glands** secrete oil around the shaft of the hair.
4 **Nerve endings** (receptors) are of various kinds.

Fat deposits, nerve cords and blood vessels are also present.

Control of body temperature

The human body is hotter than the air around it. It loses heat through conduction, convection and radiation. This heat loss is kept down partly because of the fat layers in the skin and partly by the clothes you wear. Clothing that traps a layer of air is a particularly good insulator.

Tissue respiration takes place continuously inside the body, even when you are sleeping. Heat is produced all the time. There is usually as much heat produced as lost from the body.

Human body temperature should not vary by more than a few degrees from the normal 37°C. Vigorous exercise, like playing games, produces body heat more rapidly. One reaction, controlled automatically by the nervous system, is that you sweat. The sweat passes out of the body onto the surface of the skin. Here it evaporates and takes latent heat from the body.

Nerves and hormones also bring about the dilating (widening) of the blood vessels near the surface of the body. More heat is brought to the surface and lost. You can see this by the red flushing of the skin when you get hot.

Cross section of human skin

dermis

dead cells

hair

sebaceous gland

epidermis

Meissner corpuscle

sweat duct

capillary network

erector pili muscle

sweat gland

fat layer

blood vessels

Body temperature
The body temperature is kept in balance by several factors

If you are in danger of losing heat too quickly, you may shiver. Shivering is a kind of muscle action which produces heat. Also, the blood vessels leading to the skin constrict (narrow) so that the blood does not flow into the capillaries so quickly. The blood flow may be so sluggish that the blood in the blood vessels looks blue. So the healthy body has an automatic built-in system, a kind of thermostat, to keep the temperature steady.

1 What is latent heat?
2 Why do you shiver if you come out of water and do not dry yourself right away?
3 What part of the nervous system do you think is involved in the control of blood vessels as mentioned above?

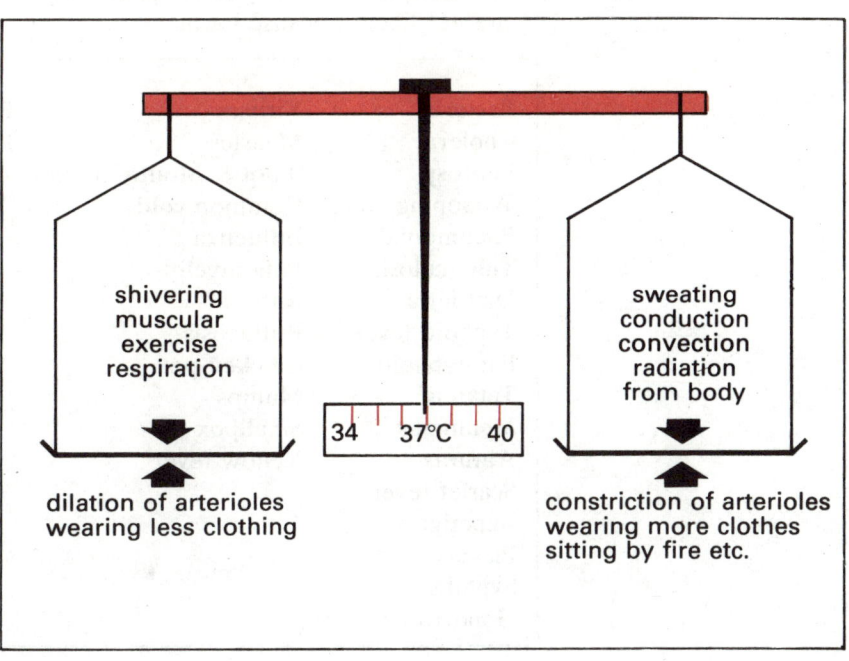

shivering
muscular
exercise
respiration

34   37°C   40

sweating
conduction
convection
radiation
from body

dilation of arterioles
wearing less clothing

constriction of arterioles
wearing more clothes
sitting by fire etc.

Disease

All plants and animals are liable to be attacked by other organisms. Man's greatest enemies (apart from other men) are very small organisms, like bacteria and viruses.

Bacteria

You have seen that some bacteria are quite essential to us (Topic 104), but many bacteria, called **pathogenic** bacteria can cause diseases.

Bacteria have different names given them according to their shapes. They are very small. When temperature and food supply are suitable bacteria increase their numbers very rapidly. If they can get inside human tissues they find ideal conditions.

The table shows some of the diseases caused by bacteria.

Viruses

These are particles smaller than bacteria. They do not reproduce outside the cells that they infect, but they disorganize these cells, causing disease. They are very large molecules or groups of molecules. Unlike bacteria, they cannot be grown in culture media.

Effect of micro-organisms

Pathogenic bacteria and viruses cannot usually be destroyed completely. They are always present in the air. They are found in drinking water and in food, but as long as they are present in only small quantities they are not likely to cause harm.

They are not likely to get into the body tissues unless there is an opening like a cut or a wound. Antiseptics should therefore be used on cuts and wounds. Even if they do invade through a cut, the body has a defence mechanism. Most proteins (**antigens**) if they enter the body and invade the blood stream, cause **antibodies** to form. These antibodies join the antigens and render them harmless. White corpuscles in the blood can now ingest them. Not only do the bacteria themselves contain protein, but they produce poisonous substances, called **toxins.** The blood produces antitoxins to remove the toxins. If the bacteria are in large numbers the blood may not be able to cope with all the toxins. The patient then develops a disease.

If the patient recovers he may have some antibodies for that disease left in his blood. So for a time he is immune and the bacteria causing that disease do not affect him.

Each disease needs its particular antitoxin. But it is possible to make a person immune without his having to suffer the disease. He can be **vaccinated.** He is given an injection of a small dose of the disease bacteria which have been killed or made harmless. The body still makes the antibodies. This is called **active immunity.**

Another method is to inject the patient with an extract of antibodies that have been made in another animal. This is called **passive immunity** because the patient himself does not make the antibodies. In either case the blood is ready to deal with the disease bacteria if they arrive.

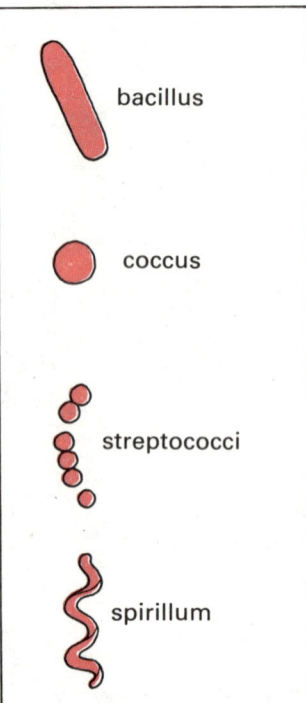

Bacterial shapes

(bacillus, coccus, streptococci, spirillum)

Bacteria	Viruses	Fungi	Protozoa	Worms
Cholera	Measles	Ringworm	Malaria	Elephantiasis
Leprosy	'Foot & Mouth' disease	Athlete's foot	Sleeping sickness	Trichinosis
Whooping-cough	Common cold		Amoebic dysentery	
Pneumonia	Influenza		Kala-azar	
Tuberculosis	Poliomyelitis			
Diptheria	Rabies			
Typhoid fever	Psittacosis			
Paratyphoid	Chicken pox			
Tetanus	Mumps			
Meningitis	Smallpox			
Anthrax	Yellow fever			
Scarlet fever				
Impetigo				
Plague				
Syphilis				
Gonorrhoea				

If prevention fails and bacteria invade the body, then chemicals called **antibiotics** can be used to kill the bacteria.

Other agents of disease

Some unicellular protozoans and some tape worms cause disease. These organisms are parasites which have a complex life-cycle. A parasite is an organism that depends on another organism for its food. It often lives inside the 'host' animal and cannot live without it.

Plasmodium for example spends part of its life cycle in a mosquito. When the female mosquito bites a man to draw blood, she may pass the tiny parasite into the man's blood. The outcome is malaria. Diseases like this can often be avoided by breaking the life-cycle at a suitable point. Malaria can be defeated by killing the mosquito larvae, which live in stagnant water.

We can fight disease by taking special care, for example: by personal hygiene, such as washing the skin with soap, and cleaning the teeth regularly. You should never handle food without first washing your hands thoroughly. You should always clean any cuts or wounds.

Bacteria should be prevented from breeding on food. Milk is pasteurised by heating it either at 60 – 65°C for 30 minutes or by heating at over 70°C for 15 seconds. This kills bacteria. The milk is then cooled quickly. Heating is also used to sterilize surgical intruments.

Some diseases result not from bacteria or other micro-organisms. They are caused by tissues of the body not working properly. Cancer, anaemia, thrombosis, heart and kidney diseases, varicose veins and diabetes are some examples.

1 People travelling abroad sometimes suffer from upset stomachs. But the local people eating the same food do not suffer. How can you explain this?
2 Name three ways of preventing the presence of large number of pathogenic bacteria in the food that you eat.
3 Can you explain why it is a serious crime for a visitor from abroad to bring an animal, such as his pet, into this country instead of leaving it for six months in a special quarantine centre?

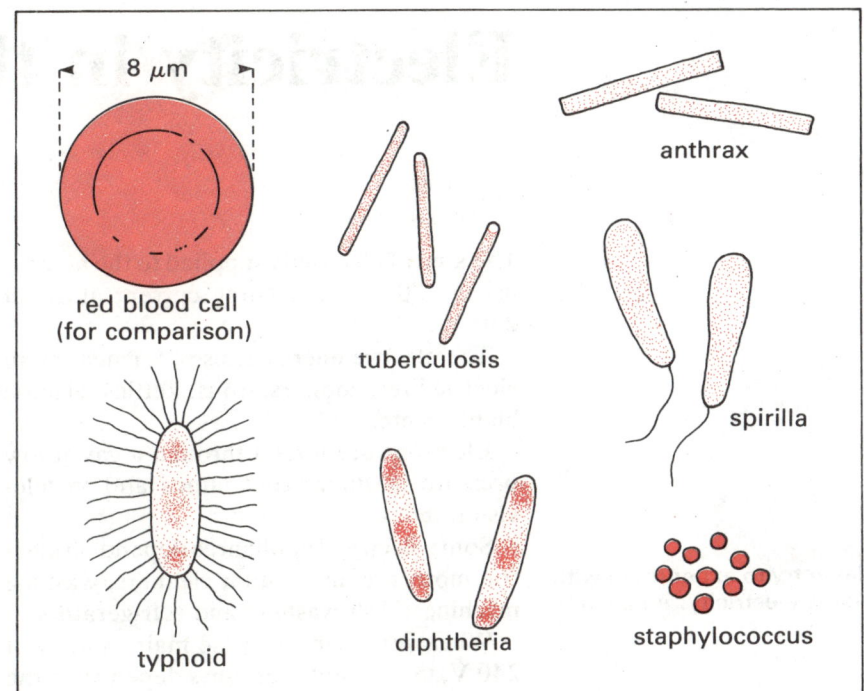

Pathogenic bacteria

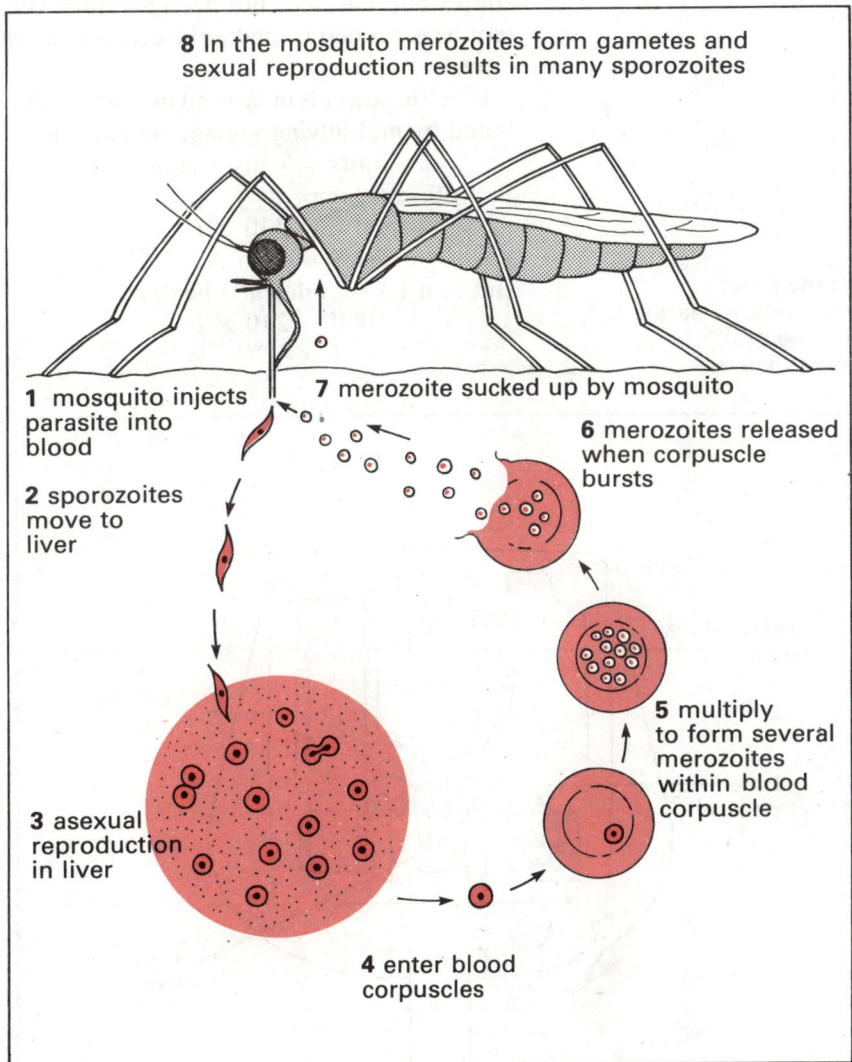

8 In the mosquito merozoites form gametes and sexual reproduction results in many sporozoites

7 merozoite sucked up by mosquito

1 mosquito injects parasite into blood

2 sporozoites move to liver

6 merozoites released when corpuscle bursts

5 multiply to form several merozoites within blood corpuscle

3 asexual reproduction in liver

4 enter blood corpuscles

Life cycle of the malarial parasite

The kind of electricity supplied to the home is usually 50 Hz alternating current at about 240 V.

The electric energy is used for heating in electric fires, cookers, irons, kettles, electric blankets etc.

Electrons are forced through a gas at low pressure in fluorescent lamps and in television tubes.

Some electrical appliances depend on electric motors as in vacuum cleaners, washing machines, dish-washers and refrigerators.

When you connect to the mains supply of 240 V, the current that flows depends on the resistance of whatever is connected. To make a very thin tungsten filament in an electric lamp white hot does not need as much current as is needed to make the element of an electric fire red hot.

Electric power is measured in watts. This is found by multiplying voltage by current:

$$watts = volts \times amps$$

In a 60 watt lamp

$$60 = 240 \times I$$
$$so\ I = 0.25\ A.$$

But in a 1 kW (kilowatt) heater

$$1000 = 240 \times I$$
$$so\ I = 4.2A$$

When you pay your electricity bill you pay according to to the number of kilowatt-hours (kWh) units used. A kilowatt-hour is the amount of energy consumed by the use of 1000 watts for an hour. So if 1 kWh unit costs 3p you can run:

a 3 kW heater for 20 minutes;

or a 6000 watt electric oven for 10 minutes;

or a 100 watt lamp for 10 hours, on the 3p of electricity.

You must be careful when you use mains electrical apparatus, because 240 V AC can be fatal. It is especially dangerous in the kitchen and bathroom. If you touch something that is live with wet hands there is a far better electrical contact made with the body. A larger current flows and this is much more dangerous.

Earth

Electrical appliances should be 'earthed'. A kettle, for example, is connected to the mains by a three-point plug. The three wires are **live** (brown); **neutral** (blue) and **earth** (green or green and yellow). Electricity passes to the heating element inside. This should be insulated from the outside body of the kettle itself. But suppose the live wire somehow comes into contact with the outside metal body of the kettle (by short-circuiting). You will have a sudden shock when you touch the metal body of the kettle. In order to prevent this from happening, the metallic framework of electric appliances should be connected to the earth wire.

Fuses

If too large a current happens to flow through an electrical appliance, it will be damaged, perhaps beyond repair. In order to guarantee that this will not happen a fuse is included in each circuit.

Suppose a 2 kilowatt heater is being used on a 250 V circuit. The current flowing will be:

$$\frac{2000}{250} = 8\ A$$

In this circuit a 15 A fuse could be included. If, due to a fault, a current greater than 15A flows, the fuse wire will melt and the circuit

Warning
Never try to experiment with mains electricity: it can kill

Electric plug
Green/yellow: earth
blue: neutral
brown: live

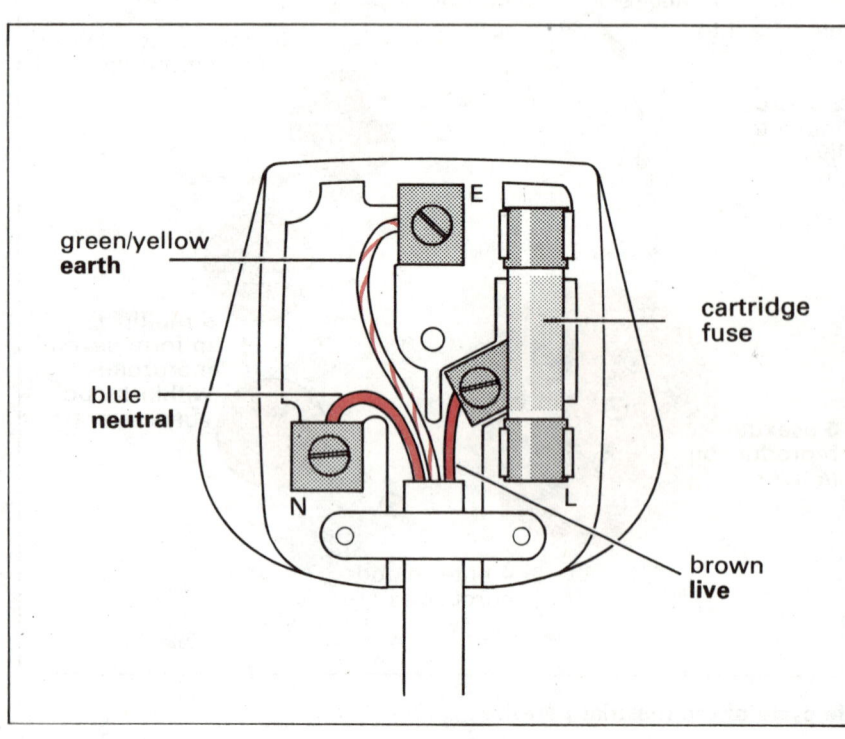

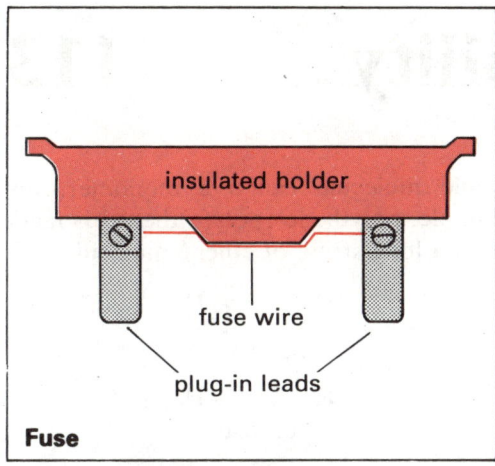

Fuse

insulated holder

fuse wire

plug-in leads

Mains supply

A ring main supply is used to supply the outlet sockets for general use (13 A fuse). A ring main saves having to connect each individual socket to the main fuse board.

A separate circuit is used for each of the heavy current items, such as cooker or washing machine (with a 30 A fuse).

Another circuit supplies the lighting points (5 A fuse). The diagram shows the position of the fuse box and the rest of the wiring.

will be broken before this high current can damage the heater or the wiring.

But what happens if two heaters are plugged in at the same time? The current now is:

$$\frac{2 \times 2000}{250} = 16 \text{ A}$$

The fuse would 'blow' because the circuit is 'overloaded'.

A fuse of 5 A is suitable for the lighting circuit.

1 The following apparatus is to be used on a 240 V mains supply: electric kettle 2640 W: vacuum cleaner 720 W: lamp 60 W: tape recorder 30 W. What current flows in each? State in each case which of the fuses available, 1, 2, 5, 13 amp, would be suitable.
2 Which two appliances in question 1 must not be used together if 13 A is the fuse used in the circuit?

Mains supply for house

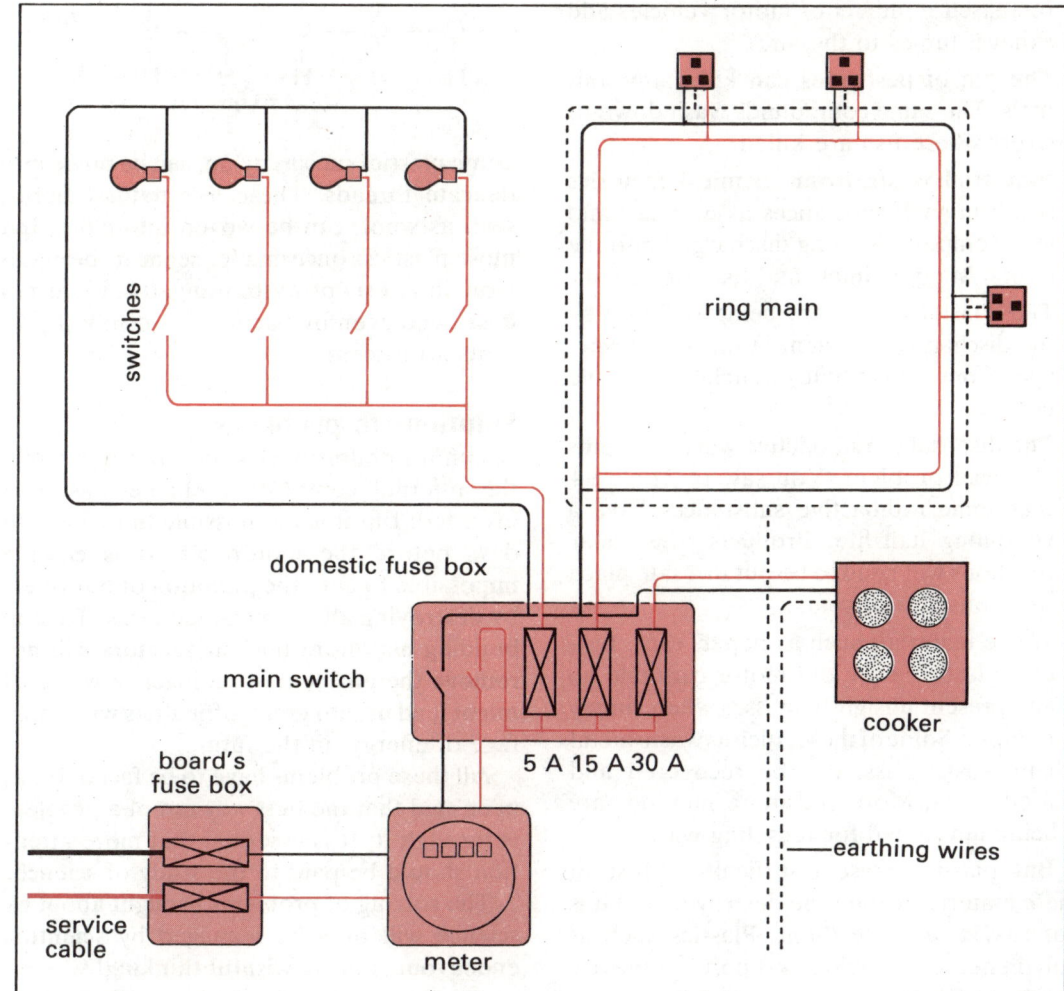

switches

ring main

domestic fuse box

main switch

board's fuse box

5 A 15 A 30 A

cooker

service cable

meter

earthing wires

In Topic 111, you saw how micro-organisms can affect man's way of life. He is also affected by many other factors such as climate and how much food is available.

Man in his turn makes his own impression on nature. His achievements have been tremendous. Not the least impressive of these are the advances brought about by the study of science and technology, for example, machines, aircraft, computers and so on.

But some of this progress has brought unpleasant by-products. Pollution is now a very serious matter. The careless way in which some people disfigure the countryside with litter is sad enough. The large scale escape of dangerous products is much more serious.

Problems of pollution

1 The atmosphere especially near industrial centres has increasing amounts of smoke and gases such as sulphur dioxide. Increased numbers of motor vehicles add exhaust fumes to the air.

2 The use of pesticides can kill many animals. The same compounds wash down to rivers where fish are killed.

3 Industrial waste from chemical factories result in such substances as mercury and lead compounds being discharged into the rivers. Many animals and fish are killed.

4 The coastal waters are being polluted by the discharge into them of untreated sewage. From time to time oil slicks pollute the beaches.

5 The disposal of radioactive waste presents a special problem. You saw in Topic 94 that some radioactive substances have a very long half-life. Products of nuclear reactions will need to be put in a safe place for thousands of years.

6 Waste materials such as paper, cans, bottles, plastics, even old motor cars pile up and present unsightly masses when simply dumped. Some of these, such as waste metal and waste glass, can be recovered and used again. More and more methods are being developed for recycling waste.

But plastics present difficulty. Most of these materials cannot be destroyed and it is not easy to recycle them. Plastics such as polythene, p.v.c., nylon and polystyrene are polymers. Their molecules are multiples of a simple molecule called a monomer. For example, polythene [poly(e)thene] is made up of a long string of ethene molecules:

$$H_2C = CH_2$$
ethene

$$- CH_2 - CH_2 - CH_2 - CH_2 -$$
polymer

Polyvinyl chloride (PVC) in the same way is a polymer of vinyl chloride:

vinyl chloride

PVC

Some plastics such as nylon, can be made into delicate threads. These like natural fibres, such as wool, can be woven into cloth, but most plastics, once made, seem to be indestructible, except by burning. They are not destroyed even by bacteria, so disposal presents a problem.

Solutions to problems

A famous philosopher said that it was a pity the internal combustion engine was ever invented. But it is not possible to go back to days before the motor car. It is equally impossible to cure the pollution of our rivers by destroying all chemical factories. To stop building any more nuclear reactors will not remove the problem of radioactive waste. It might lead us into great difficulties with shortage of energy, in the future.

All these problems have to be faced. It has been said that the best way out of a problem is through it. It is essential that more attention should be paid to the study of science.

The solving of problems brought about by science will only be managed by scientific endeavour, not by wishful thinking!

Atmospheric pollution (top)
A variety of plastic objects (left)
Recycling of glass (above)

1 What other 'pollution' in the air, not mentioned above, is caused by aircraft?

2 Why may we be in energy difficulties in the future?

3 The formula for a molecule of the monomer styrene is: Write down the formula for polystyrene.

$$\begin{array}{cc} H & C_6H_5 \\ | & | \\ C & = & C \\ | & | \\ H & H \end{array}$$

Examination Questions
X

1 Give one function of each of the following: (a) the human skin; (b) the skeleton; (c) the white corpuscles; (d) the red corpuscles; (e) the blood plasma.
Draw a diagram to illustrate the circulation of the blood in the human. WYL

2 Where in the body would you find the following?
(a) the semi-circular canals; (b) the radius; (c) the cornea; (d) the alveoli. M

3 Describe in as few words as possible an experiment to find out which part of the hand, arm or foot is most sensitive to touch. Which part do you think is the most sensitive? EM

4 Which of the following statements is correct?
In a reflex response:
(a) the brain decides the response;
(b) the eye must receive the stimulus;
(c) the message travels through only a part of the central nervous system;
(d) the message goes directly from the sense cell to the muscle;
(e) the muscles react without receiving any message. EM

5 Which of the following statements is correct?
The lens of the eye can be made less convex or more convex so that we are able:
(a) to see near and distant objects clearly;
(b) to recognize different colours;
(c) to see in good and poor light;
(d) to see moving objects clearly;
(e) to look at bright objects without damaging our eyes. Y

6 Explain possible dangers in four of the following:
(a) Prolonged excessive application of chemical fertilisers.
(b) Increased use of fossilised fuels such as coal, oil and petrol.
(c) Dumping and/or reprocessing radioactive waste.
(d) The release of heavy metals into the atmosphere and rivers.
(e) The excessive growth of water plants in lakes ('blooming'). EAS

7 Which of the following is **not** a reflex action?
(a) sneezing; (b) yawning; (c) speaking; (d) dilation of pupil of the eye. NW

8 Light from a distant object travelled into an eyeball as shown in Diagram 1.
(a) Name the eye-defect shown in the first diagram.
(b) Copy and complete the second diagram to show how this would be corrected. EM

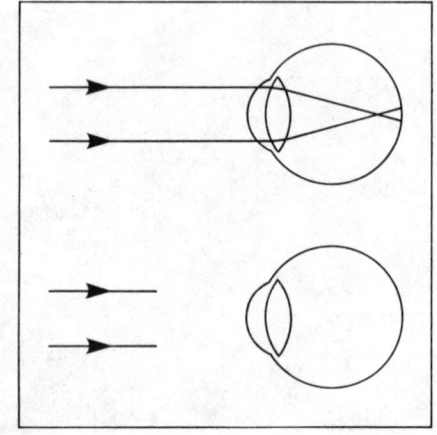

Diagram 1

9 (i) From Diagram 2, give the letter which indicates the position of the following:
pinna, eardrum, auditory nerve, eustachian tube, ossicles, cochlea, semi-circular canals.
(ii) What are the names of the three ossicles?
(iii) Give one difference between a noise and a pure note.
(iv) Explain the terms: (A) pitch; (B) frequency; (C) wavelength; (D) amplitude.
(v) By placing your ear on a screwdriver resting on a car engine it is possible to tell if the engine is misfiring. Explain why you can hear the sounds more easily through the screwdriver than through the air. SE (part)

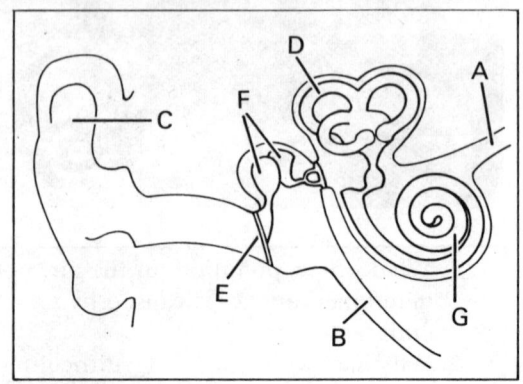

Diagram 2

INDEX

Numbers refer to Topics

The author and publisher wishes to express their thanks to the following Examining Boards for permission to use selected examination questions from the General Science papers set by these authorities:

Associated Lancashire Schools Examining Board	ALS
East Anglian Examinations Board. Mode 1: North	EAN
East Anglian Examinations Board. Mode 1: South	EAS
East Midland Regional Examinations Board	EM
Middlesex Regional Examinating Board	M
North West Regional Examinations Board	NW
The South-East Regional Examinations Board	SE
South Western Examinations Board	SW
Welsh Joint Education Committee (Cyd-Bwyllgor Addysg Cymru)	W
West Midlands Examinations Board	WM
The West Yorkshire and Lindsey Regional Examining Board	WYL
Yorkshire Regional Examinations Board	Y

Thanks are due to the following for permission to reproduce copyright material:

Camera Press (Topic 1); David Wade (3, 37, 38, 48, 55, 61, 62, 84); Institute of Geological Sciences (3, 27, 40, 46); NASA/Space Frontiers (7, 10, 88, 89); Simon Warner (8); Blue Circle Industries Limited (8); Crown Copyright, Science Museum, London (9); Spenby Limited (9); Keystone Press Agency (10, 11, 39, 60, 61, 113); British Steel Corporation (12); British Railways Board (15); Aerofilms (17); Peter Storm Limited (20); Plessey Co. Allan Clarke Research Centre (22); John Topham Picture Library (26); Dyno-Rod (27); Thames Water Authority (27); Ann Ronan Picture Library & E. P. Goldschmidt & Co. Ltd. (28); Air Products (31); BOC Limited (31); Chubb Fire Security Limited (32); Popperfoto (13, 35, 37, 50, 58, 60, 78, 82, 105); Photographed by Walter Nurnberg for the Distillers Co. Ltd. (35); Scottish Tourist Board (36); By courtesy of the Trustees of the British Museum (38); Henry Grant Photos (40, 55, 104); Mansell Collection (41); I.C.I. Mond Division and Calor Gas (41); I.C.I. Mond Division (43); I.C.I. Nobel's Explosives Co. Ltd. (43); Chloride Automative Batteries Ltd. (43); Natural History Photographic Agency (46); The Press Association Limited (47); Timothy Horne (48); Geoslides Photo Library (49); British Aluminium Co. Ltd. (54); Hoover Ltd. (57); Central Electricity Generating Board (58); United Kingdom Atomic Energy Authority (58, 94); Photographed by courtesy of A.E.R.E. Harwell (60); All England Netball Association (61); Donald Normal (63); National Coal Board (69); Boosey & Hawkes (71); Phillips Small Appliances (72); Topham/Scowen (73); Thorn Lighting Ltd. (76); Geoslides Photo Library (77); Perkins Engines Ltd. (79); Biophoto Associates (82, 85, 97); Andrew Wiard, Report (85); John Watney Photo Library (85, 106); Nuffield Radio Astronomy Laboratories (87); Royal Greenwich Observatory (89); G.E.C. Machines Limited (92); St. Mary's Hospital London (94); Royal Society for the Protection of Birds (98); Griffin & George Ltd. Gerrard Biological Centre (100, 102); Jack Cohen, Philip Harris Biological Ltd. (100); John Sturrock, Report (100); Philip Harris Biological Ltd. (102); Biophoto Associates (104); LBC (113); A Shell photograph (113); Picturepoint (front cover); UKAEA (back cover).

The publishers have made every effort to trace copyright holders, but if they have inadvertently overlooked any, they will be pleased to make the necessary arrangement at the first opportunity.